PURE INTELLIGENCE

synthesis

A series in the history of chemistry, broadly construed, edited by Angela N. H. Creager, Ann Johnson, John E. Lesch, Lawrence M. Principe, Alan Rocke, E. C. Spary, and Audra J. Wolfe, in partnership with the Chemical Heritage Foundation

PURE INTELLIGENCE

The Life of William Hyde Wollaston

MELVYN C. USSELMAN

THE UNIVERSITY OF CHICAGO PRESS
CHICAGO AND LONDON

Melvyn C. Usselman is professor emeritus in the Department of Chemistry at Western University in London, Ontario.

The University of Chicago Press, Chicago 60637
The University of Chicago Press, Ltd., London

Printed in the United States of America

24 23 22 21 20 19 18 17 16 15   1 2 3 4 5

ISBN-13: 978-0-226-24573-7 (cloth)
ISBN-13: 978-0-226-24587-4 (e-book)
DOI: 10.7208/chicago/9780226245874.001.0001

Usselman, Melvyn C., author.
Pure intelligence : the life of William Hyde Wollaston / Melvyn C. Usselman.
pages cm — (Synthesis)
ISBN 978-0-226-24573-7 (cloth : alk. paper) — ISBN 978-0-226-24587-4 (e-book)
1. Wollaston, William Hyde, 1766–1828. 2. Scientists—England—Biography.
I. Title. II. Series: Synthesis (University of Chicago Press)
Q143.W795U87 2015
509.2—dc23
[B]

2014039432

⊗ This paper meets the requirements of ANSI/NISO Z39.48-1992 (Permanence of Paper).

DEDICATED TO MY WIFE TRIXIE

AND OUR CHILDREN JASPER, CHARLOTTE,

RICHARD, AND DAVID

CONTENTS

ACKNOWLEDGMENTS

I first encountered William Hyde Wollaston while preparing a history of chemistry course for chemistry majors at the University of Western Ontario in the mid 1970s. Following his trail through the science of the early nineteenth century has directed much of my historical research ever since. I owe much of my intellectual development to Paul de Mayo and F. Larry Holmes. Paul de Mayo demonstrated for me on a near-daily basis the power of critical thinking, and Larry Holmes helped me complete my transition from chemistry to historical research by proving to me many times over the intellectual rewards of seeking out the merits of divergent viewpoints. Discussions with both of them instilled in me the belief that modern science owes as much to the personalities and cultural context of the past as it does to the content base of its many disciplines. The life and science of Wollaston that I present in this book illustrates, I hope, how superficial (and historically impoverished) it would be to investigate scientific discovery without consideration of enabling societal factors.

The richness of the Wollaston material—scientific, artifactual, social, familial, and anecdotal—and its distance from my university location in London, Canada, explains (in part) the long gestation of this biography and (in toto) my great debt to the many librarians, archivists, and curators who spent so many tedious hours carefully locating and copying the thousands of pages I have required to recover his life. I have also benefited enormously from the endless goodwill and encouragement of historical and scientific colleagues around the world, and I am pleased to acknowledge their specific contributions in relevant footnotes. Some people, many of whom are no longer around to read and critique this biography, merit special mention. Leslie Hunt of Johnson Matthey played a pivotal role in my early studies of Wollaston's platinum researches, and Peta Buchanan hunted down for me details of London life with the tenacity of a bulldog. Lionel Felix Gilbert collected all the Wollaston materials now in the D. M. S. Watson Library of University College London, and David Goodman wrote an excellent D. Phil. thesis on Wollaston, which I have mined for much secondary literature. Mrs. Vaire Solandt (née Wollaston), a fellow Ontarian, has been a most gracious and enthusiastic conduit to the several Wollaston descendants who have so willingly aided my study of their illustrious ancestor.

I owe a great debt also to Geoffrey Cantor and Jed Buchwald and to my chemistry colleagues Dick Puddephatt, Peter Guthrie, and Edgar Warnhoff, all of whom read one or more draft chapters and helped improve every one. Special thanks are due to Bill Brock, David Knight, Trevor Levere, Rob Lipson, Alan Rocke, and Willemina Sennema, each of whom read an early draft of the entire manuscript and gave many wise suggestions for its improvement. My study of Wollaston could not have occurred without funding support from the Canada Council (1976), the Social Sciences and Humanities Research Council (1978 and 1982), the Hannah Institute for the History of Medicine (1987), and the Department of Chemistry of the University of Western Ontario. I am also most grateful to all those at University of Chicago Press who have helped bring this biography to press. And finally, I thank my wife, Trixie, for her continuing cheerful support, editorial assistance, and, especially when needed, encouragement.

NOTE TO READER

To minimize anachronistic interpretation, I have used early nineteenth-century terminology in descriptions of Wollaston's science and technical innovations. I have, however, placed modern terms and formulations in brackets whenever I think such a clarification will aid a reader who wishes to bridge the two centuries between his time and ours.

Footnote citations give short form versions of the full references in the Bibliography. Footnotes to papers by William Hyde Wollaston include the dates of publication because some of his research results are grouped by theme instead of chronology.

ABBREVIATIONS

Gilbert Collection	L. F. Gilbert Collection, D. M. S. Watson Library, University College London.
Wollaston Mss.	Wollaston Manuscripts, Cambridge University Library, Mss. Add. 7736.
DSB	*Dictionary of Scientific Biography.* New York: Charles Scribner's Sons, 1970–1978.
oDNB	*Oxford Dictionary of National Biography.* Oxford: Oxford University Press, 2004.
TNA	The National Archives of the UK: Public Record Office.

PROLOGUE

April 1801

> I remember Dr. [William] Whewell remarking to me once, just after a return from London where he had conversed with Dr. Wollaston, that "it was like talking to pure intelligence."[1]

Wednesday April 15, 1801, was another in a succession of cool breezy spring days in London, England, but it was to be a seminal one in the history of one of the world's most valuable metals. John Dowse, a young man in his early twenties, had risen early to stoke the coal fires that warmed the living quarters of the house at 18 Cecil Street before proceeding into the back room that had been converted into a chemical laboratory. There he tended to the furnace that kept the vessels on its surface heated day and night, for his duties had become more and more those of a laboratory assistant than of a manservant to the young physician, about ten years his senior, who employed him.

John had begun working for Dr. William Hyde Wollaston in early 1800. Since that time, he had witnessed a steady decline in the number of patients that showed up for treatment at the short street in Westminster that ran from the banks of the Thames to the prosperous and bustling Strand running parallel to the river. Surprisingly, the doctor appeared to welcome the freedom a declining number of patients afforded him. Although, as a relatively new member of the Royal College of Physicians, Wollaston spent much of his idle time reading the texts, pamphlets, and journals in his extensive medical library, his passion did not lie in service to his patients. It manifested itself instead in the observation, measurement, and comprehension of the beauty, intricacy, and rational design of the natural world around him. The doctor's home contained telescopes, microscopes, and a variety of electrical and mechanical devices. One after another was used to observe natural phenomena, and the results were often recorded late into the night. John had little personal involvement with those scientific pursuits, but he had been charged with maintaining the room in which the doctor's chemical experiments were conducted. He ordered coals for the furnaces and candles for lighting, and also took delivery of chemicals and apparatus. Most importantly, he was responsible for maintaining the temperatures of the reaction vessels near the values specified by the doctor, often for days in succession. This April morning was a special one, however, for Wollaston would on this day bring a critical series of experiments to a successful conclusion, and lay the foundation for a career that would establish him as one of the leading scientific figures of the early nineteenth century.

The plan for the day appeared to be a relatively simple one. Wollaston wished to convert a few ounces of a powdery, grey metal into a solid mass. Ominously, however, he knew that many before him had tried to do the same thing, and none had been able to attain consistent results. The stakes were high, for Wollaston recognized that success promised to change his life forever. Consolidation of the metallic powder into a malleable solid would permit him to bring a new and valuable metal into widespread commercial use. Moreover, profits from its sale had the potential to liberate him from the oppressive medical career that had brought him more unhappiness than financial reward. Although months of pioneering chemistry had been required to obtain the pure, powdery metal, he reckoned that it would take little more than a day to perform the key, carefully-planned, metallurgical step. That day had now arrived. In mid-morning, Wollaston entered his home laboratory and asked John to bring the furnace to full heat. Then he moved to the bench where several samples of the grey metallic powder awaited him.

The powdery metal had an interesting history of its own. It originated in the gravels of the meandering rivers of coastal New Granada (now Colombia), then a South American territory under Spanish control. It had been deposited with gold in the river beds as heavy silvery grains of a metallic ore termed platina and was a troublesome byproduct of the gold-mining operations of the Spanish conquerors. Like the gold it accompanied, alluvial platina became a controlled material belonging to the king of Spain, and only small amounts were made available for scientific study. In the middle of the eighteenth century, chemists had discovered that the major component of crude platina ore was a new noble metal (and chemical element), then known generally as platina or platinum, which possessed properties similar to gold. The metal had proven to be hard to purify because its melting point was too high to allow it to be liquefied. Moreover, its hardness made it nearly impossible to fashion it into useful objects. A few chemists had been able to sporadically obtain platinum pure enough to make it into things such as laboratory crucibles, but the procedures gave inconsistent results, which continually frustrated the experimenters. Nearly all of them had given up in despair, and the crude ore had become little more than a chemical curiosity, available principally from miserly Spanish sources.

In December 1800, hoping that valuable platinum could be extracted from the alluvial deposits, Wollaston and a business partner spent nearly £800 (about $150,000 today) to purchase several thousand ounces of the crude ore. This large quantity of platina had been smuggled out of New Granada and delivered to traders at Kingston in the British colony of Jamaica. They in turn sold it to a dealer working on Wollaston's behalf. Even before the ore reached London, Wollaston had begun to experiment with a few ounces of locally available crude platina and had worked out a novel chemical process for isolating purified platinum in powder form. Then, after a portion of the Jamaican purchase arrived in London in early 1801, he began to purify greater amounts of the ore and had been able to accumulate about 60 oz. of platinum powder by mid April. His goal this day was to convert the powder into solid metal ingots suitable for commercial applications. To do this, he planned to compress the powder into a compact mass, heat the fragile solid to the highest temperature possible, and then attempt to hammer the hot plug of metal into a solid, high-density mass of malleable platinum. Much depended on the results of the planned forging trials, both financially and professionally.

Wollaston closed one end of a hollow cylinder of iron, which he referred to as a (gun) barrel, with an iron plug held in place by blotting paper and set the cylinder, closed-end-down, upright in a jug of water. Then he filled the barrel with water and added the first sample of platinum powder. After the heavy metal sank to the bottom he placed a circle of blotting paper and a wooden plug on top of the wet mud, took the barrel out of the water and placed it in a specially designed screw press. He then compressed the wet metal to expel as much water as possible and thereafter pushed the loosely compacted platinum plug out of the barrel. Next, he heated the plug to redness for several minutes on a charcoal fire, then covered and heated it for another twenty minutes at the highest heat attainable with a coal fired wind furnace. Finally the hot metal, now in the form of a cylinder about 2 inches long and 0.7 inches in diameter, was set vertically atop an anvil for the final, critical step. Wollaston struck the ingot on the top with a heavy hammer, slowly increasing the force of the blows while trying as best he could to keep the sides parallel as the ingot became shorter and denser. And then, as he neared the end of the hammering procedure with success only a few hefty hammer blows away, the ingot broke into two pieces. With obvious disappointment at the failed consolidation, he tersely wrote in his laboratory notebook that the metallic ingot had "snapped."

Moving next to the second sample of powdered platinum, Wollaston repeated his consolidation procedure and again began to hammer the hot metal into a compact mass, maybe faster, maybe slower, maybe more carefully, maybe after stronger heating, but this time the ingot cohered well enough to withstand the most powerful hammer blows. This second consolidation trial was a resounding success. With satisfaction, and undoubtedly great relief, he entered in his notebook only a succinct comment, the second ingot, "rolled well"; he was not one to convey emotion in records of experimental observations. These two words, however, mark the beginning of our modern platinum industry. Wollaston had found a way to prepare the metal in malleable form. Buoyed by this one successful result, Wollaston continued on to attempt the consolidation of the third sample of the day. Unfortunately, whatever operational insights he had gained from the successful forging of ingot number 2 were not enough. During the final hammering, the third ingot also broke apart, and Wollaston disappointedly entered "split" into his notebook. That concluded the forging trials of April 15, with but one success and two failures. Nonetheless, the one malleable ingot was the first encouraging sign that his technique for purifying and consolidating platinum could yield a commercially-valuable product. It ultimately proved to be so, and malleable platinum was to make him a wealthy man years later. Not surprisingly he proudly saved a sample of his first malleable ingot. That souvenir, the first platinum produced by Wollaston's unique chemical and metallurgical process, marks a milestone in the evolution of our modern global platinum industry. A portion of the first ingot, rolled flat and shaped into a rounded blade, remained with his descendants for over a century before it was placed on loan with London's Science Museum, where it is now occasionally placed on display. All things considered, that brisk spring day in April 1801 was a portentous one for Wollaston, chemistry, and the platinum industry.

The foregoing account has been composed from information in primary and secondary historical sources and is intended to capture some of the events, and drama, in the home laboratory of William Hyde Wollaston on one auspicious day in April 1801. There is no reason to doubt the factual content of the discovery process, for there are several source documents from which the pertinent information can be extracted. The weather details, for example, are taken from readings made on the premises of the Royal Society of London, a short distance from Wollaston's Cecil Street home. Some other relevant information, however, is taken from sources compiled days, months, or even years later. For commercial reasons, Wollaston did not publish a complete description of his platinum process until late in 1828. Several details of the April trials are taken from this later account, and it is certain that the forging process in 1801 differed to a greater or lesser degree from the perfected one published twenty-seven years later. But such easily accessible information provides only part of the story. The rest, and the best, comes from original laboratory notebooks that contain succinct entries on nearly every aspect of Wollaston's research career, including crucial details of his 1801 discovery. Unfortunately the notebooks went missing in the nineteenth century and did not become available for study until they resurfaced in the middle of the twentieth century in the holdings of the Department of Mineralogy and Petrology of the University of Cambridge. The collection consists of nineteen notebooks covering a wide range of scientific investigations and financial accounts, together with medical lecture notes, numerous letters to and from Wollaston, superficial accounts of Wollaston's Cambridge years by a fellow student (and failed biographer), and even a copy of a post mortem examination carried out by Wollaston's doctors. The entire collection has since been transferred to the archives of Cambridge University Library, and it has provided much of the new information I have included in this biography. However, a reader cannot fail to notice that the discovery narrative in this prologue supplements reliable source material with generous portions of circumstantial evidence and biographer's intuition; it is unlikely to be an entirely accurate description.

Intuitive statements about historical events do not by themselves weaken descriptions of those events, but a reader should be given the means to differentiate reliably between facts and opinions. An honest historical account, and a compelling one, should maintain a thorough and consistent distinction between well documented events and an author's informed judgments. So the following chapters will expand on the format of this prologue by inclusion of references to the supporting literature and textual clues to personal judgment and opinion. Speculative judgments will enrich the narrative, but they will be acknowledged, and the reader can contemplate alternatives when so desired.

Details of experiment, participants and context aside, there are other cautionary remarks that one should make about discovery tales like the one just presented. The first is the heroic aura placed around the successful forging and the implicit "eureka" moment of discovery. Wollaston's research notebooks give ample evidence of enabling observations and discoveries in the days, weeks, and months prior to production of the first malleable ingot. Also, numerous unanticipated problems (and even a few serendipitous discoveries) were to appear in his platinum work over the years following the first successful consolidation. In fact, Wollaston's discovery of a novel method for the purification and consolidation of platinum is impossible to confine to a single momentous event on a singularly portentous day. It evolved irregularly over time and reached its final form only when his researches on the metal ultimately came to an end. For any scientist in any age, significant discoveries seldom occur in a flash of inspiration or in a single experiment.

An additional concern with such a heroic discovery account is the implicit value given to it by an unchecked presentist perspective. The process for making malleable platinum was to become Wollaston's greatest achievement, but he could not have known that at the time of the crucial breakthrough. Simultaneously, he was attempting to synthesize high-value organic compounds from wine dregs. He was also involved in optical researches that would lead to patents on two optical devices. Each of these research interests was pursued to generate marketable goods, and Wollaston could not know with any certainty which, if any, of his scientific quests would bear fruit. Even the most lucrative product, malleable platinum, was to become so only after unforeseen technological applications emerged several years after the 1801 forgings. Yet, although Wollaston clearly had high hopes for the commercial potential of his product (and kept a sample of his first success as a commemorative artifact), he could have had no conception of the enormous impact his discovery would have on the future emergence of a global platinum industry. It is instructive from a presentist vantage point to seek the origins of our core science and technology, and to use accumulated knowledge to enrich our understanding of past events. But it is a mistake to evaluate the importance of past events solely by their fit with, or impact on, modern beliefs and materials. Scientific work, like all creative endeavors, is best comprehended when convincingly located within its appropriate contextual environment.

Finally, Wollaston's discovery as dramatized above is too simplistically self-contained. Its enabling condition was a novel chemical purification of crude platina, and its crucial feature was powder consolidation brought about by strong compression and hot forging. These improvements were all designed and executed by Wollaston, but he was not a lone intellect. We can trace his interests in platina to his Cambridge undergraduate days and his interactions then and later with his business partner, Smithson Tennant. There were published papers in several journals known to Wollaston that contained variants of purification techniques and powder compaction. Perhaps even more importantly, he had the financial resources through an inheritance and a brother's generosity, the driving ambition to become financially independent, and the personal connections and entrepreneurial acumen to establish a virtual monopoly in platina purchases and products. Furthermore he was active when a happy confluence of politics, technology, and economics allowed his discoveries to gain traction and flourish. His discoveries have synergistic intellectual and social components.

As we will see in the following pages, Wollaston was highly intelligent, endlessly curious, minutely observant, and a dogged worker. He was averse to any but the most careful theorizing and impatient with superficial thinking. He was aloof in unfamiliar and unselected company, bluntly intolerant of pretense and pomposity, but compassionate and engaging to family and friends, male and female, young and old. He lived in tumultuous times at the intellectual center of one of the world's great cities. He made significant contributions to a wide range of scientific specialties and was consulted by natural philosophers, businessmen, entrepreneurs, medical men, politicians, churchmen, and military leaders. Nonetheless, he has been overlooked by historians, in large part because much of the primary source material disappeared upon the death of his delinquent biographer. The reappearance of that material has made this biography possible, and, as we shall see, the many interests, accomplishments, and legacies of William Hyde Wollaston in late Georgian England make interesting and instructive reading, even without the dramatic flourishes employed at the outset of this prologue.

The following account, then, is the first comprehensive biography of one of England's greatest scientists, and a man whose legacy was succinctly described by a famous contemporary interpreter of human nature, the Irish novelist Maria Edgeworth:

> Wollaston was in truth consistently great and good, living and dying. Esteemed, beloved, admired, how rare that union of sentiments for one object! Yet I believe it was a union felt towards Wollaston by all who knew him, whom he ever admitted to his regard, who were ever near enough to appreciate his character.[2]

CHAPTER 1

The Making of a Physician 1766–1795

> A numerous family made it necessary for Wollaston's father to prepare each of his sons for acquiring a livelyhood by his own exertions; and on sending his son to Cambridge, had already determined that Wollaston should practice Physic.[1]

In the summer of 1758, Francis Wollaston, a twenty-seven-year-old Anglican priest delivering the Sunday morning sermons at St. Anne's Soho, married nineteen-year-old Althea Hyde. The families of bride and groom were well known to each other as friends and neighbors in Charterhouse Square, London. The marriage was to be a most fecund one, for over the following twenty years Althea would give birth to seventeen children, fifteen of whom survived childhood. In honor of the mother and her family, all were given the second name Hyde, including the seventh child and third son, William Hyde, the subject of this biography.

Soon after the marriage, Francis was appointed to the church living at Dengey in Essex, northeast of London just inland from the North Sea. Because there was no house available there, he and his young wife chose to reside in Richmond, Surrey, on the western outskirts of London and near the summer home of Francis's father.[2] After five years there, Francis and Althea moved in early 1763 to East Dereham, Norfolk, a small town close to the bustling and prosperous East Anglian city of Norwich. There they settled into the newly renovated vicarage with their three young children: Mary (b. 1760), Althea (b. 1760), and Francis John (b. 1762).[3] And more were on the way. Althea, pregnant during the move, gave birth to Charlotte in 1763, Katherine in 1764, George in 1765, William in 1766, and Henrietta in 1767. Nine more births followed in Chislehurst, Kent, where the family relocated in 1769. This Wollaston family was a remarkable one in both quantity and quality. Many of the children were to achieve scientific and business success, but the third son was to make the greatest mark on the world.

WOLLASTON'S EARLY YEARS

William Hyde Wollaston was born at 6:30 AM on August 6, 1766, and baptized two days later.[4] On the reverse of the piece of paper giving William's birth and baptism details is a listing of his childhood illnesses, which reminds us of the perils to life that were common before modern medicines and vaccination programs. The entries are:

• Inoculated	Sept. 27	1766
• Worms		1768
• Water-pox	Jul	1769
• Measles	Oct	1772
• Whooping cough	May	1774
• Scarlet fever	Dec	1777
• Typhus	March	1791

The first entry is notable, for it reveals that William was inoculated for smallpox when about six weeks old. Inoculation had been introduced to England in the 1720s and involved the introduction of a minuscule amount of smallpox serum taken from a pustule of an infected person into an incision in the skin of the person being inoculated. The process usually caused only a mild form of the disease and generally bestowed lifelong immunity on the recipient. It had become fairly common among upper-class families by mid-century, but it was still viewed by many as a risky procedure. Such inoculation with active smallpox serum did occasionally lead to serious disease, unfortunately, and the procedure was ultimately made illegal in England in 1840. By that time, it had been replaced by Jenner's much safer technique of vaccination with nearly equally effective but less virulent cowpox serum.

There is good reason to believe that inoculation was recommended to the Wollaston family by the eminent English physician William Heberden, who had married Francis's elder sister Mary in 1760. In 1755 Heberden had signed a resolution by the Royal College of Physicians, of which he was a leading member, in support of smallpox inoculation.[5] Moreover, at the request of Benjamin Franklin (then in England representing the Pennsylvania legislature), he had written a pamphlet with instructions for the procedure that also encouraged parents to inoculate their children.[6] On the assumption that all Wollaston children were inoculated for smallpox, it is less surprising that so many of them survived to adulthood, although they still had to survive the gauntlet of childhood diseases endured by William—water pox (elsewhere known as chicken pox), measles, whooping cough, and scarlet fever, among others. All had the potential for serious illness and death, especially among the poor and undernourished, but fatalities were rare in better-off families like the Wollastons. Typhus, which William contracted in 1791 while attending patients at London hospitals, was more serious. But again he survived, giving evidence of a reasonably strong constitution and casting doubt on the claim of a contemporary that William's "constitution was naturally feeble."[7]

In 1769, when William was three years old, the family moved to Chislehurst in Kent, a small town southeast of London, when Francis became rector of St. Nicholas Church.[8] There Francis, near again to his aging parents still residing in Charterhouse Square, London, found contentment. He continued ministering to the parish until his death in 1815. At Chislehurst, the remaining nine Wollaston children were born: Anna (1769), Frederick (1770), Louisa (1771, died in 1772), Charles (1772), Henry (1774, died the same year), Amelia (1775), Henry Septimus (1776), Sophia (1777), and Louisa Decima (1778). A few years later the seventeen surviving family members had their images captured for a group silhouette, which remains in the possession of their descendants (Figure 1.1).[9] Father Francis is on the

Fig. 1.1. Silhouettes of the Family of Francis and Althea Wollaston, ca. 1783.

upper-left playing chess with William, and mother Althea is shown serving them tea. It is fitting that William is seated at the chess table, for he was an avid player throughout his life. His surviving notebooks and letters contain several references to chess invitations, games played, and his habit of playing the game by memory while traveling.

Francis took seriously the role of educating his family, and he makes several references to their tutelage in his memoir *Secret History of a Private Man*. This focus on home education runs strongly through the distinguished Wollaston lineage, beginning with an earlier William.

THE WOLLASTON LINEAGE

The ancestry of the Wollaston line that runs through William Hyde begins with forebears living near the town of Wollaston in Staffordshire in the fifteenth century, but the first ancestor of import is his great-grandfather, also William, known for his moral philosophy. This ancestral William was born into a family of moderate means in 1659.[10] With financial support from a cousin of his father, he was admitted to Sidney Sussex College, Cambridge, in 1674 and obtained his BA and MA. After leaving university, he took up a post at a school near Birmingham and was ordained as a priest. In 1688, after receiving a substantial inheritance from the same relative who had supported his university education, great-grandfather William moved to London. He and his wife moved into a house in Charterhouse Square and had eleven children, nine of whom survived childhood. William was a reclusive scholar who read widely and wrote extensively on philology, ecclesiastical topics, and morality. His one major work, which established his reputation as a free thinker, was *The Religion of Nature Delineated*. First printed privately in 1724, it is said to have sold over 10,000 copies through several editions, a huge selling in an age when literacy in England was not high. The seventh edition published in 1750 was a favorite of Queen Caroline, who had a bust of the author placed in the royal garden at Richmond. The book strove to equate the moralities of good and evil with what human reason understood to be naturally right or wrong, without recourse to divine revelation. William Wollaston's intellectual morality appealed to many in the eighteenth century, especially those who saw the Creator's handiwork in nature's beauty and complexity. His views were certainly judged as unorthodox by some, and his proclivity for the primacy of rational thinking, even when in conflict with religious tenets, re-emerged from time to time in his descendants. His third son Francis, grandfather of William Hyde and his baptism sponsor, acquired the Charterhouse home and was father to three sons, all to become Fellows of the Royal Society, and a daughter who married one.

Grandfather Francis, born in 1694, was educated at home by his father and then proceeded to a degree at Sidney Sussex College, Cambridge, the preferred college of the Wollaston males.[11] In 1723 he was elected a Fellow of the Royal Society of London, the first of several Wollastons to become a member. The Royal Society, founded in 1660, was a locus for England's leading natural philosophers and for those of eminence who wished to be associated with them. Scientific attainments were not to become an essential requirement for admission until well into the nineteenth century, and grandfather Francis's admission was based more on his interest in scientific knowledge than on his own contributions to its acquisition.

In 1728, Francis married Mary Fauquier, the same year that his older brother William married her sister Elizabeth. The Fauquier women were daughters of John Francis Fauquier, a wealthy Huguenot immigrant who had become deputy to Isaac Newton at the Royal Mint and a director of the Bank of England. The union between the Wollaston and Fauquier families enhanced both the financial base and the scientific gene pool of Francis's and Mary's children. Financial acumen thereafter united with natural philosophy and religion as cornerstones of their intellectual development.

Francis and Mary had four children, all of whom had some impact on William Hyde. The first child, Mary, married William Heberden in 1760, six years after the death of his first wife. Heberden, elected to a fellowship of St. John's College in 1731, delivered a popular course of lectures entitled "Introduction to the Study of Physic" at Cambridge from 1740 to 1748. He then relocated to London, where he established himself as a popular and influential physician, and became a member of both the Royal College of Physicians and the Royal Society.[12] From 1748 to 1769, he ran his practice from a house in Cecil Street, London, the same street where William Hyde was to begin his London career three decades later. In the years following his marriage, Heberden published several papers on medical topics in the *Medical Transactions* of the Royal College of Physicians, a journal he had been instrumental in founding. In the last years of his life he completed a collection of medical case histories, entitled *Commentaries on the History and Cure of Diseases*. These were published a year after his death in 1801, and cemented his reputation as one of the leading clinical physicians of the eighteenth century.

The second child of Francis and Mary was Charlton, who graduated from Sidney Sussex College, Cambridge, with an MD in 1758, having become a Fellow of the Royal Society in 1756.[13] He began his practice in Bury

St. Edmunds, where William Hyde later established his own medical career. Charlton moved to London in 1762 when he accepted a position as a physician to Guy's Hospital. In 1763, he became physician to the Queen's household but died a year later from an infection initiated by the dissection of a mummy. The couple's third child was Francis, to whose career I shall return below. The fourth was George, who graduated from Sidney Sussex College with a BA, MA, and, lastly, Doctor of Divinity in 1774. He was for a time mathematical lecturer at Sidney Sussex and helped edit an edition of excerpts from Newton's *Principia*. He was elected a Fellow of the Royal Society in 1763. He was an orthodox Anglican priest who was presented to Dengie in Essex in 1762 when his brother Francis vacated the position for one at East Dereham. He later relocated to a church in London and moved to Richmond, Surrey, where he died in 1826.

Francis, father to William Hyde, was born in 1737. In his autobiographical *Secret History of a Private Man*, he acknowledged his early home education under the tutelage of his father, with its "particular bias towards religion, and philosophy, and scientific pursuits."[14] In 1748, he and brother Charlton went together to Sydney Sussex College, where Francis embarked on the study of law. He graduated LLB in 1754 after having been admitted in 1750 to Lincoln's Inn, one of London's four Inns of Court for barristers. But the practice of law was "a Profession ill suited to his disposition, . . . [because] the idea of [holding] himself ready, to defend either side of any question, as clients should happen to retain him, he could not digest."[15] Here we catch a glimpse in Francis of that intellectual independence, bordering on obstinacy, so characteristic of his lineage. After some soul-searching he decided to leave the law and take Holy Orders, in the optimistic belief that in church matters "he thought himself sure, of never being [required] to defend a position, which he did not fully in his heart, and from conviction, judge to be the truth."[16] He was ordained priest in 1755, but remained at Cambridge for another year to assist his younger brother George during his first year there. In 1756 Francis returned to London to preach at St. Anne's Soho and, two years later, to marry Althea Hyde. Francis's career invites more scrutiny, for its evolution is important for understanding William Hyde's life trajectory.

Francis entered the church as an orthodox believer, with the intent "to embrace Truth wherever he could find it, and to follow whithersoever it should lead him."[17] This rational bent bore great similarity to his grandfather's as expressed in *The Religion of Nature Delineated*, and soon led him to question some tenets of his faith. Especially troublesome was the Athanasian Creed, which enshrined the concept of God as a tripartite entity, the

Trinity: the Father, the Son and the Holy Spirit, and condemned all who did not accept the doctrine to eternal damnation. Although not critical of the Trinity itself, Francis came to believe that no true Christian could accept the damnatory clause, and he refused to read the offending clause to his congregation. So much for his earlier hope that religion would provide a safe haven from conflicting interpretation. Moreover, he opposed traditional thinking on the faith requirements for admission to the two major English universities. He wrote and circulated a pamphlet in support of a 1772 parliamentary initiative to replace the regulation requiring Oxford and Cambridge students to subscribe to the Church of England's Thirty-Nine Articles with one that required only a pledge of faith in the scriptures. Passage of the bill would have allowed those in the dissenting faiths, who could not subscribe to the mandatory Articles without abandoning their own beliefs, to obtain degrees from England's two universities. The bill, nonetheless, was defeated by a large majority in the Commons. Undeterred, Francis in 1773 sought to have parliament pass a simpler law exempting dissenters from the subscription requirement, and he circulated a pamphlet to all members of parliament in support of his proposal. This initiative also came to naught. Finally in 1774 he published another pamphlet, which contained a renewed attack on the Athanasian Creed. Not surprisingly, his dissident views were stridently criticized by more orthodox Churchmen, causing him to withdraw in frustration from religious and political activism. Writing in the third person, he explained his subsequent transition to astronomical studies:

> In Astronomy he trusted, that he should be at a distance from any of the jealousies, any misrepresentations of narrow-minded bigots. . . . There he could allow his thoughts to range, without fear of giving offence. He could look up to the heavens, and adore his Maker, and admire His works, without presuming to pry in to His Essence.[18]

Francis's battles, both with fellow clergymen and his own conscience, occurred during the youthful years of his children, and it would be very unlikely indeed if they were not fully aware of their impact upon their father.

Francis, elected to the Royal Society in 1769, became an accomplished astronomer. He published several scientific papers on instruments and observations in the *Philosophical Transactions* between 1769 and 1793. From an observatory near his home in Chislehurst, Francis observed the surface of Jupiter, and published details of its belts and great spot in 1772. Several years later, in 1789, he published a general astronomical catalogue of stars to aid the search for small stellar motions. The catalogue was much used by astronomers including William Herschel, the discoverer of Uranus and a frequent correspondent of Francis. In 1800, Francis published a catalogue of circumpolar stars and in 1811 a set of ten plates showing naked-eye views of the heavens. All this science was done while tending to the needs of his parishioners and his large family.

In the closing paragraph of the *Secret History*, Francis sums up his experiences and expresses his hopes for his offspring.

> His *Ambition* has been; to render himself as useful in the World, . . . and his *endeavours* have been; to educate a very large family in those sentiments . . . ; his hope and his aim of late years have been, to get the several branches of his family rewarded; for that deep sense of religion, that steadfast loyalty, and that indefatigable attention to their respective occupations, which he feels the satisfaction of having instilled into them with success; wherby they *may deserve* the notice of the Public; and, whether noticed or not, they certainly *will secure* the BLESSING OF GOD, and their own Comfort.[19]

We will see that William Hyde had a very close relationship with his father, frequently sought his counsel, and worked hard to embody the characteristics so valued and encouraged in the *Secret History.*

THE ROUTE TO A MEDICAL DEGREE

At five years of age, William began his formal education on Lady Day (March 25) 1772 at Lewisham, then a small town about midway between Chislehurst and London.[20] There he attended Colfe's grammar school, whose founder had given the minister of Chislehurst the privilege of sending his sons, one at a time, to the school.[21] William remained there for only two years, before moving in June 1774 to Charterhouse school in London.

Charterhouse was a public grammar school endowed, together with a chapel and hospital, by Thomas Sutton in 1611.[22] It occupied about eleven acres near the center of the city, including much open space. Three acres, known grandiloquently as the Wilderness, were laid out in grass and gravel walks, while another square of three acres, the Green, was available for recreational activities such as cricket. The school accepted three classes of students: scholars, boarders, and day students. The forty scholars, admitted when between ten and fourteen years of age, were fully funded by the foundation.[23] Continued funding support, known as exhibitions, for study at Oxford or Cambridge was given to those on the foundation who successfully completed their course of education and demonstrated their abilities in December examinations. The funds provided by the exhibitions were substantial—£80 per annum for each of the first four years and a further £100 for each of another four years if the recipient remained at university. The other two classes of students were not so favored. The boarders lived in residences at the school and paid a fee of about £80 per year. The day students lived away from the school and paid an annual fee under £20.

The Wollaston family had close connections to Charterhouse, and all of their six boys went there. The eldest, Francis John, was already there when George and William arrived together in 1774, and the three younger brothers attended a few years later. William first began at age eight as a day scholar and is listed in the school Register as a Berdmore student. This refers to his being under the supervision of Samuel Berdmore, the schoolmaster at the time who provided boarding for several students.[24] In June 1778, at twelve years of age, William was admitted to the foundation of the school, and became one of the resident "Gown Boys." William continued at the school until May 1783, when he graduated as an exhibitioner.[25] There is not much specific information available on the details of his education other than the Charterhouse declaration that students received instruction in classical learning, writing, and arithmetic, including Latin and Greek grammar and literature. Nor is there anything to indicate that William's academic performance was exceptional. He may, in fact, have been overshadowed by his older brother Francis John who became an "orator" of the school in his final year and preceded William as an exhibitioner.

After their Charterhouse days, the six Wollaston boys followed different career paths. Francis John, destined as the eldest to inherit the bulk of the father's estate, followed family tradition by moving on to Sidney Sussex College, Cambridge, to become a churchman. George left Charterhouse in 1778 to attend the "Academy" of Clapham Common and became a merchant and banker in Genoa before returning to England in 1796. Two younger brothers, Henry and Frederick, pursued business vocations after Charterhouse. The third, Charles, entered Sidney Sussex College as an exhibitioner and he, too, sought ordination as a priest. William, perhaps under the added influence of his famous uncle William Heberden and alone among his siblings, decided to become a physician.

Thus, in 1783 and seventeen years of age, William set out to prepare for a career in medicine. He had the advantages of support from a well-connected family, an intellectual inheritance steeped in duty, religion, and natural philosophy—all abetted by a sound classical education. Such benefits of birth and family undoubtedly generated high expectations for the young man, both from within and without, and it would have been with a mixture of excitement and trepidation that William headed north to Cambridge.

In the last quarter of the eighteenth century, the University of Cambridge was a collection of sixteen small, largely independent colleges situated within a curve of the river Cam in the town of the same name. The then-small university served primarily as the ecclesiastical training school for Anglican clergymen, and all graduates had to subscribe to the articles of faith of the Church of England.[26] Even so, attendance at the university brought a student to an intellectual center of English learning, and a degree from Cambridge (or Oxford) opened the doors to a number of career options other than those in the church. Most good public and private schools required university degrees for their senior personnel, as did many professional societies. One was the Royal College of Physicians in London, which restricted its fellowships to graduates of Cambridge and Oxford. Alternatively, top students could hope for a coveted fellowship at one of the constituent colleges, which entitled them to a yearly income for little mandated university labor, as long as they remained unmarried. Thus students who showed academic promise in their primary schooling, who could get the needed financial support, and who wished to gain the credentials for a learned profession other than the clergy, made up a good proportion of the university registrants.

When at Cambridge, students enrolled in one of the constituent colleges, each of which had its own intellectual and social character. Because of his desire to become a physician, William Hyde registered at Gonville and Caius College, the favored college of many medical aspirants. From the time of its founding Caius (the common shortened name for the college) drew most of its students from the nearby counties of East Anglia, Norfolk, and Suffolk, perhaps because one of its founders, John Caius, was a Norwich native. One estimate gives the fraction of East Anglian students at Caius in the late eighteenth century at seven-eighths.[27] Scholarships and fellowships were predominantly awarded to students from this favored region, and, because the total number of Caius students was not large, chances for success were high. Three of the twenty foundation fellowships at Caius were explicitly reserved for men from Norfolk, others were preferentially given to Norfolk and Suffolk residents, and two or three were held for physicians.[28] In addition, while most colleges required their fellows to take Holy Orders within a few years after election to a fellowship, the majority of Caius fellowships could be held by laymen. It is unlikely that a teenaged William would have foreseen these opportunities, but his well-connected father would certainly have weighed the possibilities, anticipated a positive outcome, and guided his son there. The break with family tradition ultimately turned out as planned, for Wollaston did indeed win financial support as a Caius scholar during his undergraduate years. Moreover, after graduation he was immediately elected to a fellowship, which he continued to hold until his death in 1828.[29] The Caius scholarship money, together with the funding available to him as a Charterhouse exhibitioner, would have allowed William to obtain his degree with little additional financial support from his father.

The Caius College records indicate that William was first admitted as a pensioner, a fee-paying student, on July 6, 1782, prior to his last year at Charterhouse. His registration status was changed to that of a Tancred scholar beginning with the Michaelmas term in October 1782. There were four such scholarships available for medical students at Caius College, but competition for them was not severe. William was the only medical student of the six Caius registrants beginning in his year and possibly the only one eligible for that specific scholarship. He continued to hold the scholarship until he obtained his Bachelor of Medicine degree in early 1788.[30] Although registered at Caius for the academic year 1782–1783, William spent much of that period completing his preparatory education at Charterhouse. It appears as if he moved between the schools when circumstances allowed, for steward's records at Caius indicate that William resided there for a few days each term, perhaps during Charterhouse vacations.[31] He finally became a full-time Caius student in October 1783, at seventeen years of age.[32] Once there, he would have benefitted from the guidance of older brother Francis John, who had been on scholarship at Sidney Sussex and graduated BA in 1783, achieving honors as the top graduating student in the university, known as the senior wrangler. Francis John later became Jacksonian Professor of Natural Philosophy at Cambridge in 1792, and soon began lecturing on both natural philosophy and chemistry. It would have been natural for the younger Wollaston to wonder if he would be able to match the achievements of his older brother.

University statutes specified few formal requirements for a Bachelor of Medicine (MB) degree. Surprisingly to modern eyes, there was no structured medical curriculum. A candidate at Wollaston's time had only to "keep his name on the books of a college [pay fees] for five years, to reside for nine terms, to witness two dissections, and keep one act [an oral defense of a medical proposition]."[33] The students were encouraged, but not required, to attend relevant lectures. Consequently they were free to acquire knowledge of any type, in any way they wished, and to whatever degree of expertise they were inclined. An MB from Cambridge in the late eighteenth century clearly was no signal that its holder knew much about medicine. Those who wished to obtain the advanced MD degree required of a licensed physician generally obtained the bulk of their specialized education elsewhere, as Wollaston was later to do. But it is unlikely that William, given his upbringing and intellect, wasted his time at Caius. The little reliable information available about his university studies comes mainly from Warburton's draft notes on the topic.

> From this danger [of intellectual idleness], however, [Wollaston] was saved by lessons impressed upon him by his parents, . . . and by the circumstance that his eldest brother was then a resident at Cambridge, . . . William Wollaston directly on his arrival at Cambridge, applied himself to the study of the Principia of Newton, and the mathematical branches of Natural Philosophy.[34]

There is some other evidence that William pursued his mathematical studies with diligence. L. F. Gilbert, who collected much Wollaston material in the early twentieth century, reported that he occasionally crafted mathematical propositions for use by his friends in their required undergraduate disputations, and "was more than once gratified by having his brother [Francis John] as Moderator commend a disputant for the novelty and ingenuity of an Argument which he himself had supplied."[35] Another account by a later, close friend illuminates Wollaston's intellectual curiosity, which manifested itself during his undergraduate years.

> [Wollaston] had been speaking about some mathematical matters, . . . on the *Ninth section of Newton*—the motion of the Apsides. . . . Upon my expressing my surprise at *his* being acquainted with it (as he had graduated in *medicine*); "to tell you the truth," said he "a man of our college was going to keep an act upon it; I suspected that he did not know much upon it, so I determined to ascertain the point, and read up the question."[36]

Warburton elsewhere mentions that Wollaston began to study chemistry in 1786, his third full year at Cambridge, by attending the lectures of Isaac Milner. Milner had been appointed the first Jacksonian Professor of Natural Philosophy in 1782, and he presented alternating yearly courses in mechanics and chemistry, lecturing from 8 until 10 each morning. Unlike many of his contemporaries, he taught his courses well and included many demonstration devices and experiments. Warburton wrote that Milner "possessed in a singular degree the power of engaging the attention of his pupils, by appearing to view with wonder and delight the results of the experiments he exhibited. He was, as Wollaston said, a First-rate Showman."[37]

Warburton also states that, near the end of his undergraduate years, Wollaston became interested in astronomy. He befriended John Pond, then a student at Trinity College, who was also much interested in chemistry at the time. Pond, who was several years later to become England's Astronomer Royal, allowed William to use several of his astronomical instruments. Wollaston soon became skilled in the use of Halley's sextant and became a "most perfect observer,"[38] much like his father.

Another observational and more accessible science also caught Wollaston's interest. Warburton writes, "Botany at that time was much attended to by several of the Fellows of Caius College, but he [Wollaston] soon outstretched them all, by his accuracy of observation, at discerning the number of rare plants."[39] Botany had become fashionable in the middle of the eighteenth century due largely to the novel plant classification system and binomial system of nomenclature popularized by the Swedish naturalist Linnaeus. Another close friend attests to Wollaston's notable powers of observation in all realms of natural philosophy, demonstrated a few years after his Cambridge period.

> [Wollaston exhibited] always the same quickness and keenness of observation; he was fond of *Botany*, and soon knew the habitat of every rarer plant of which in this neighbourhood [Bury St. Edmunds] there are several. Nothing escaped his eye. When we were crossing a heath at a smart trot, I remember his suddenly pulling up, and exclaiming "there's the Linum radiola," a plant well known, but so *minute* that his companion, when alighting from his horse, and looking close to the ground, could scarcely at first descry it.[40]

These examples of Wollaston's intellectual interests give valuable insights into his Cambridge years. There is no mention of special interest in any medical topics, nor regular attendance at medicinal or anatomical lectures. But there is much evidence that William sought out and acquired advanced knowledge in many fields of natural philosophy, including astronomy, chemistry, botany, and Newtonian mathematics and mechanics. It is hard not to conclude that science held much greater appeal for him than did medicine. And this proclivity was strengthened by his emerging friendship with a man who was later to become his business partner, Smithson Tennant.

THE ECCENTRIC SMITHSON TENNANT

Smithson Tennant was born in 1761 in Selby, Yorkshire, the only child of Calvert and Mary Tennant.[41] After the death of his father in 1772 and mother in 1781, Smithson inherited the family estates and became financially independent. He then went to the University of Edinburgh, where he attended the chemistry lectures of the famed Joseph Black and conducted chemistry experiments with him. Upon becoming of age to receive his inheritance in 1782, Tennant transferred to Cambridge and became a fellow commoner (a socially privileged student who paid higher fees) of Christ's College, with the intention of proceeding to a medical degree. But his primary interest was chemistry, a science then in the throes of experimental and theoretical turmoil. He toured the Low Countries in the summer of 1783 and one year later traveled for several months in Sweden, Denmark, and Germany.[42] In Sweden Tennant met two of that country's best chemists, Carl Wilhelm Scheele and Johan Gahn. Gahn taught Tennant the techniques of small-scale mineral analysis by use of a simple device known as a chemical blowpipe. In Germany, Tennant attempted unsuccessfully to purify platina with Lorenz Crell. Following his return to England he was elected a Fellow of the Royal Society in January 1785. Later that year he was off to the continent once again, this time visiting France and Switzerland and making contact with chemists there.

In France he met Claude Louis Berthollet and Guyton de Morveau. From them and others, it is likely he would have learned more of the multifaceted revolution in chemical thinking then being formulated by Antoine Lavoisier (there is no evidence the two met), considered by many to be the key figure in the development of modern chemistry. Lavoisier's reformulation of chemical theory and nomenclature appeared in near final form in 1789 in his *Traité élémentaire de chimie* (an English translation entitled *Elements of Chemistry* was printed in 1790). It is impossible to know how many of, and to what extent, Lavoisier's early ideas were assimilated by Tennant in the mid 1780s but one of his contemporaries states that, while still at Cambridge, Tennant had "entirely satisfied himself as to the truth of this doctrine [Lavoisier's oxygen, or antiphlogistic, theory of combustion]."[43] This would have placed Tennant among a small group of British philosophers, most of whom were in the more open-minded Scottish universities, who were quick to accept the new French theory of combustion.[44] It seems plausible to conclude that Tennant, while still an undergraduate, was one of the best-informed chemists at Cambridge, and certainly one with a very broad theoretical view of the subject, including an appreciation for the novel concepts being promoted in Paris.

Warburton's preliminary drafts of Wollaston's life portray Tennant as a person of great intellect, wide-ranging interests, and an engaging personality, an assessment echoed by many others who were associated with him. But there was another aspect of his personality which had the potential to both charm and frustrate his acquaintances. Tennant's friend, John Whishaw, provides some details.

> Although [at Cambridge Tennant was] incessantly employed, there was a singular air of carelessness and indifference in his habits and mode of life; and his manners, appearance, and conversation, were the most remote from those of a professed student. His College rooms exhibited a strange disorderly appearance of books, papers, and implements of chemistry, piled up in heaps, or thrown in confusion together. He had no fixed hours or established habits of private study; but his time seemed to be at the disposal of his friends; and he was always ready either for books or philosophical experiments, or for the pleasures of literary society, as inclination or accident might determine.[45]

As such observations reveal, Wollaston and Tennant could not have been more different in background, experience, personality, and mode of conduct. One was from a large and influential family, the other was an only and orphaned child. One was continually mentored by relatives and shepherded through traditional educational opportunities, the other left to his own devices from an early age. One's world encompassed only Cambridge and London, the other was from Northern England and a continental traveler. And one was quiet, reserved and disciplined, while the other was none of these. But a shared interest in natural philosophy, especially chemistry, drew them together, and their lives remained closely interconnected until Tennant's death in 1815. Warburton provides the only known insight into their Cambridge association.

> A meeting at the rooms of a common friend was the commencement of the acquaintance of these two men. . . . Wollaston held Tennant's knowledge as a Chemist in profound admiration, and a year or two after this period, expressed his despair of ever becoming Tennant's

> equal . . . Wollaston cultivated Tennant's acquaintance; but disparity of age, which tells at this period of life, . . . was too great to admit of the closest intimacy. . . .
>
> With desire of emulating his distinguished friend he [Wollaston] applied sedulously to chemistry, not making many experiments in his own rooms, but availing himself of a Laboratory which his brother Francis had fitted up. Platina even at that time engaged his attention; he made persevering attempts to fuse it in a Blacksmith's forge, aided by Dr. Pemberton, then of the same College.[46]

There can be little doubt that Tennant played a key role in sparking interest in chemistry in the younger Wollaston and had a great influence on his future career choices, including the fateful decision to focus on platinum refining. But Wollaston's fascination with chemistry could not last for long, and as he neared the end of his undergraduate years, he began to apply himself more assiduously to the career his father had planned for him.

In 1786 he began to attend Addenbrooke's Hospital in Cambridge, and the following year he sat in on the anatomy and physiology lectures of Busick Harwood, the professor of anatomy.[47] In that, the final year of his undergraduate tenure, Wollaston was chosen to give the Thruston speech in July and the Tancred speech in October.[48] The Thruston lecturer was chosen by the Master and Fellows of Caius from the medical graduates, and the speech, in Latin, was to be on a medical subject. The Tancred speech, also in Latin, was to be given by one of the Tancred students on the progress of medicine since the time of Dr. Caius. Finally on October 31, 1787, Caius College passed a grace (statement of approval) to Wollaston for the degree of Bachelor in Physic and he was officially created MB by the University of Cambridge in early 1788.[49] Shortly after, he was elected and sworn in as a Senior Fellow on the Caius Foundation, at the young age of twenty-two.

Foundation Fellows received a yearly stipend from their college, which varied from year to year depending on the balance sheet of the college's endowment funds, and they were generally expected to reside in their college to devote themselves to study and to participate in the education of undergraduates. However, exceptions to the residency regulations were normally granted in the eighteenth century to those Fellows whose careers required them to be absent.[50] Such was certainly the case for Wollaston, who wished to proceed to an MD degree, and needed to acquire the requisite medical training elsewhere. He decided to attend the hospitals in London, but before moving there later in 1788, he remained at Caius and attended the regular meetings of the College Seniority from January to August.[51]

BECOMING A DOCTOR

Regulations for a Cambridge MD degree were even less demanding than those for the preliminary undergraduate degree. The principal requirement was to pay registration fees to a college for five years after the MB, but residency was unnecessary. The academic requisites were simply attendance at two dissections, plus the keeping of two acts (oral defenses of propositions) and one opponency (successfully-argued objection to a proposition).[52] These requirements would not have been a barrier to any aspiring physician, but the responsibility of providing care to one's patients meant that most doctoral students spent their five years conscientiously studying medicine. So, late in 1788, Wollaston moved into his grandparents' home in Charterhouse Square, where he began to assemble a more complete medical library and to gain clinical information by visiting hospitals.[53]

Notebook evidence suggests that Wollaston was directed to the relevant medical literature by his uncle William Heberden. In support of his medical lectures at St. John's College, Heberden in 1741 completed a manuscript entitled *An Introduction to the Study of Physic* for transcription by interested students. There is a copy in the Wollaston collection dated 1786.[54] Although the copy is not in Wollaston's handwriting, it does contain a few marginal comments in his hand, so it is certain that he read and studied the work. The manuscript gives directions to Cambridge students on how to become "Masters of their Profession."[55] The bulk of the *Introduction* concerns the sequence, and duration of study, of topics to be mastered by the aspiring physician. Two subjects, botany and chemistry, were recommended by Heberden for study at the outset of a physician's education, but only to the extent to which they might aid patient care. Study of these two disciplines, plus *materia medica*, pharmacy, anatomy, and physiology, was to be completed in the first four years of medical studies. The fifth year "must be employed among the practical authors, the writers of Surgery & in attending an hospital."[56] The manuscript ends with a recommended list of books on each of the topics covered in the *Introduction*. Nearly all of the 114 Latin, Greek, French and English titles have check marks beside them, indicating that Wollaston had acquired them for his library. At the end of Wollaston's copy is a more complete list of books in his possession, with the purchase price of most of them.

Once settled in London Wollaston began making hospital rounds in the company of Christopher Pemberton, a friend and Caius medical graduate of 1789. An entry in his daybook links the names of Abernethy and Dr. Pitcairn with mention of attendance at hospitals.[57] Both had posts at

London's oldest hospital, St. Bartholomew's, which was situated very near Charterhouse Square. David Pitcairn was physician, and John Abernethy assistant surgeon, to the hospital. In contrast to the feeble medical training provided by Cambridge, clinical instruction available to fee-paying students in both private and hospital settings in London was perhaps the best in the world.[58] By paying fees to Pitcairn and Abernethy, Wollaston could accompany them during their rounds of the hospital wards, visit patients independently, observe operations, witness dissections, and even carry out an occasional autopsy. It is also possible that Wollaston attended some of the private lectures offered by London practitioners, such as Abernethy's anatomy lectures given in a house near St. Bartholomew's. The Wollaston manuscripts at Cambridge include copies of George Fordyce's lecture notes on *Materia Medica* as well as *On Chronic and Acute Diseases* and William Osborne and John Clarke's *On Midwifery*, all with addenda that appear to be in Wollaston's handwriting. One entry in his daybook, for example, mentions his diligence in studying Fordyce's *Practice of Physic*.[59] Overall it is probable that Wollaston gave his full attention to furthering his medical education in the manner recommended by Heberden during the London years prior to his MD degree. Although details of his intellectual development are lacking, we do know one consequence of Wollaston's contact with patients at hospitals. This was his contraction of typhus in 1791.

Wollaston did not set aside all his other interests while he readied himself for medical practice, for his book collection contains several titles on chemistry and a few others on botany, mineralogy, and natural history. He was also observing the stars at night, for Warburton mentions an astronomical study begun by Wollaston about this time, the first of his scientific studies on which we have information.

> In 1791 accompanied by Dr. Brinkley and Mr. Vince, he [Wollaston] visited Dr. [William] Herschel at Slough, on which occasion he exhibited to his companions a method which he had devised for measuring the intensity of different lights by comparing their shadows with that of a candle; a method which subsequently formed the principles of the Photometer described by Count Rumford.[60]

Wollaston worked on his technique and the accompanying celestial observations on and off for over three decades before finally compiling his results in 1828 and publishing them posthumously.

In the summer of 1792, as the five-year mandated period between an MB and an MD degree was drawing to a close, Wollaston left London and moved to Huntington, a small town about fifteen miles northwest of Cambridge. There he began his practice as a country physician. From this point of his life onwards, Wollaston began to record his daily activities in a commonplace book, and to maintain a record of his medical practice. We know, for example, that on August 13 he treated his first patient, a man aged about fifty who was diagnosed with upper respiratory difficulties. His second patient, a fifteen-year-old girl, was diagnosed with goiter two weeks later. In both cases, Wollaston placed a plus sign in a column beside their names indicating that he judged them to have improved under his care.[61] There were patients who Wollaston judged to have gotten no better or no worse (indicted by a zero) after treatment, such as a man with a cough, one with tuberculosis, and another with dropsy. And, inevitably, there were some patients whose health deteriorated (indicated by a minus sign) after treatment. Wollaston was unable to help a woman with typhus, another with suspected stomach cancer, and a third with that scourge of childbirth, puerperal fever. There was very little in the doctor's arsenal in the late eighteenth century that could arrest the more serious medical conditions, including aggressive bacterial (unrecognized at the time) infections. Up to January 11, 1793, when the Huntington record ends, Wollaston had treated only nineteen patients. Of those, nine improved under his care, four deteriorated and the other four showed no change. Unfortunately Wollaston does not record how he treated any of the patients in his list.

The brief Huntington practice was not especially lucrative, for Wollaston collected fees of only £27.16.6 in 1792 (for fifteen patients), which nonetheless meant each patient paid nearly £2 for his services, a fee that would have prevented all but the well-off from seeking the care of a physician. On the debit side, he calculated his expenses as £105.[62] Perhaps to save him money, or maybe just for sibling companionship, his eldest sister Mary came up in October to assist William with housekeeping. That same month, Caius College approved Wollaston's doctorate in physic and on January 29, 1793, Wollaston was created MD.[63] With the coveted degree gained, Wollaston sold his Huntington practice to a successor for the surprisingly high sum of £270. This payment made his short stay in Huntington a profitable one.

The gaining of his MD quickly set other plans in motion. On February 7, a week after his degree was approved, Wollaston's application to become a Fellow of the Royal Society of London was presented to the Society and his certificate of election was placed on display for the required ten weeks. The illustrious supporting names on Wollaston's certificate virtually guaranteed a positive outcome: William Heberden, Henry Cavendish, Francis

Wollaston, George Wollaston, Smithson Tennant, William Herschel, and seven other astronomers, physicians, and antiquaries.[64] The membership vote was successful and, on May 9, William was elected a Fellow of the Royal Society, more for the potential to become a good natural philosopher than on demonstrated talents, for he had yet to publish a scientific paper. Wollaston paid his admission fee shortly after (through an intermediary) plus the usual sum that was accepted in lieu of ongoing annual payments, but he did not attend a Society meeting for formal admittance until he was next in London on March 6, 1794.

Election to the Royal Society obviously went as planned, but an unrelated problem was developing that could have interfered with Wollaston's official acceptance of his medical degree in Cambridge's Senate House in the summer of 1793. Although Wollaston was created MD in January, he had to complete his required two acts before the degree could be conferred. He completed his first act successfully but the proposal for his second, the potential of poisonous substances to act as medicines, met with opposition from Isaac Pennington, the Regius Professor of Physic. After receiving a letter of protest from Wollaston, Pennington acquiesced and Wollaston did complete his second act successfully.[65] Consequently, he received his degree in the Senate House in July 1793.

LIFE AS A COUNTRY PHYSICIAN

In April 1793, Wollaston left Huntington and moved to Bury St. Edmunds, an ancient and thriving market town in Suffolk about twenty-five miles east of Cambridge. A daybook entry reveals that "he was sent to Huntington & Bury because his uncle Dr. Charlton Wollaston had pursued that course."[66] This entry, attributed to "G.W.," most likely William's older brother George, gives further evidence that William's development had been mapped out in advance. In fact, William's entire progression through the educational system, quick election to the Royal Society and establishment as a country physician appears to have followed a script. If so, the script had run its course, and, after establishing himself in Bury St. Edmunds, William needed to make his own way in life.

Wollaston was a great list maker, and this valuable gift to historians allows us to see the care he put into setting up his household. He rented a house in Bury, brought furniture from Huntington, and added kitchen utensils, linens and china sufficient for hosting several guests, including a corkscrew, two wine decanters, and a dozen wine glasses.[67] Larger expenses were for a horse, harness, and stable fittings. A good horse was essential for a country practitioner to visit and treat patients in their homes, often several miles distant, and at all hours of the day. He also hired a young liveryman, and outfitted him with a livery coat and a frock coat, both with accompanying waistcoat, breeches, and boots.

Wollaston treated his first two Bury patients on April 3, 1793, a one-year-old girl with smallpox and a sixty-year-old woman with dropsy.[68] He judged both to have improved after treatment. By the end of the month Wollaston had seen ten patients, and his practice continued to grow thereafter. He continued to keep patient records until late in 1794, recording the name, age, address, diagnosis, and outcome for each. The cases were recorded over a period of 725 days, which means that Wollaston saw a patient, on average, about once every second day. This afforded him much free time, a portion of which was spent compiling lists of letters sent and received, books purchased and read, and events attended. There is record of one incident which suggests that Wollaston, not unusually for a physician, found the emotional demands of treating his patients sometimes overwhelming. Henry Hasted tells of the attendance upon his father.

> Soon after he came to Bury he was called in to attend a relative of the narrator in what was thought a serious case, and asked immediately to give his opinion on it; he replied "You must consider I am a young man, I see nothing to be alarmed at, but you cannot expect me to speak at once decidedly," and he burst into tears. He was right in the opinion which he afterwards gave, as well as in his treatment of the case; but the circumstance shewed even in early days what he suffered when having a patient seriously ill under his care.[69]

Wollaston's internal empathy for his patients was one manifestation of a personality trait that became characteristic: heartfelt sympathy for the misfortunes of others and a personal aversion to causing harm to anyone, even in casual conversation. He was quite reluctant to discuss his own and others' personal matters with others, even family members. He preferred instead to limit his conversations to matters of observed fact and dispassionate interpretation. Not surprisingly, such a guarded persona prompted many to judge him as aloof and reserved, even cold-hearted at times. Emotional detachment is important for a doctor expected to rationally diagnose a patient's malady, but a lack of visible empathy would have been disconcerting to most patients. Perhaps Wollaston's bedside manner hindered the growth of his practice.

Wollaston ceased keeping records of his patients in 1794, but records of income earned suggest his practice continued much the same way for the rest of his stay in Bury. His medical income was never enough to do much more than cover his basic expenses, but his financial standing was helped by the Caius fellowship monies, which ranged between £50 and £80 per annum. Fortunately Wollaston acquired a new source of revenue. On May 22, 1795, his uncle General West Hyde gave him title to seventeen acres of land at Peckham, a short distance southwest of London, which delivered a yearly rent of close to £50.[70] Even though his annual income amounted to between £250–£300 per year (normally more than enough to support a comfortable lifestyle), Wollaston still occasionally fell short of balancing his budget. When he could not, his father appeared willing to make up the difference, for one entry in his daybook dated May 1795 mentions receipt of a paternal allowance of £100.[71]

In 1794, several months after the move to Bury and a year after becoming FRS, Wollaston applied for membership in the Royal College of Physicians of London. In March and April, he appeared before the board of censors, who examined him in each of physiology, pathology, and therapeutics.[72] He passed all three exams, and on April 14 he was elected as a Candidate of the Society, with his uncle William Heberden in attendance. The following year, he returned to London once again and on March 30, 1795, was elected a Fellow of the Society.[73]

Membership in both prestigious London societies, the Royal College of Physicians and the Royal Society, placed Wollaston firmly among the intellectual and social elite of Georgian England. He was ideally positioned to ascend ever higher in the medical world, perhaps even with the potential to become physician to the royal household, as his uncle Charlton had done. Nonetheless, he chose to celebrate his election rather modestly, by having dinner with his brother George and then returned to Bury to continue his practice as a country physician.

CHAPTER 2

Early Medical and Scientific Interests 1792–1800

Indeed it was scarcely possible to be in Dr. Wollaston's company half an hour without learning something; without hearing some new fact, or having some old one put in a new light, almost incidentally, without effort or design.[1]

Life as a country physician left Wollaston many hours of free time between patients, and he used them productively. He spent much time outdoors as an avid naturalist, observing and recording details of soils, crops, waters, plants, and animals, often in the company of Henry Hasted, whose father was an apothecary in the town. The two young men became close friends, and, after Wollaston left Bury, they exchanged letters until Wollaston's death. The surviving letters are a valuable source of information about several aspects of Wollaston's personal life, as they contain candid, often light-hearted, exchanges of ideas and opinions. More importantly, suspecting that Warburton's biography would never see the light of day, Hasted produced his own short life of Wollaston in 1849. In it he described his friend's wide-ranging interests and idiosyncratic methods of information exchange.

Having a similarity of tastes in many things we were very frequently together, either riding, walking, or talking on them; . . . every notable spring, or mineral, or tree, in the neighbourhood was known; experiments made on them in his little study with a few small phials, tests, and watchglasses; the time of leafing and flowering of plants, the notes and scales of birds, the habits of animals, the motion and velocity of the clouds and winds; there were to him "sermons in stones and *food* in everything." And when the day was gone, the stars were looked at with an artificial horizon of quicksilver in a saucer. . . .

> There was a kindness in the manner of communicating [his findings]; but if any great error was asserted, with a single look, or a single question, he would convince the assertor that he was wrong.[2]

Hasted's observations of Wollaston's interests and insights into his character are sound, and are the more compelling because Hasted had considerable intellectual ability himself, having graduated BA in 1793 from Christ's College, Cambridge, as sixth wrangler. In contrast to the more reserved nature of his friend, Hasted displayed "constant cheerfulness, unclouded good humour, and universal benevolence and kindliness. . . . It was always sunshine with him."[3] Hasted was the kind of person (as was Tennant) that Wollaston forged relationships with throughout his life—intelligent, curious, amiable, outgoing, and most importantly, modest and self-effacing. Those who overestimated their own cleverness, or assumed an unwarranted degree of intimacy or familiarity, were rarely able to penetrate Wollaston's reserved and occasionally aloof personality.

EARLY SCIENTIFIC INTERESTS

Numerous entries in Wollaston's notebooks give substance to Hasted's mention of his eclectic interests. The first scientific entry is dated September 6, 1792, a few weeks after he treated his initial patients at Huntington.[4] It gave the temperatures of water in a well and a nearby spring, and analyses of water from wells surrounding Huntington and Bury. Elsewhere, there are miscellaneous notes on veterinary problems and their treatment, oddities such as the amount of blood ingested by leeches, and rates of growth of fingernails, together with a careful listing of the flowering and greening dates of plants and trees, and the first spring sighting of birds, covering the period from 1794 to 1797.

Wollaston's analyses of spring and well waters reveal an attention to detail that would become a hallmark of his later researches. In the spring water he collected from a source near Huntington he was able to detect in the solids left after evaporation selenite (calcium sulfate), aerated lime (calcium carbonate), and common salt (sodium chloride).[5] These and other analyses of solids reveal interests and techniques that were to be refined over many years, and were to underpin several of his later analytical discoveries. Wollaston was primarily interested in the identification of the individual substances that could be found in solid or liquid mixtures and solutions (now known as qualitative analysis) and was less concerned with their exact proportion by weight (quantitative analysis). Like most of his peers, Wollaston characterized individual substances by their known physical and chemical properties, but he excelled in his ability to observe those characteristics on amounts just visible to the naked eye. And, most importantly, he combined his qualitative analyses with crystallographic data, uniting information on the chemical composition of substances with the physical parameters of their very pure crystals. This allowed him to make a positive identification with confidence. Thus, for example, one analysis of the solid residue of evaporated well water gave "common salt which crystallized entirely in cubes without any tendency to deliquesce."[6] The chemical details are without novelty in these observations, but the methodology is noteworthy. Wollaston's correlation of crystallographic details with chemical ones was to become the crucial factor in his later discovery of two new chemical elements. Since substances normally crystallize with a very low degree of contamination, well-formed crystals provide a criterion of purity that is reliable and easily perceived. He was not the only one to recognize the value of distinguishing pure substances by this combination of techniques, but Wollaston was to become a master of those methods carried out on microscale quantities. One striking example of his expertise was the identification, made many years later, of very small amounts of potassium in sea water.

When his good friend Alexander Marcet, a London physician and chemist, carried out an extensive study of sea waters about 1818, Wollaston obtained samples taken far from land to determine if potash (from the presumed decay of land plants) could be detected in them.[7] He evaporated samples of sea water to about one-eighth of their original volume and precipitated a yellow triple muriate of platina and potash, (potassium chloroplatinate, K_2PtCl_6), a salt well known to him from his earlier platinum researches. He then converted the precipitate to well-defined "crystals of nitrate of potash [KNO_3]," giving incontrovertible proof of the presence of potash in sea water. From the weight of crystals isolated, Wollaston could estimate the proportion of potassium in sea water as somewhat less than one part in two thousand (500 parts per million, ppm), not far off the modern estimate of 400 ppm.

At Bury he also became interested in the rings of darkened grass and barren soil commonly found in country pastures known, then and now, as fairy rings. Wollaston began collecting data in October 1792 with measurements on three rings in a Huntington common.[8] After five years of study, he determined that at least five different species of fungi produced fairy rings in the same general manner. Mushroom spawn, he concluded, grew out from a point of origin in pasture lands, consuming essential soil nutriments in the process. Continued growth then proceeded only on the outer circumference of the exhausted soil, often leaving behind a narrow, barren ring of soil where mushroom growth had prevented the growth of grass. The inner circumference of the barren ring was marked generally by a band of rich grass growth, fertilized by the decaying spawn of the outwardly expanding mushroom circle. Wollaston waited several years before publishing what he announced as his "clear and satisfactory" conclusions.[9] The paper is the only one of Wollaston's publications that can be classified as botanical in content, and its conclusions remain valid today.[10]

Entries in Wollaston's notebooks relating to his personal life and social interactions during the Bury years reveal him to be a man with broad interests outside of science and one who sought out, and enjoyed, the company of others. For the entire Bury period, from May 1793 to August 1797, Wollaston kept a daily log of social engagements. He dined most nights with others, either in his own home or as guest in theirs, and regularly attended card assemblies, a variety of balls and plays, and the periodic traveling criminal courts known as the assizes. His brothers, sisters (especially), and parents visited him regularly, and his sisters when visiting individually or in pairs often stayed for several days. Although he was never to marry, he enjoyed the company of women and children, as Hasted recalled.

> His presence was courted by all; even in female society, it was remarked, "we are always glad to have William Wollaston to join our circle, for he always suggests something or other about our work, or what happens to be before us, which we were not aware of before;" and amongst the young, those at least who had any mind, or any desire to learn, he entered into all their views and cheerfulness.[11]

Wollaston's great enjoyment from interaction with children remained with him throughout his life, and was often commented upon by friends and relatives. One example is contained in a letter written by his niece Caroline after the 1867 death of her father Henry Septimus Hyde Wollaston.

> I should rather like to tell of the way his [William's] great mind condescended to the capacity of a child under 10 years of age. Often he devoted an hour every morning to giving me a lesson in Chemistry, when he was staying at Chislehurst. I well remember seeing him melt a tube of glass and form it into a vessel in which to show me the expansion of water when boiling. And some years afterwards when my father had a young family [with a later wife] I have seen him on the floor playing with the children.[12]

It is likely that Wollaston was content in Bury but in early 1797 his uncle West Hyde died and the future took on a different prospect.

William's mother Althea was one of six Hyde children, four girls and two boys. The large Hyde family fortune had become concentrated in the hands of her older brother, West, a colonel of the 20th Foot Guards. Unmarried himself and without heirs, West bequeathed his considerable estate to the second son of Francis and Althea, on the traditional belief that the father had the responsibility of providing for the eldest son. Upon Hyde's death on March 2, 1797, William's elder brother George became sole beneficiary of his lands in England and America, ships, stock holdings, and London house. Although it took two years for the estate to pass into his hands, George subsequently gave William £8000 in 3 percent reduced annuities and a share of Bank of England stock and/or 4 percent annuities to the annual value of £50.[13] Ledger entries in the archives of the Bank of England confirm William's receipt in March 1799 of this substantial gift and reveal more of the destination of some of the estate.[14] George also simultaneously transferred to the joint account of William and his younger brother Henry £8000 in Bank of England stock. This amount was subsequently dispersed from 1801 to 1808 to all of the surviving Wollaston siblings (£800 for each brother and £600 for each sister; excepting only Frederick who left England in 1796 under some cloud of parental disapproval). The uniquely large transfer to William had special significance for his future because, shortly after learning of the bequest in 1797, he decided to relocate to London.

STUDIES ON HUMAN CALCULI

The year 1797 prior to the move was important to Wollaston in another way, for it marked the publication of his first scientific paper, "On Gouty and Urinary Concretions."[15] The pioneering biochemical work described in the paper was carried out at Bury and integrated Wollaston's talents at chemical analysis with a long-standing medical problem–the nature of urinary stones generally called calculi, and gouty deposits.[16] Gout was marked by acute pain in joints and tendons caused by the deposition of a chalky white deposit. Wollaston correctly determined that gouty deposits were composed of a water-insoluble salt of mineral alkali (sodium) and lithic acid (uric acid, $C_5H_4N_4O_3$). He identified lithic acid, so named earlier by Scheele as the principle component of bladder stones, by precipitating it in crystalline form from samples of gouty solid. The straightforward chemical processes used to identify both components of gouty matter exemplify

Wollaston's reliance on crystal structure as a proof of identity for qualitative analysis. He neither sought to weigh the component substances or calculate their specific combining proportions. Instead, by use of common reactions to convert the constituent substances of an unknown compound to known crystals whose composition was undisputed, he provided a readily comprehensible and compelling result, the kind that his colleagues would increasingly come to label as characteristically ingenious, code then and now for "why didn't I think of that?"

The paper then turned to the analysis of four different types of urinary calculi, generally called bladder stones, which we understand today as a variety of minerals that precipitate from urine. The first type was termed "fusible calculus" because Smithson Tennant had found that it melted under the heat of a blowpipe to form an opaque white glassy material. Wollaston found that some perfectly white specimens of the calculus had the appearance of sparkling crystals, which had the form of "a short trilateral prism, having one angle a right angle, and the other two equal, terminated by a pyramid of three or six sides."[17] He knew that such well-defined crystals had to be those of a pure salt, which he found by analysis to contain phosphoric acid, magnesia, and volatile alkali (the crystalline compound is magnesium ammonium phosphate hexahydrate, $MgNH_4PO_4 \cdot 6H_2O$, a type of stone now known to be caused by urinary tract infections). Wollaston determined that the triple salt made up the principal part of the fusible calculus, the remainder being phosphorated lime (likely calcium hydrogen phosphate dihydrate, $CaHPO_4 \cdot 2H_2O$), and some lithic acid, the same substance involved in gout. Wollaston determined that the three ingredients mixed in various proportions in different stones, and that all three were required to make the calculus fusible upon strong heating.

The second type of bladder stones Wollaston investigated were those generally known as "mulberry calculus" because of their irregularly knotted surface and dark color, somewhat similar to the appearance of mulberries. The smoother varieties he found to be a mixture of phosphorated lime and a new salt composed of lime and acid of sugar (calcium oxalate monohydrate, $CaC_2O_4 \cdot H_2O$). As usual he identified the component acid of sugar by isolating it in crystalline form. The rougher-surfaced varieties generally contained some lithic acid, which caused the surface irregularities.

The third category of stone, called the "bone-earth calculus," tended to be deposited in the form of concentric spheres, each composed of crystallized fibers. Wollaston found this type of calculus to be entirely phosphorated lime, which could be fully fused by heating with a blowpipe. He then gave a simple chemical procedure for converting the phosphate in bone to that in bone-earth calculus, and he was able to obtain the latter in crystalline form, once again establishing its existence as a chemical compound of fixed composition.

The fourth type of stone examined by Wollaston was a prostate gland calculus. Wollaston's description of his analytical technique for this type of stone well illustrates his small-scale procedures.

> A small fragment being put into a drop of marine acid [HCl], on a piece of glass over a candle, was soon dissolved; and upon evaporation of the acid, crystallized in needles, making angles of about 60° and 120° with each other. . . .
>
> This crystallization from marine acid is so delicate a test of the neutral phosphorated lime, that I have been enabled by that means to detect the formation of it, although the quantities were very minute.[18]

This stone, like the bone-earth calculus previously described, contains a salt of lime and phosphoric acid (calcium hypophosphate hydrate, $Ca_2P_2O_6{\bullet}2H_2O$), but one that is perfectly neutral from an acid/base standpoint.

Wollaston's acquired expertise in the analysis of fusible calculi and bone was later to find application in a much different context, as so often happens in science. In 1821, the geologist William Buckland investigated a large collection of bones, teeth, and other remains of "pre-diluvial" animals found in a cave in Yorkshire. He sent samples of preserved dung to Wollaston, who found in them "the ingredients that might be expected in faecal matter derived from bones."[19] This finding supported Buckland's conclusion that the dung was from a species of hyena which had in ancient times used the cave as a den.

A year after Wollaston's paper, George Pearson, chief physician at St. George's Hospital in London, published a tedious and much inferior paper on the same topic in the same journal.[20] In a brief introductory section Pearson mentioned the contributions of Scheele and the French chemist Fourcroy, but inexplicably made no reference to Wollaston's results. Another more substantial study of urinary stones was published in 1799 by the French chemists Fourcroy and Vauquelin.[21] By analysis of over 600 calculi, they recognized 12 main types, including those previously identified by Wollaston. But they, too, made no mention of Wollaston, although Pearson's paper was cited.

Wollaston's novel researches on calculi were finally given their proper due two decades later by Alexander Marcet, one of Wollaston's closest

London friends. Much impressed by the small-scale chemical and analytical techniques perfected by Wollaston, Marcet set about to extend the chemical understanding of calculi begun by his mentor. In 1817, he published the results of his long and thorough study of calculi in a book on the subject.[22] The book was dedicated to Wollaston for both personal and professional reasons. The personal was to Wollaston's profound influence on Marcet's investigations, and the professional was to connect his name to the chemical study of calculi, correcting the oversight of Fourcroy on the subject. Marcet was determined to set the record straight.

> It is the more desirable, that his [Wollaston's] claims in this respect should be placed in the clearest point of view, as the late celebrated M. FOURCROY, . . . has, in a most unaccountable manner, entirely overlooked Dr. WOLLASTON's labours, and in describing results exactly similar to those previously obtained and published by the English chemist, has claimed them as his own discoveries.[23]

Marcet was unwilling to excuse Fourcroy on the grounds of ignorance, for Marcet noted that Fourcroy did cite the works of Pearson in his publications and, as we have seen, Pearson's paper followed Wollaston's by a year in the *Philosophical Transactions*. Whether or not Wollaston shared his friend's sense of injustice is unknown, for nowhere does he comment on the issue.

Wollaston's first paper on calculi was based on research carried out in Bury, but he retained interest in the subject for the rest of his life and published a second important paper on the subject in 1810. It described the analysis of yet another new type of stone, quite different from previous types.[24] The stone had been taken years earlier from the young brother of a Norwich doctor, who gave Wollaston a portion for analysis. He found that it shared chemical properties common to animal substances like uric (lithic) acid, but it combined readily with both acids and bases and on heating emitted a distinct odor unlike any other Wollaston had smelled. He was able to obtain a very few crystals in the form of flat hexagonal flakes, but investigation of another stone of the same type (the only other example known to Wollaston) held by Guy's Hospital exhibited crystals on its surface that were nearly cubic. Wollaston was able to confirm that the substance in the stone contained oxygen, and since it had been removed from a person's bladder, he named it cystic oxide. Marcet accepted this name in his book on calculus disorders but the great Swedish chemist Jons Jacob Berzelius wrote to his good friend Marcet in 1818 to state his disapproval of naming the substance an oxide because most animal substances contained oxygen which had no predictable affect on their properties.[25] Shortly afterward Marcet replied to Berzelius stating that he and Wollaston agreed that cystic oxide was not a good name for the substance, but they preferred a bad name to a changed one.[26] Berzelius demurred for a while but in the 1833 edition of his *Treatise on Chemistry,* he introduced the name cystine for the organic substance, a name it retains today. Cystine is an amino acid, the second of the amino acids found in proteins to be discovered (the first, asparagine, was isolated from asparagus juice by Vauquelin and Robiquet in 1806, but was not named until later). Wollaston's discovery has been confirmed in the twentieth century by scientists who used modern techniques to examine the same cystine stone in Guy's Hospital.[27] Other authors date the beginning of studies on inborn errors of metabolism, one of which is responsible for cystine stones, to Wollaston's discovery of the first such stone.[28]

At the end of his cystine paper, Wollaston reports a related study that he believed might be of use in the prevention of gout and uric (lithic) acid–containing calculi. He analyzed the excreta of birds that fed on varying proportions of plant and animal matter, from the dung of geese that feed entirely on grass to the urine of gannets that feed entirely on fish. He found that the amount of uric acid excreted increased in proportion to the amount of meat and fish in the diet. Thus he concluded that "it would appear, that persons subject to calculi, consisting of uric acid, as well as gouty persons, in whom there is always a redundance of the same matter, have much reason to prefer vegetable diet, but that the preference usually given to fish above other kinds of animal food, is probably erroneous."[29] Wollaston's supposition has proven correct, and modern medical treatment for persons prone to uric acid deposits includes decreased consumption of foods rich in purines (compounds necessary for the formation of uric acid), such as meats, meat products, and fish.

This summary of Wollaston's pioneering contributions to a chemical understanding of gouty stones and urinary calculi has extended much beyond his first paper, and we return now to 1797 when, with his paper accepted by the Royal Society for publication, Wollaston began planning his relocation from Bury to London. Most of July was spent traveling around East Anglia, ending up at the family home in Chislehurst. In August, Wollaston embarked on his first extended holiday, traveling to Weymouth, Bath, Bristol, and Oxford, among other places.[30] Such August tours would become a regular event, and allowed Wollaston to explore almost the entire length and breadth of England, Scotland, Ireland, and, after the cessation of war with France, much of western Europe.

On September 20, Wollaston moved into his new house at No. 18, Cecil Street, just off the Strand and very close to the Thames. While waiting for his medical practice to take root, Wollaston read intensively, mostly medical tracts but also books on a wide range of topics, such as geology, chemistry, gunnery, the national debt, foreign countries, music, and the natural history of religion. As well, he had to adjust his eating habits, as the normal London dining time was a couple of hours later than the 3:30 time he had become accustomed to in Bury.[31]

ACTIVITIES WITH THE ROYAL COLLEGE OF PHYSICIANS

As a London resident, Wollaston was expected to attend all meetings of the Royal College of Physicians, which he did after attending his first meeting on September 30, 1797.[32] In 1798, he was elected for a one-year term as one of the College's four censors. One of the censors' major tasks was the examination, in Latin, of all candidates for election as Fellows (medical graduates of Oxford or Cambridge) and Licentiates (graduates from other institutions, primarily the Scottish universities) in the three categories of physiology, pathology, and therapeutics. One of the first men to be examined by Wollaston and his fellow censors was Alexander Marcet, who passed the exams and was admitted as a Licentiate in June 1799.

Marcet had obtained his MD in 1797 from the University of Edinburgh and moved on to become a physician at Guy's Hospital from 1804 to 1819.[33] In 1799, he married the daughter of a wealthy Swiss merchant, Jane Halliмand, who was later to write a popular series of instructional books on chemistry, botany, economics, and religion.[34] Marcet and Wollaston shared many scientific interests in general and animal chemistry, as their work on calculi exemplifies. They also became close friends and dining companions and, in later years, enjoyed many sporting excursions together.

A second duty of the censors of the College of Physicians was the annual visitation to, and assessment of, London's apothecary shops. Wollaston and his three fellow censors visited seventy-seven shops over three days in July and August 1799. In each they assessed the quality of ten to fifteen substances and the overall appearance of the shop. Poor shops vastly outnumbered the good ones. The censors prepared a report on the results of their visitations and the College decided to print five hundred copies to be sent to Fellows of the College and the Company of Apothecaries. I have not been able to locate a record of this report, but a censors' report of 1802 was read into the minutes of a College meeting, and it mentions the precedence of the 1799 report in lamenting the sorry conditions of many apothecary shops and the lack of power to force improvement.[35] But, like its 1792 predecessor and 1802 successor, the report of Wollaston and fellow censors failed to initiate any meaningful action.[36] Professional tensions between physicians and apothecaries were finally resolved by the passage of the Apothecaries' Act of 1815, which granted the Society of Apothecaries rights to license and regulate its own members.

At the same meeting of the College of Physicians in October 1798, at which he was elected as a censor, Wollaston was also chosen to be one of the five Madhouse Commissioners for the following year. The Commissioners were delegated to visit madhouses, collect license fees, and take notes on the condition of patients.[37] On six days over the period November 1798 to June 1799, the Commissioners visited a total of thirty-four houses within a seven-mile radius of London and Westminster.[38] In the following year, Wollaston was chosen for a second and last term, and he and his fellow Commissioners visited forty houses over four days from October 1799 to May 1800. The Commissioners of the College of Physicians, however, had no power to effect changes in madhouse care, other than refusing a license to keepers who refused them entry. There was no other penalty that could be applied to houses in which ill treatment or neglect of patients was observed, so the visitations were little more than a formality. His two years as a Madhouse Commissioner marked the end of Wollaston's active role in the Royal College of Physicians, except for his advisory role related to one of the eighteenth century's most revolutionary medical discoveries.

The country surgeon Edward Jenner published a book in 1798 announcing his observations that deliberate infection of a person with material taken from cowpox lesions, a process he dubbed vaccination, caused only a mild disease in the recipient but provided him or her with lifelong immunity both to cowpox and the related, but more deadly, smallpox. Wollaston, as we have seen, had been inoculated as a baby with a weakened form of smallpox and had been rendered immune to subsequent infection, a procedure that was generally effective if done properly. But inoculation with active smallpox serum occasionally led to serious illness, transmission of the disease to others, and the odd fatality. Vaccination with cowpox agent was believed by its supporters to provide all the benefits of inoculation, but without the consequent risks. Wollaston studied Jenner's publications and quickly became a strong supporter of vaccination. He read widely on the subject, studied firsthand the development of inoculated smallpox pustules and even exchanged letters with Jenner.

Not everyone was as certain as Wollaston about the validity of all of Jenner's claims. William Woodville, for example, the physician to the London

Smallpox and Inoculation Hospital in London, was a strong proponent of inoculation with weakened smallpox serum. After publication of Jenner's book, he took advantage of a cowpox outbreak at a London dairy to obtain cowpox serum and conduct a large-scale trial of vaccination. His results, published in 1799, confirmed the most important of Jenner's claims, but also revealed that some patients developed symptoms of smallpox.[39] Wollaston studied Woodville's results closely, compiled a tabular summary of the two hundred vaccination trials and concluded that the results were best explained by contamination of cowpox serum by smallpox agent circulating within the hospital at the same time.[40] Wollaston's re-interpretation of Woodville's vaccination data was well known to members of the College of Physicians. When the College was asked by a committee of the House of Commons in 1802 to comment upon the safety, utility, and benefits of vaccination, the letter of reply stated that, on the evidence available, vaccination was safe when properly conducted, and of great advantage to the public.[41] The House of Commons must have taken note of the College's approbation, for it acted quickly to award Jenner a grant of £10,000 in 1802 (and an additional £20,000 in 1807), quite generous disbursements at a time when war funding was draining the national treasury.

FAMILIAL AND POLITICAL STRESSES

On June 8, 1798, Wollaston's mother Althea died at fifty-nine years of age, twenty years after bearing the last of her seventeen children. Her death stunned Wollaston, who confided to Hasted how attempts by others to console him were often ineffective.

> I must confess that the total silence of some friends has appeared ambiguous, while the mere manner of others seems feelingly affectionate. The majority have no conception of the harshness of empty sounds. Happy beings, they never felt the loss of either friend or relation, & think the customary jargon of politeness excessively consoling.[42]

The letter exposes the sensitivity of Wollaston to nuances of spoken words, especially on personal matters. But he appears not to have appreciated the dilemma facing those wishing to extend well-intentioned sympathy. This failure to appreciate the emotional context of comments directed his way made it challenging for others to know how to speak naturally and empathetically with him after intellectual exchanges had run their course. Throughout his life, Wollaston maintained a similar emotional distance from the great majority of his acquaintances, forming very close personal friendships with only a select few, like Hasted.

Other letters to Hasted about this time contain intermittent comments on politics, especially local consequences of the war with France that had begun in 1793. Wollaston followed the conduct of the war closely, but with a curious blend of both patriotic concern and detachment. After the sensational, but ill-fated, landing of a disorganized French force in Wales (known as the battle of Fishguard) had led to the creation of several local militias and debate about one's political loyalties, Wollaston wrote:

> Neither the horror of a French invasion, nor the dread of being quizzed, which I find to be equally prevalent & equally efficacious have yet prompted me either to sport a red coat or to engage in any military association. The fear of approaching danger does not rouse me, to what appearance of courage, to what deeds of rashness the actual presence of democracy may irritate me, I will not pretend to answer till it arrives. I shall probably face it come when it will.[43]

As the fortunes of war waxed and waned, Wollaston's irritation with his government's increasing need for money grew. In early 1799, he opined, "I am far from sanguine in my expectation of peace. Our purse strings are untied and Pitt can take out ad libitum."[44] The reference to Pitt and purse strings refers to the income tax introduced by the prime minister in December 1798. Wollaston's distaste for a tax on income was shared by nearly everyone to whom it was applied (incomes over £60 per year), and he found the examination of his personal finances to be especially aggravating. In a series of letters to Hasted, we learn of Wollaston's antipathy to the tax, expressed in a caustic tone rare in his correspondence. In 1801 he wrote

> but to lose a summer, . . . in this solitude of London, to be boiled into a fever of extreme indignation against the insolence of the Commissioners of Income, to be kept 6 weeks in suspense for their arbitrary decision, & to doubt whether enough of what once was England remains to leave me a hope of justice by any appeal are not conducive to [my] wellbeing.[45]

This first implementation of an income tax in England was rescinded with the short peace of 1802, a decision that mollified Wollaston, at least temporarily (the tax was reinstated in slightly different form in 1803), as he confided to Hasted:

> you may form some judgement of the gratification I have lately experienced to find that all persons begin to view that detestable inquisition in the same light, that the whole country seemed at once roused from their lethargy to express their abhorrence of it. . . . One begins to feel that 2 and 2 are really 4 again, & to hope that the profits of industry may be enjoyed in security without fear of confession.[46]

Wollaston had become a chemical entrepreneur before he wrote this letter, so his comments on retaining the profits of industry are particularly relevant.

In contrast to what we know of Wollaston's activities in the College of Physicians and the advancement of his professional medical career in his early London years, details of his private practice in Cecil St. are almost nonexistent. Wollaston's notebooks contain no entries on patient care, and letters to Hasted are silent on the subject. The last entry in the daybook listing income from medical practice was made in December 1800, showing that Wollaston received medical fees from October 1797 to the end of 1800 totaling only £188, suggesting that he let his practice wither away once settled in the city.[47] Certainly the dividends on his bank stock more than offset the reduced medical income, but contemporaneous events suggest that the retreat from medicine was deliberate.

A CHEMICAL PARTNERSHIP

After his move to London, Wollaston was able to expand his association with Smithson Tennant, and sometime in 1800 the two decided to combine resources in a joint chemical business. After obtaining his MD degree in 1796, Tennant took up residence at Garden Court, Temple, a short walk from Wollaston's home in Cecil Street. Financially independent and without any desire to practice medicine, Tennant lived a busy but unfocused life following his intellectual interests wherever they led him. And he continued to be mercurial, innovative, and astonishingly undisciplined. For example, in the middle of one experiment in 1797 with Wollaston at his side, "suddenly recollecting that his hour for [horseback] riding was come, he left the completion of the process to Dr. Wollaston, and went out as usual to take his ride."[48]

Although Tennant and Wollaston were frequently seen together, they led independent personal lives and there is no contemporary reference to any kind of a shared endeavor, even in their own publications and letters. Knowledge of the partnership only came to light in the twentieth century through information in the Cambridge collection of their manuscripts.[49]

Documentary evidence of the partnership consists of entries in the account books of each man, the first of which was in December 1800. Those entries were expenditures for a large purchase of crude platina, which marked the opening of the business. Each man maintained a running account to keep a tally of debits and credits so that profits and losses could be equally shared. The entries provide a rich record of the varied chemical ventures undertaken by the two men.[50] The partnership, which continued until Tennant's death in 1815, was to generate financial and intellectual rewards far greater than either man could initially have hoped for. Nonetheless, one is led to wonder why Wollaston would contemplate linking his financial future to a partnership with such an eccentric and unreliable character. I believe three contributing factors merit consideration. The first, and most important, was the financial security that Wollaston had gained by the acquisition of some of West Hyde's estate, as has already been discussed. The second was Wollaston's own passion for natural philosophy in general, particularly chemistry. And the third was his growing unhappiness with medical practice. I will discuss the last two in turn.

At Cecil Street, Wollaston continued to use the time not needed for medical practice to pursue a wide range of scientific interests. After his first winter there, he wrote Hasted to tell him of the rich intellectual life of the capital, where "I have grasped at & with the most undistinguishing appetite devoured everything till I am disgusted at the thought of any mental food."[51] Much of that intellectual sustenance was obtained at the 8 o'clock Thursday evening meetings of the Royal Society, which became a regular part of his schedule after the move to London. Fellows could bring one or two guests to the meetings, which consisted of a business portion followed by the reading of a scientific paper by its author or one of the two secretaries.[52] Most of the papers read to the Society were subsequently published in the *Philosophical Transactions*, so regular attendance at the meetings would expose one to a wide range of the best English science of the time. Wollaston often brought guests to the meetings, including his younger brothers, yet another indication of the strong sibling connections among the Wollaston family and the mutual interest in natural philosophy.[53] Another frequent guest at the meetings was his Bury friend and lifelong correspondent Henry Hasted, who had the background to comprehend much of the science presented there.

NEW SCIENTIFIC INTERESTS

In addition to learning about the researches of others, Wollaston continued to be very active with studies of his own. In 1799 he began a long series of experiments on ink formulations.[54] None of this work was ever published but the best of the inks were used in his notebooks and correspondence. He developed his best recipe, the twenty-seventh in a series, in 1811. It contained a carefully measured combination of coarsely powdered galls, powdered gum arabic, green sulphate of iron and powdered cloves. These ingredients yielded a good black ink that flowed freely and was resistant to mould.[55] He continued to make ink according to this recipe right up to his death in 1828. The hours spent preparing inks, and testing their quality, while of no great scientific importance, give us some insight into Wollaston's motivations in this research work. His studies were fueled equally by a fundamental intellectual curiosity and a desire to gain firsthand practical knowledge from making and testing inks himself. Such a hands-on approach to knowledge acquisition and application is a common element in all of Wollaston's scientific work. He wanted to understand scientific principles on a fundamental and practical level, one that was internally consistent and made sense to him, and one that would enable him to speak confidently on the principles so learned. In short, he wished to satisfy fully his own curiosity. If someone were to ask him about the qualities of a good ink, Wollaston could give an answer based on a personal investigation of its composition, formulation, and application. And his answer would be thorough and convincing. Such confidence in his acquired knowledge was the character trait most noted by his contemporaries.

Wollaston's ink studies were little more than a private recreational pursuit, but a more substantial research interest at this time led to the first of a notable series of publications involving optical phenomena, which would make him one of England's leading experts on the subject.

STUDIES ON THE REFRACTION OF LIGHT

In August 1799 Wollaston traveled overland through Surrey to Portsmouth and toured the neighboring region including the Isle of Wight. One calm, bright morning, he noticed a striking optical phenomenon, the complete inversion of a coastal image viewed over water from a distance.[56] Such refraction of images near the horizon was a well-known optical phenomenon to astronomers and seamen. It was generally recognized that light rays were bent by the atmosphere in a way that caused them to follow somewhat the curvature of the earth. Consequently objects such as stars sighted a few degrees above the horizon could in actuality lie below it. Mariners also knew that the horizon visible at sea was not an accurate indication of its real position and the true horizon was lower than the observed one. But

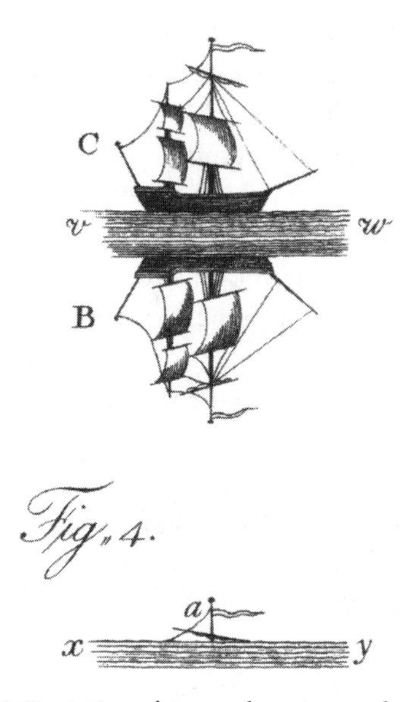

Fig. 2.1. Vince's Depiction of Anomalous Atmospheric Refraction.

refraction also caused other unusual effects, such as the one noted by Wollaston. Some objects sighted near the horizon could display additional images, elevated or depressed, even sometimes upside down, relative to their real position. Samuel Vince, the Plumian Professor of Astronomy and Natural Philosophy at Cambridge, presented a Bakerian Lecture to the Royal Society on the topic in 1798, and included several examples of unusual refractions seen over water.

One, shown here as Figure 2.1, was of a distant boat viewed on a hot August day with a telescope. The boat's mast *a* was just visible over the horizon *xy*. Refracted secondary images of the boat appeared above the mast in both inverted, *B*, and erect, *C*, forms, with an image of water, **vw**, between them.[57] Vince proposed that the refracted images resulted from different paths taken by light rays moving through an atmosphere of variable refractive power dependent upon its density, composition, and moisture content, although he provided no causal mechanism.

Wollaston knew of Vince's paper and, after observing the optical illusions for himself, began an investigation designed to provide a theoretical explanation for unusual refractions, and to model the phenomena with fluids of different densities. His resulting paper began by claiming that a fluid containing a continuum of "parallel indefinitely thin strata" could have a density gradient containing a "point of contrary flexure."[58] His argument

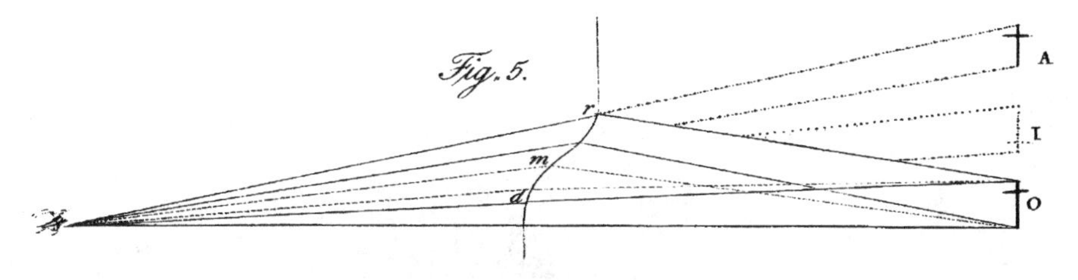

Fig. 2.2. Refraction of Light Rays Passing through a Density Gradient Containing a Point of Contrary Flexure.

was based on the geometrical construction shown in Figure 2.2. He postulated that the density gradient of atmospheric air could vary in the vertical direction in a way represented by the line *dmr*. Then, the eye of an observer on the left viewing the object *O* shown on the right through an intervening fluid whose density curve has the necessary point of contrary flexure, *m*, will see refracted images (by tracing the light rays striking the eye back to their perceived point of origin) that correspond with all of those described by Vince. The object will simultaneously be seen upright at position **O**, raised and inverted at position **I**, and elevated even more but upright again at position **A**.[59]

Because the density differences between infinitely thin layers of differing density and refracting power established in air at different temperatures are extremely small, the secondary refracted images only appear if the light rays pass through a long length of such layers, and only when the layers remain undisturbed. These factors made controlled study of them nearly impossible, so Wollaston devised a model system using liquids of different densities which could give the predicted secondary refracted images. He placed into a square glass phial, without mixing, a bottom layer of syrup, a middle layer of water, and a top layer of rectified spirit of wine (ethanol). After allowing the phial to stand undisturbed for a short time, he viewed words on a printed card through the regions where the different liquids intermingled to produce zones of varying density. The results are shown in Figure 2.3. At the syrup/water interface (the refractive index of water is lower than that of syrup), three images of the word syrup appeared: one erect in its normally refracted place, a second elevated and inverted, and a third even more elevated and erect. At the water/spirit interface (the refractive index of rectified spirit is greater than that of water), three images of the word spirit appeared: one erect in its normally refracted place, one depressed and inverted, and a third more greatly depressed and erect. The images remained stable for hours and sometimes, if other more immiscible liquids were used,

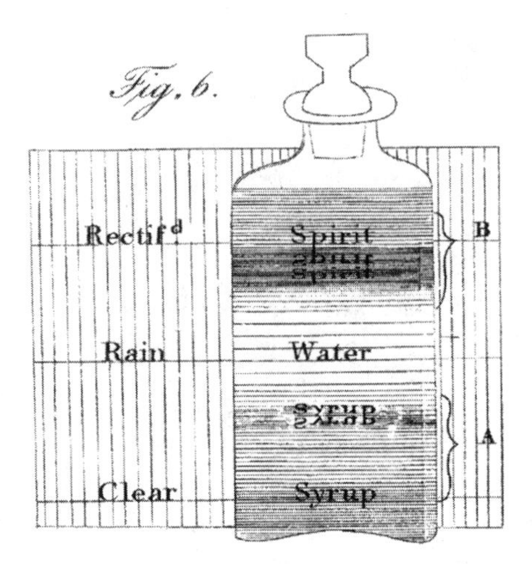

Fig. 2.3. Refracted Images Seen through the Contact Zones of Different Liquids.

for days. In short, Wollaston created a simple, ingenious model system that displayed all the features of atmospheric refraction, thereby transferring study of the phenomenon from a rarely observed and transient occurrence over the sea to a table-top display. His model system could serve both as an instructional aide and a basis for further experimentation and study.

Wollaston advanced his model even further by carefully depositing a layer of hot water over cold to generate similar refracted images, showing that the phenomena were a consequence only of a density gradient in a fluid medium (hot water has a lower refractive index than cold water). It was not necessary to have two different fluids. He also presented evidence in the paper to show that secondary refracted images could be produced by temperature variations in a column of air surrounding a red-hot fireplace poker. These observations supported his conclusion that temperature gradients in the lower layers of the atmosphere, especially common over open bodies of water, were sufficient to cause the observed refractive phenomena.

Overall, Wollaston's paper on atmospheric refraction in 1800 made a rare natural phenomenon previously observed and contemplated by only a few into a bench-top display accessible to all. This was its greatest contribution. Although his explanation for the means by which a density gradient could yield a refraction curve containing a point of contrary flexure was vague and incomplete, his conclusion that variations in the density of the refracting medium, produced mainly by temperature or, perhaps, by humidity acting independently or in concert, were responsible for secondary refracted images was generally accepted at the time. However, subsequent studies established that humidity gradients could not, by themselves, generate secondary refracted images. It is now known that temperature gradients alone create the atmospheric lensing actions that result in secondary images.[60]

The paper was Wollaston's first on a nonmedical topic and marked him as a natural philosopher of talent in the field of linear optics. His work was most appreciated in France, where theoretical studies of light were more advanced than in Britain. For example, the French physicist Jean-Baptiste Biot, in his thorough analysis of atmospheric refraction published in 1810, praised Wollaston's fundamental contribution to the subject.

> [Wollaston] imitates these [atmospheric] phenomena and magnifies them by observing them on heated bodies, across liquids of unequal density, and even on the surface of red-hot iron. . . . These results are very precious for verifying the mathematical theory of the phenomenon, . . . [although] Wollaston has not given it. . . . But in relation to physics, his work leaves nothing to be desired, it is full of experiments, conceived with all sagacity, and performed with all the skill, which characterize this able philosopher.[61]

The study of atmospheric refraction was the first of Wollaston's investigations into optical phenomena, and he was later to consider other applications of refraction, two of which led to patented devices. These will be discussed in a later chapter. But some nonscientific events in 1800 were more portentous for Wollaston.

THE END OF DOCTORING

In March 1800 the prominent physician Matthew Baillie relinquished his appointment to St. George's hospital, and Wollaston added his name to the list of candidates for the position. Hospital appointments, although unsalaried, were highly sought after, and Oxbridge doctors were often preferred over those with degrees from elsewhere.[62] Wollaston was a highly qualified candidate with several years of patient care, professional status, personal connections, and a medical publication. Nonetheless, he lost the election. In a vote by the hospital's governors on March 13, Charles Nevinson, an inferior candidate in many ways, received 111 votes versus 78 for Wollaston.[63] Whatever the political backdrop to the election, the result must have rankled Wollaston, who had not previously suffered any setbacks in his career. Although it is likely that Wollaston was both disappointed and angered by the election defeat, the result may not have had a profound impact on him. We have no comments on the episode from Wollaston himself, nor did he ever mention the election defeat in correspondence with Hasted. Perhaps the loss to Nevinson only confirmed what Wollaston had already suspected—medicine was not the career for him. We know that he had ceased attending to patients in 1800, had no dire need for a medical income, was planning on forming a chemical partnership with Smithson Tennant that very year, and was very active in scientific research. The possibility that Wollaston turned with more conviction to a career trajectory that had already been set in motion gains some credence from the details of a second election at St. George's, which took place one month after the election of Nevinson. Wollaston chose not to place his name in competition for the new position, even though he might well have been the leading candidate by then. One suspects that Wollaston, if he really had been committed to a future in medicine, would have pursued a hospital position at the second opportunity.

Wollaston continued to attend meetings of the College of Physicians during 1800, and completed his second term as madhouse commissioner in May. These were, however, the last of his medical activities. Back in London in the fall of the year, he plunged into a wide range of scientific investigations, as he contemplated forming a partnership with Tennant. Near the end of the year, the two men finalized their agreement and began to keep records of their shared expenses. Then, in late December, shortly after he had paid £484 for his share of the purchase price of 5,959 oz. of crude platina ore and had attended his last meeting of the Royal College of Physicians, Wollaston wrote in his daybook (slightly adapting Jean d'Alembert's lament of the philosopher), "Farewell to Physic 'la médiocrité des desires est ma fortune; l'indépendence de tout, excepté des devoirs, est mon ambition.' "[64] Obviously, d'Alembert's words resonated with Wollaston: he had no desire for professional success or acclaim and was happiest when pursuing his own interests, on his own schedule. He had, he realized, the temperament of a natural philosopher, not a doctor.

Wollaston's abandonment of medicine was more fully explained in a heartfelt letter to Hasted.

> I cannot help thinking that I have at various times given you reason to think with me that the practice of physic is not calculated to make me happy. I am now so fully convinced of it & have so well satisfied those who are more interested for my welfare & whom I thought it most pru-

> dent to consult that I have fully determined & now declare that I have done with it. What I shall do instead I do not yet know. I feel no doubt of finding employment & turning my time to account in some way or other less irksome to me, for even if I turn waiter at a tavern ready to say "Yes Sir" to everyone that calls at any hour of the day or night, I cannot be a greater slave. . . .
>
> At present I certainly am far happier than I have been for many months, (I might say years)—although I have to encounter the decided censure of 9/10 of those I meet. I know best what is most for my happiness & those who best know me blame me least (if not approve). Would you submit to be flogged like a slave for any compensation?—then do allow me to decline that mental flagellation termed anxiety, compared with which the loss of thousands £ is a mere flea-bite.[65]

Hasted must have been shocked to read that Wollaston compared treating patients to slavishly waiting on patrons at a tavern, and mental anxiety over medical problems to being flogged like a slave. Such misery must have been further exacerbated by a feeling of personal and familial failure, for Wollaston had to accept that he was abandoning a career he had worked very hard to establish, and one that his father had planned and supported. Interestingly, the letter makes no mention of the new partnership with Tennant, and Wollaston coyly says only that he is uncertain about his future prospects. Secrecy, it has to be assumed, was viewed by Wollaston and Tennant as essential to the commercial success of their endeavor. Of course, very few of Wollaston's other acquaintances knew why he chose to leave doctoring behind, and most attributed the surprising change of direction to the election defeat. For example, Thomas Thomson, a chemist and historian who knew Wollaston well in later years, superficially blamed the incident for the career change.

> A vacancy occurring in St. George's Hospital, [Wollaston] offered himself for the place of physician to that institution; but another individual, whom he considered his inferior in knowledge and science, having been preferred before him, he threw up the profession of medicine altogether, and devoted the rest of his life to scientific pursuits.[66]

Of course, Thomson had no knowledge of Wollaston's financial windfall or his partnership with Tennant, so his conclusion did little more than reflect faulty contemporary opinion. The letter to Hasted, however, brings clarity to Wollaston's decision. He disliked the loss of freedom that attendance on patients required, and he was unable to ameliorate the anxiety he experienced at being unable to do much to relieve the suffering of patients. Failure to obtain a hospital post is not mentioned as a contributing factor, and circumstantial evidence suggests it probably wasn't. One gets the sense that, even if the election had gone his way, Wollaston would not have lasted long as physician to St. George's, and his departure from physic would only have been delayed, not avoided. But what was to be his new career? How would it generate an income? Could it repay his family's faith in him? The answers to these questions would become clearer in 1801, as researches on platinum and organic chemicals would move toward the production of commercial goods, but the sale of products was several years even further into the future. Hard times still lay ahead.

CHAPTER 3

Early Years as a Natural Philosopher 1800–1802

> He was remarkable, too, for the caution, with which he advanced from facts to general conclusions; a caution which, if it sometimes prevented him from reaching at once to the most sublime truths, yet rendered every step of his ascent a secure station, from which it was easy to rise to higher and more enlarged inductions.[1]

The years 1800–1802 were the most critical, and spectacularly successful, of Wollaston's life. His retreat from medicine in the first of those years was accompanied by an intense engagement in practical and commercial scientific endeavors. He had his own intellectual interests in science to pursue, as well as several chemical initiatives in partnership with Tennant. He recognized that a great deal of work had to be done to establish himself as a competent natural philosopher and chemical entrepreneur. Consequently, he immersed himself fully in a wide range of investigations, nearly every one of which was to lead to novel discoveries and critical acclaim. To better understand the content and impact of Wollaston's new lines of research, it is best to look at his major discoveries in electricity, optics, platinum metallurgy, and chemical element isolation in sequential chapters, even though each line of research began in 1800 and extended several years beyond. The topics will be covered in the order in which they were made known to the public. The first was a fundamental study of electrical phenomena.

RESEARCH ON ELECTRICITY

During the last years of the eighteenth century the Italian Alessandro Volta had developed two prototypes of a novel electrical device capable of producing a sustained electric current. On March 20, 1800, he wrote the first part of a two-part letter to the president of the Royal Society, Joseph Banks, in which he described the construction and properties of his electricity-producing apparatus. Volta asked that his results be communicated to the Royal Society, and, after the second letter reached Banks, the results were read to the Society on June 26 and published in the *Philosophical Transactions* shortly thereafter. However, soon after Banks received the first letter in April he showed it to members of the Society, some of whom immediately set about to confirm Volta's astonishing results. Prior to Volta's discovery, only static electricity (also known as common electricity), produced by an electrostatic generator and stored in a glass and metal vessel known as a Leyden jar, was known. The current produced by an electrostatic generator was very feeble, and although a Leyden jar could store opposite electrical charges at high potential, it could only deliver its energy in sudden bursts or discharges. In contrast, Volta's pile, as it became known, delivered an almost constant electrical current for a long period of time, from an easily constructed stack of metallic discs. It was the most sensational discovery of its time, and news of it spread like wildfire. By the end of April, Anthony Carlisle had constructed an electrical device consisting of seventeen pairs of silver and zinc discs separated by pasteboard that had been soaked in salt water, and verified Volta's results. In early May, Carlisle and William Nicholson found that, when the pile discharged electrical current through water, hydrogen and oxygen were produced in a ratio of 2:1, the same ratio in which they had decades earlier been combined to synthesize water. Although they withheld publication of their results until after Volta's paper appeared, word of their discoveries spread quickly and several persons began to assemble improved and more powerful Voltaic piles.[2]

Wollaston, too, was caught up in the excitement of this new electrical device, and he initiated his studies with simple versions of the battery, as he informed Hasted soon after learning of Volta's discovery.

> I too have been dabbling with so curious a subject & can tell you a few facts. 40 pieces are not necessary—3 shillings are sufficient with similar pieces of zinc and pasteboard to decompose water, & if the pasteboard is wetted with salt-water even two pieces will do it.[3]

One week later when he set out on a tour of the Lakes District with Hasted, Wollaston had already made an even smaller display device to illustrate the phenomenon.[4] Wollaston never published a description of his simple electrolytic cell, but the Swiss natural philosopher Marc-Auguste Pictet was present at the Royal Institution in the summer of 1801 when it was used,

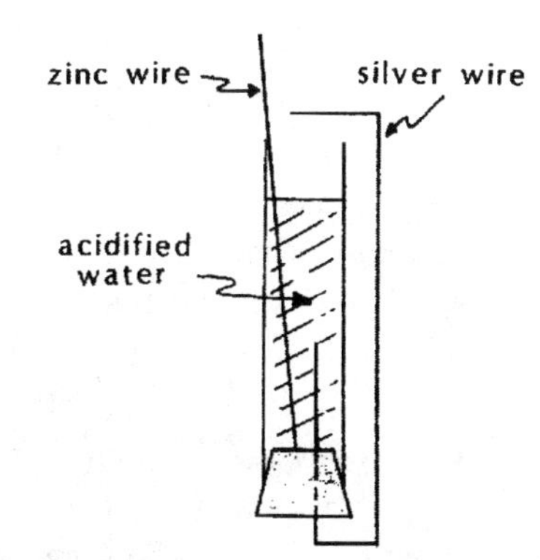

Fig. 3.1. Wollaston's Miniature Galvanic Cell. Reprinted with permission from Melvyn C. Usselman, "Wollaston's Microtechniques for the Electrolysis of Water and Electrochemical Incandescence," in *Electrochemistry, Past and Present,* eds. John Stock and Mary Orna. Copyright 1989, American Chemical Society.

perhaps by Wollaston himself, to demonstrate galvanism. Pictet described what he saw.

> I also witnessed the [production of hydrogen from water] effected by the following apparatus, which is of an admirable simplicity. It consists of a small glass tube, full of water slightly acidified by a common acid. In this tube is a silver wire, which passes through a cork from above, bends immediately, and extends upward along the outside of the tube until it bends again above the tube so that it can be made to touch, or not, the end of a zinc wire (itself protruding). As the two metals touch outside the tube, a stream of small ascending bubbles appears at once at the end of the silver wire; when they are separated, the chemical action in the liquid on the silver ends, and the stream ceases.[5]

From Pictet's description I have made a replica of the device, as shown in Figure 3.1; it works just as described.[6] Wollaston's motivation in designing his simple electrochemical device was largely pedagogical: he wanted to show the basic principles of water electrolysis (as he referred to it) in a manner as free from confounding principles as possible.

Wollaston also carried out a broader investigation of chemical phenomena associated with the operation of the voltaic pile, and published the results in 1801.[7] His research sought answers to two questions raised by studies of voltaic electricity. The first was the source of the electricity. Did the electrical current originate in the simple contact of the two dissimilar metals of the pile, as Volta believed, with chemical changes as a consequence? Or did the electricity arise from the chemical reaction of one of the metals with the liquid on its surface, with current flow to the second metal as a consequence? Through use of a cell similar (if not identical) to the one shown to Hasted, Wollaston studied the behavior of cells made with different metallic wires. He observed that metals with greater affinity for acid were the ones that dissolved in voltaic cells, and that hydrogen gas always formed on the surface of the metal of lower affinity. He therefore concluded that oxidation of a metal in acid gives rise to the electricity produced, which subsequently travels to the non-oxidized metal to produce hydrogen.

Humphry Davy independently came to the same conclusion after discovering that oxidizing conditions were essential to the operation of a cell, and his opinion was published in late 1800.[8] But these and other arguments for the chemical primacy of galvanism did not win the day, for they could not provide an explanation for the separate production of oxygen, when it occurred, at the electrode that did not produce hydrogen (Wollaston did not address the issue).

The second question that Wollaston sought to answer was whether or not voltaic electricity was identical to common, or static, electricity. If, he reasoned, he could obtain results using common electricity similar to those he observed with a voltaic cell, then it would be reasonable to conclude that both types of electricity were the same. Of course, common electricity could only be generated in much smaller intensity, so the experiments presented difficulties. But micro techniques were his specialty, and he again rose to the challenge. His goal was to repeat the deposition of copper on silver by common electricity, as he had observed in his voltaic experiments. By using an electrostatic generator to pass a small current through two narrow silver wires immersed in a copper sulphate solution, he was able to demonstrate that common electricity produced the same deposition phenomena as the electricity generated by a voltaic cell.

Wollaston then tried to decompose water with the feeble current produced by his electrical machine. Although others had done so using the powerful discharges of large Leyden jars, he believed he could do the same with an ordinary electrical machine if he could reduce the surface of the metal exposed to the water, thus concentrating the electrical effect on a very small area. Using specially-prepared gold wires of diameters estimated at 1/700 inch, or even finer ones of 1/1500 inch, electrical current generated by turns of his electrical machine exceeded his expectations by producing "a current of small bubbles of air."[9] Encouraged by this result, he continued to miniaturize his wires. With great creative imagination, he filled a capillary glass tube with an aqua regia solution of gold, and heated the tube to evaporate the acid and deposit a film of gold on the inner surface. He next melted the glass capillary to give an extremely fine gold thread coated in glass. When current from an electrical machine was passed through this exceedingly small gold tip immersed in water and placed next to a conducting wire, Wollaston "found that the mere current of electricity would occasion a stream of very small bubbles to rise from the extremity of the gold."[10] Thus, even the small current from an electrical machine turned by hand could decompose water. When he tried next to mimic the voltaic production of both hydrogen and oxygen by connecting to the electrical machine two microscopic gold tips immersed in water, he observed bubbles of air from each tip, but "in every way in which I have tried it, I observed that each wire gave both oxygen and hydrogen gas, instead of their being formed separately, as by the electric pile."[11] He attributed the different actions of the two electricity sources to the greater "intensity" (voltage, in modern terms) of the common electricity delivered by the electrostatic generator. In other experiments, Wollaston demonstrated that the passage of both common and voltaic electricity induced the same color changes in the natural acid-base indicator litmus. He was thus led to conclude that the similar actions of the two types of electricity (on the decomposition of water and coloration of litmus) provided compelling evidence of their identity.[12] With his questions on the nature of galvanism answered, Wollaston did no more fundamental work on the phenomenon. His curiosity had been sated, and the great advances in voltaics that occurred in the early nineteenth century were made by others. But his pioneering contributions were noted by his contemporaries.

Pictet, after witnessing Wollaston's experiments at the Royal Institution in 1801 commented that they "reveal more on Galvanism than has been discovered to date on the subject . . . [by them] is Galvanism fully reconciled with common electricity, and one might say, identified with it."[13] Davy acknowledged in 1810 that Wollaston was the first to demonstrate the identical chemical effects of common and voltaic electricity.[14] Michael Donovan, in his comprehensive 1816 *Essay on Galvanism*, made a similar assessment.[15] Even later, after Arago had shown in 1820 that both common and voltaic electricities could induce a magnetic field in steel, the editor of the *Annals de Chimie* reprinted Wollaston's paper with the comment that "it would be beneficial for investigators to study in detail the ingenious experiments by which Wollaston demonstrated the identical chemical effects of common and voltaic electricities."[16] And the brilliant experimentalist Michael Faraday was one who did so. In a sweeping and comprehensive study of electricity, he improved upon several of Wollaston's results and distinguished more clearly between intensity and quantity of electrical current. Faraday admitted that "there cannot be now a doubt that Dr. Wollaston was right in his general conclusion; and that voltaic and common electricity have powers of chemical decomposition, alike in their nature, and governed by the same law of arrangement."[17]

Despite his departure from voltaic investigations, Wollaston remained interested in the subject and produced another uniquely clever device about a decade later. We learn of it from Davy's biographer, John Paris, who wrote

> Shortly after he [Wollaston] had inspected [in 1813] the grand galvanic battery constructed by Mr. Children, . . . he accidentally met a brother chemist in the street, and seizing his button, (his constant habit when speaking on any subject of interest,) he led him into a secluded corner; when taking from his waistcoat pocket a tailor's thimble, which contained a galvanic arrangement, and pouring into it the contents of a small vial, he instantly heated a platinum wire to a white heat.[18]

The "brother chemist" was probably Thomas Thomson, for he persuaded Wollaston to publish a description of the small voltaic cell in his journal *Annals of Philosophy.* There Wollaston reported that he had first constructed the thimble battery in 1812, shortly after he had perfected the method of making very fine platinum wires.[19] The device, shown schematically in Figure 3.2, consisted of a plate of zinc about one inch square fastened inside a topless, flattened copper thimble by sealing wax.[20] Attached to the metals were two sturdy, parallel platinum wires, bridged by a

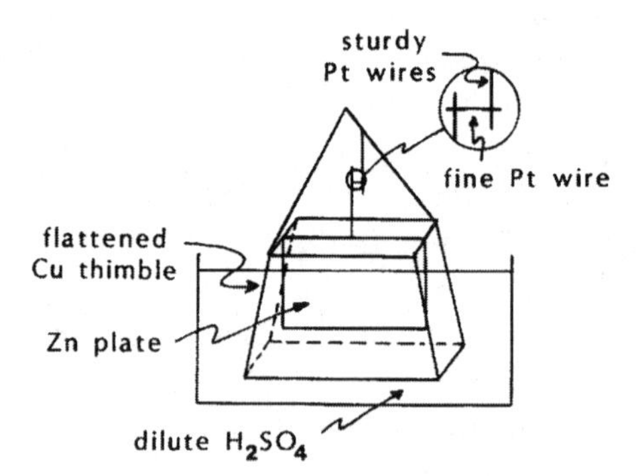

Fig. 3.2. Wollaston's Thimble Battery. Reprinted with permission from Melvyn C. Usselman, "Wollaston's Microtechniques for the Electrolysis of Water and Electrochemical Incandescence," in *Electrochemistry, Past and Present,* eds. John Stock and Mary Orna. Copyright 1989, American Chemical Society.

very short length of exceedingly thin platinum wire (about 1/3000 inch in diameter). When the battery was lowered into a dilute solution of sulfuric acid, the resulting electric current heated the fine platinum wire to incandescence for a brief, but spectacular, display. This thimble battery might be judged, in hindsight, as a prototype of an electric light. In fact, Thomas Edison's first patent for an incandescent glow lamp in 1878 incorporated a fine platinum wire encased in a partially evacuated glass globe.

MOVE TO BUCKINGHAM STREET

In 1801 Wollaston directed most of his efforts to the purification and consolidation of platinum, an experimental investigation that will be described in the next chapter. Since no results from those investigations were made public until 1803, attention will now be directed to other events in Wollaston's life that followed his withdrawal from medicine. First, he decided that he could not carry out chemical research effectively at his cramped Cecil Street location, so he purchased a large Georgian-style house with a side entry on the northwest edge of Westminster at 14 Buckingham Street, near Regent's Park. He moved there in September.[21] (On July 4, 1934, a commemorative plaque to Wollaston sponsored by six of England's scientific societies was placed on the Buckingham St. house.[22] Unfortunately, the building was destroyed by bombing in 1945 and the plaque was placed in the archives of the Geological Society, where it now remains.) He continued to live there until 1825, when he no longer needed laboratory space and moved a bit further west to 1 Dorset St, Fitzroy Square.[23]

A combination of time-consuming laboratory work and planning for a move kept Wollaston in London that summer, one of the only years when he did not leave the city for a summer excursion. In the fall, shortly after settling into his new home, Wollaston wrote to Hasted of his hopes for the future.

> I am partial to Chemistry; I have here room for a laboratory, & tho' many have spent fortunes in such amusements more have made fortunes by the same processes differently conducted. Is it impossible to mix the utile dulci [the useful with the agreeable] —if it be I have erred egregiously & may be ruined, but I have no fears at present. When I quitted the terra firma of Physic this was my sheet anchor, . . . , & I now hope that I am fixed for life. If I make £40 it is as good as most curacies & I may be content excepting that there need then be no obstacle to my making 10 times as much.[24]

The letter was typically circumspect on the matter of personal finances. It is noteworthy, however, for its mention of chemistry as his new vocation and commercial venture, although the partnership with Tennant remained undisclosed. The commitment to a chemical business is made even more forcefully in a following letter.

> My business must be Chemistry. It is late to be beginning life entirely anew, & as efforts that I am not equal to may be requisite, I must not play with botany or think of chemistry at present. . . .
>
> After putting one's hand to the plough there may be "no looking back" for a prudent man but I have done it & must abide by the consequences, come what will.[25]

Wollaston appears in this letter to differentiate between the business of chemistry, by which he must mean production of commercial products for profit, and intellectual or curiosity-driven chemistry, what he refers to as thinking about chemistry. Subsequent events would, however, demonstrate that the pursuit of applied chemistry (a modern term) did not put an end to his pure chemistry (another modern label). In fact, the two were inextricably linked, even synergistic. Nonetheless, his distinction between them was to lead him into a public controversy in 1803.

As Wollaston entered into his post-physician life, he began to shed several of his past obligations. He attended his last meeting of the Royal College of Physicians in December, 1800, and encouraged his friends thereafter to address him as "no more Dr but plain Mr I beg."[26] The request went largely unheeded, for he continued to be known and addressed most commonly as Dr. Wollaston. In time he must have grown accustomed to the title, for he did nothing to suppress its use in later years. In January 1801, he wound down another commitment by traveling to a meeting of the seniority of Caius College for the last time. He never did resign his fellowship, though, and continued to receive an annual stipend for it until his death. As Wollaston shed these ties to his first profession and his Cambridge college, he became increasingly active in the affairs of the Royal Society, the home of England's pre-eminent natural philosophers.

ROYAL SOCIETY INVOLVEMENT

During his London years, Wollaston became a regular attendee at the Royal Society's Thursday evening meetings and soon began to lend his support to others who wished to become Fellows. He frequently added his name to the certificates of admission of medical candidates, hoping no doubt to advance the cause of science in the diagnosis and treatment of disease.[27] But of even more importance for its long-term impact was Wollaston's entry into the dining club of Royal Society Fellows, a more select, and convivial, collection of scientifically-minded philosophers.

Members of the Royal Society and some of their like-minded friends had begun dining together in a convenient tavern prior to the 8 o'clock meetings as soon as the Society received its royal charter in 1662.[28] By the end of the eighteenth century the diners had organized themselves more formally into the Royal Society Club, consisting of forty elected members of the Society, together with its principal officers as ex officio members. Each member had the privilege of bringing one visitor to the dinners, and the names of all guests and their hosts were entered in the minute book.[29] The atmosphere at Club dinners was a congenial and lively one, with a free exchange of ideas among intelligent men of diverse interests and talents. Following their meal, most of the diners would move on to the Royal Society for its business meeting and the reading of papers. Although the Royal Society adjourned its scientific meetings for the summer holiday months, the Club continued its Thursday dinners all year round and, in Wollaston's time, met at 6 o'clock at the Crown & Anchor tavern on the Strand.

Wollaston was invited to dine with the club for the first time on March 19, 1801, at the invitation of Henry Cavendish, the famed English natural philosopher and social recluse.[30] Why would he invite Wollaston, thirty-five years his junior and generally known as a physician? One salient fact was that Cavendish and Francis Wollaston, William's father, were contemporaries at Cambridge in the early 1750s. In addition Cavendish and William Heberden were long time, influential colleagues at the Royal Society, and Cavendish could well have shared Heberden's interest in advancing the career of Wollaston. Whatever the reasons, the pathologically shy Cavendish introduced Wollaston to the Club and invited him a second time in 1803. In marked contrast to Cavendish, Wollaston's shyness dissipated with closer acquaintance, and he welcomed interaction with others. While most contemporaries accepted Cavendish's social anxiety, Wollaston recognized "the way to talk to Cavendish is never to look at him, but to talk as it were into vacancy, and then it is not unlikely but you may set him going."[31]

Wollaston was invited to the Club on other occasions by the well-known surgeon and anatomist Everard Home, but only became a regular attendee after he was appointed a secretary of the Society in November 1804. From that time forward Wollaston dined with the Club many times each year, and became a very popular member. Secretary to the Admiralty John Barrow, who was also a regular attendee at the dinners, commented on Wollaston's contributions.

> In fact it was impossible to be in company with Dr. Wollaston without acquiring new information, on whatever subject might be under discussion, more especially when at the Club of the Royal Society, where the members were intimately known to each other, and great freedom of speech prevailed. Here it was not unusual to start a subject for no other purpose than to draw out Wollaston's opinion, or remarks upon it. I must plead guilty of having not unfrequently done this.[32]

Wollaston's discussions at Club dinners will arise again later in this biography, as the participants and topics of debate become relevant to his ongoing research interests. Then, as now, many of the arguments over theoretical and experimental issues in science received their most candid and probing discussions during such social engagements.

LIFELONG COLLEAGUES: HUMPHRY DAVY AND THOMAS YOUNG

The year 1801 marks the first meeting between Wollaston and his soon-to-be famous contemporary, the chemist Humphry Davy. Davy, twelve years younger than Wollaston, had arrived in London in March to take up a position as Director of the Laboratory and Assistant Lecturer in Chemistry at the recently founded Royal Institution.[33] A few weeks after settling into his rooms at the Institution, Davy began a short and very popular course of lectures on galvanism, which was attended by several of London's leading natural philosophers. It is not known if Wollaston was at any of the lectures but he certainly was aware of them. Perhaps as part of that course, or at least as a likely consequence of it, Wollaston's two small galvanic demonstration devices (the small galvanic cell shown in Figure 3.1, and the decomposition of water effected by the electric current from an electrostatic machine passing through very fine wires) were used in the summer of 1801, as described by Marc-Auguste Pictet. Davy also carried with him from previous work at the Pneumatic Institution in Bristol an interest in nitrous oxide (N_2O, also known as laughing gas), and encouraged his new London colleagues to try its effects. Pictet was at one of these laughing gas trials in June 1801, and described Wollaston's participation.

> [Wollaston] experienced sensations very similar to those I felt; only furthermore, the same sensation in both hands one gets by rubbing the ends of all fingers in succession against the thumbs; he moved them so solemnly during the height of the effect without embarrassment at our bursts of laughter. His pulse was exceedingly irregular.[34]

This interesting anecdote, besides revealing what great risks early nineteenth-century experimenters took with their health, demonstrates Wollaston's interest in all new phenomena and his powers of concentration when mentally recording sensory input. He did not allow the laughter of those near him to affect the acuity of his observations.

Wollaston and Davy were to be associated in a wide variety of endeavors as they went on to become the two most influential chemists of Regency England. They were in 1820 to compete for the presidency of the Royal Society and, later, to travel and fish together. But the two men were quite different in temperament and research style, and occasionally friction developed between them. Not surprisingly, their colleagues often compared them and emphasized the greater impact of one or the other, depending on which scientific traits the assessor wished to see in a man of science. Davy was generally viewed as a man of great imagination, bold speculation, and social ambition while Wollaston was judged to embody the characteristics of close observation, careful induction, and personal discretion. Fortunately we have illustrative examples of Wollaston's cautious attitude toward hypothesis and theory formation in letters of this time to the third member of London's elite trio of natural philosophers, Thomas Young.

Thomas Young was born in 1773 in Milverton, Somerset, the eldest of ten children in a Quaker family.[35] After preliminary training to become a doctor, he moved to London in 1793 and attended lecture courses of the leading medical men. A precocious intellect, he published a paper on the accommodating power of the lens in eyes in the *Philosophical Transactions* in 1793 and was elected as a fellow of the Royal Society a year later. He received his MB degree in 1803 and his MD, together with fellowship in the Royal College of Physicians, in 1808. Like Wollaston, whose early career had many parallels, Young spent much of his free time pursuing interests in natural philosophy, especially the phenomena exhibited by sound and light. He published an important paper on the topic in the 1800 *Philosophical Transactions* in which he suggested that light, similarly to sound, might be propagated as longitudinal vibrations through a medium of transmission. In 1801 he was appointed Professor of Natural Philosophy at the Royal Institution and began a comprehensive course of lectures on natural philosophy in 1802.[36]

Wollaston, Young, and Davy all began to make their marks in the activities of the Royal Society, and the scientific literature, in the early years of the nineteenth century. A telling indicator of their influence is the fact that all three were chosen multiple times to give the Bakerian Lecture to the Royal Society. Over the thirteen-year period from 1800 to 1812, Davy gave the lecture six times, Young and Wollaston three times each. All three of the great men continued to play a major role in British science until their deaths (within a six month period) in 1828–1829. However, Wollaston's future career was to become even more entwined with Young than with Davy. The two served together for many years as secretaries of the Royal Society, on various scientific committees, and on the Board of Longitude in the 1820s. Their dealings with one another were almost entirely professional and cordial, and although they rarely interacted socially, they found themselves in each other's company nearly weekly. Undoubtedly they probed one another for facts and opinions. A recent biographer of Young provocatively titled his book *The Last Man Who Knew Everything*, a bit of hyperbole that has some basis in contemporary opinion.[37] Leonard Blomefield, for example, commenting on the life of Wollaston (his second cousin), mentioned a question posed at one of Joseph Bank's Sunday evening soirées to someone who replied, "I cannot answer your question myself, but there stand Young and Wollaston, and between them both they know everything."[38] This comment suggests that the knowledge bases of the two men were impressively complementary. Both were studying the effects of light in the early years of the nineteenth century, though, and a short series of letters from Wollaston to Young from this time reveals their quite different commitments to theory.

THE PRIMACY OF OBSERVATION

The varied properties of light, such as its propagation in straight lines at constant velocity, reflection and refraction, spectrum of colors, polarization, and double refraction, had presented investigators with formidable explanatory challenges. Some characteristics of light invited comparison with vibrational phenomena such as sound and water waves. Others were more easily explained by the straight-line propagation of particles. Newton, whose work was fundamental to all later English natural philosophers, favored a corpuscular interpretation, in which all visible objects emitted a high-speed, linear stream of small particles. Newton's view had become the generally accepted one when Young began his researches on light and vision.[39] Young's studies of light, especially its interference properties, led him to resurrect the controversial vibratory theory, and he was eager to draw others to his cause. In the summer of 1800, Young was investigating color vision and the mechanism of the eye, prior to his presentation on the topic in his first Bakerian Lecture. Young, knowing of Wollaston's expertise in the laws and manifestations of refraction, wrote to him seeking input on his own nascent ideas.

In 1800, just a few months after Wollaston's paper on atmospheric refractions was read to the Royal Society, Young quizzed him on the differing refractions of red and blue light rays and how they might be at odds with Newton's ideas. Wollaston replied that he was unwilling to deviate from his habit of "communicating conclusions with freedom but conjectures with reluctance."[40] But Young was unwilling to yield and tried by return mail to get more concrete answers to his questions. Wollaston responded a day later, reiterating his aversion to what he interpreted as overhasty speculation.

> You must pardon me; the matter of fact is not a question with me . . . I decline being led any farther dance after phantoms. It is rather for you who guess that inconsistencies might be found [in Newton's arguments], to spend what time you please in seeking them. Would it not be more correct to say you wish it, than that you think it, . . .[41]

Clearly, Wollaston was not going to let himself be provoked into overgeneralization. One year later, Young was continuing to develop his vibratory theory of light before his second Bakerian lecture on light and color and, knowing that Wollaston was carrying out further experiments on refraction, he again solicited his colleague's opinion on the dispersion of refracted rays. Once more, Wollaston returned a statement of his reliance on the primacy of observation.

> The fact is that however ready to communicate any decisive observation or decided opinion, I am certainly at all times very reluctant to utter & still more so to publish any mere conjecture. You may try to pump me but you cannot exhaust beyond a vacuum.[42]

It appears that Wollaston in this letter is resisting Young's prompting to endorse the vibrational theory of light, a theory that Wollaston still considered to be conjectural. But Young's marshaling of evidence consistent with the theory was beginning to have an effect. Shortly after the above reply was sent, Wollaston wrote to Hasted.

> Young (phaenomenon) finds that these [colours observed by light refraction through thin glass plates] as well as many other facts yet unexplained become very intelligible upon the old hypothesis of ethereal vibrations & I am inclined to think he will nearly prove that to be the true doctrine.[43]

There is another, undated letter (likely from 1802), in which Wollaston comments on a draft article sent him for review by Young, in which he more jocularly reaffirms his commitment to the primacy of experimental fact over theory.

> If you have the evidence of facts you may stand still to eternity but if you stir a step beyond you may set one foot in y[ou]r grave, & it is an even chance that a physician compleats your destruction, for tho' . . . yours most sincerely, W. H. Wollaston . . . magis amica veritas [truth is a better friend].[44]

We will see later in this chapter why Wollaston thought that he, the "physician," might have results troublesome to Young's conclusions.

The statements quoted above in letters to Young are the best primary evidence available on Wollaston's approach to natural philosophy, consequences of which permeate his scientific legacy. Comprehensive theories are constructed by bold thinkers through induction from selected experimental and observational data. This is very difficult, if not impossible, to do in a strictly objective way. If a person waits until the data set becomes convincing enough to make the inductive result obvious, that person will probably not be the first to introduce the theory or law to science. Someone bolder, hastier, more ambitious, less fearful of error, or just more reckless will announce the inductive result sooner, with less supporting data. And the reward system of science, together with much historical reconstruction, champions those who get to the destination first. Those who aspire to be game changers must be prepared to construct their theories on preliminary evidence and hope to see their inductions subsequently justified. Such a scientific innovator was what Young and Davy aspired to become (both were largely successful), but not so Wollaston. He preferred to build up the database on which inductive laws are constructed, to avoid advancing from conjecture to conclusion until the result was inescapable, and to recommend caution to those who were less methodical.

The caution with which Wollaston proceeded from experiment to theory was one of his defining characteristics from the beginning of his career until the end. His stubbornly careful reasoning was lauded by some and lamented by others. For example, Charles Babbage, the future inventor of the analytical engine who came to know Wollaston well in the last decade of his life, assessed his talents in the following way:

> Caution and precision were the predominant features of the character of Wollaston, and those who are disposed to reduce the number of principles, would perhaps justly trace the precision which adorned his philosophical, to the extreme caution which pervaded his moral character. . . .
>
> Dr. Wollaston [in comparison to Davy] appreciated more truly the rarity of the inventive faculty; and, undeterred by the fear of being anticipated, when he had contrived a new instrument, or detected a new principle, he brought all the information that he could collect from others, or which arose from his own reflection, to bear upon it for years, before he delivered it to the world.[45]

PIONEERING STUDIES ON THE REFRACTION AND DISPERSION OF LIGHT

Wollaston's letters to Young were written when both were investigating the refractive properties of light, and we now turn to Wollaston's important studies on the topic, which were reported in two papers. The first, a study of refractive phenomena, was read to the Royal Society on June 24, 1802, and published shortly thereafter in the *Philosophical Transactions*.[46]

The relationship between the incident and refracted rays of light striking a transparent medium such as a glass prism had been independently discovered in the seventeenth century by the Dutch mathematician Willebrord Snellius and the French philosopher René Descartes. They derived a simple equation that related the angle of the two rays to the refractive indexes of the two mediums through which the light passed, as shown in Figure 3.3. That equation is now generally known as Snell's law. When a light ray passes from a medium of greater refractive index $\mathbf{n}_2$ at an angle Φ_2 into one of lesser refractive index $\mathbf{n}_1$ such that the refracted angle Φ_1 is 90°, that ray is said to have undergone "total internal reflection." At that or any larger angle of Φ_2, nothing can be seen beyond the surface of the substance with refractive index $\mathbf{n}_2$ because the light ray has been totally reflected at its inner surface. By designing a device that allowed the measurement of Φ_2 at known values of Φ_1 (90°) and $\mathbf{n}_1$, Wollaston could apply Snell's law to give the refractive index, $\mathbf{n}_2$, of any unknown substance.

It was the method Wollaston invented to measure the desired angle that marked his ingenuity in this study. To do so he invented the simple

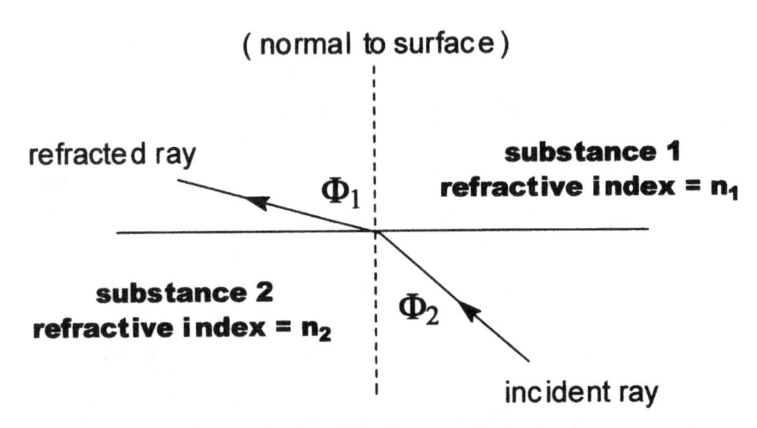

Fig. 3.3. Angles of Incidence and Refraction.

device shown in Figure 3.4. The base of the instrument is the flat piece of wood ***ab***. The substance whose refractive index is to be measured (not shown in the diagram) is applied to the underside of flint glass prism ***P***, which rests upon another flat piece of wood ***cd***, which has a cavity in its middle to accommodate the substance. Attached at ***d*** by a hinge is a 10 inch length of wood ***ed*** with 2 plane sights at each extremity. End ***e*** is hinged to a 15.83 inch length of wood ***ef***, which in turn is hinged at *f* to wooden arm ***fg***, which has a scale marked off in hundredths of an inch (Wollaston referred to it as a vernier). At the midpoint of ***ef*** is attached the length of wood ***ig***, which is equal in length to the radius of the circle shown by the dotted line. A line drawn from ***e*** to **g** thus forms a right angle with ***fg***. The entire hinged wooden construction ***defg*** changes shape when ***de*** is elevated to achieve alignment with the prismatic ray at the angle of total internal reflection. By setting the length of ***ed*** proportional to the refractive index of air, 10.00 inches, and the length of ***ef*** proportional to the refractive index of the flint glass prism, 15.83 inches, the length ***fg***, in inches, will be equal in numerical value to the refractive power of the substance attached to the prism. The instrument yields the numerical value of the desired refractive index simply and directly without the need for any calculations. As such, it was a vast improvement over previous methods.

Wollaston's total internal reflection refractometer, the predecessor of all later instruments built upon the same principle, has many advantages. It allows measurement of the refractive index of small samples because all that is needed is a very small plane surface visible through the sights on the alignment arm. Liquids, and easily melted solids, are readily measured because they can be placed in direct contact with the square prism. Nonfus-

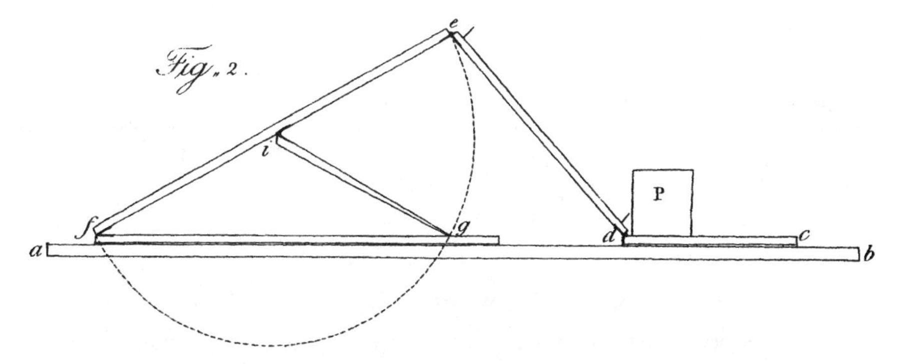

Fig. 3.4. Wollaston's Instrument for Measuring Refractive Index.

ible solid samples can also be measured if they are first polished to a flat surface and glued to the prism by a transparent cement of higher refractive index than the sample under examination. Even opaque solids could be measured, which was not possible with earlier techniques. Further, the refractive indexes of two substances can be directly compared with each other by cementing each of them side by side to the prism. This was a useful technique to distinguish between two substances of nearly identical refractive indexes. And finally, the varying refractive index of a non-uniform material, such as the crystalline lens of an eye, can be measured over its whole range by placing it in full contact with the prism. The device could not, however, yield results for a substance whose refractive index was greater than that of flint glass because such substances do not cause total internal reflection in the glass prism.

Wollaston used his instrument to measure the refractive power of sixty-two different substances, and used older techniques to obtain values for sixteen others whose refractive indexes exceeded that of flint glass. He presented all his results in a table of eighty substances. The refractive powers spanned the range from diamond = 2.44 to atmospheric air = 1.00032. Many of the substances were commercial oils, for Wollaston believed that his instrument could be used to measure their relative purity (in fact, refractive indexes have since been used for such purposes). Wollaston reports his values to three decimal places, as given by the vernier on his measuring arm. By comparison with modern values, his values are astonishingly good for substances which are likely to be of equivalent purity. Table 3.1 gives some representative results. Clearly, the easily-constructed instrument was capable of yielding accurate results, at least in the hands of a careful experimenter and keen observer.

TABLE 3.1. Refractive Indexes

Compound	Amber	Sugar	Oil of Cloves	Camphor	Lemon Oil	Alcohol	Water
Wollaston value	1.547	1.535	1.535	1.487	1.476	1.37	1.336
Modern value	1.539 to 1.545	1.56	1.535	1.485	1.481	1.359	1.333 (at 20°C)

In the second section of the paper, Wollaston extended the phenomenon of total internal reflection into an investigation of the dispersion of differently colored light rays. Newton had, of course, discovered over a century earlier that a triangular prism could disperse white light into a spectrum of refracted colors. Violet rays were refracted the most by passage through a glass prism and red rays the least. Wollaston noted in his measurements of refractive indexes that the rays of light at the angle of total internal reflection had colored fringes. He then realized he could use total internal reflection to study the dispersion properties of various liquids. He briefly described his technique in the following words.

> When a glass prism is placed in contact with water, and brought near the eye, in such a position that it reflects the light from a window, the extent of perfect reflection is seen to be bounded by a fringe of the prismatic colours, in the order of their refrangibility. The violet rays, being in this case the most refrangible, appear strongest and lowest.[47]

Wollaston does not describe exactly how he placed the prism to obtain this result but Figure 3.5 fits with his observations. A few drops of the liquid under study are deposited on the upper plane surface of a prism (the liquid is shown as a rectangle atop a triangular prism in the diagram). The refracted light at the angle of total internal refraction is viewed where it exits the triangular prism nearly perpendicular to the surface, so that further refraction at the prism/air interface is minimized. The diagram shows the refraction spectrum when water rests on the prism, in which the violet rays are refracted more than the red rays, the most well-known, so-called normal dispersion. Some liquids refracted the red rays at one end of the visible spectrum only little differently than the violet rays at the other end; they were said to have low dispersive power. Other liquids with high dispersive power separated the spectrum into a broad band. And some liquids

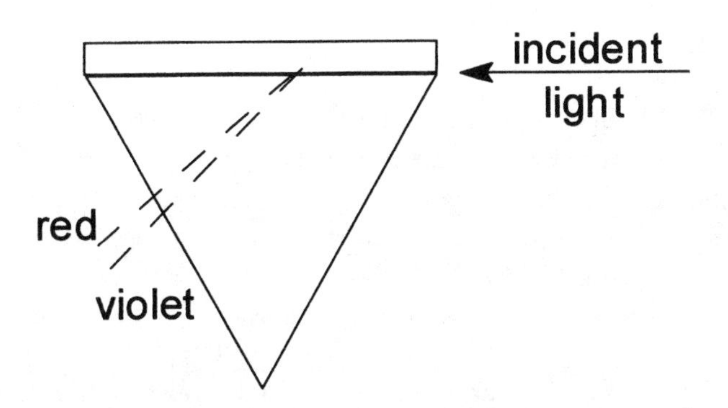

Fig. 3.5. Wollaston's Prismatic Observation of Dispersion.

even reversed the usual order of the refracted colors when combined with a glass prism, such as oil of turpentine in contact with crown glass.[48]

Wollaston found that the dispersion properties of a substance did not bear a simple relationship to its measured refractive power. Different combinations of liquid and prism material could yield any of a normal dispersion pattern, an inverted one of narrower or broader widths, or little dispersion at all. He observed that the angle of the refracted light in the prism depended on the ratio of the two refractive indexes, not on any one individually, and it varied with the color of the light ray. In substances with high dispersive power, violet rays were widely separated from the red ones whereas in substances of low dispersive power the separation was much less. He recognized three different possibilities for the net refraction of violet and red rays when light passed through two different media under the condition of total internal reflection, even though violet rays were refracted more strongly than red rays in both media.

1. When the dispersion of a substance is less than that of the prism material, violet rays are refracted through the prism at a smaller angle to the normal than the red rays, resulting in a normal dispersion pattern, as seen in Figure 3.5 when water is in contact with the prism.
2. When the dispersion of a substance is equal to that of the prism material, or very nearly so, the violet rays are refracted through the prism at the same angle as the red rays. Such refraction without dispersion of colors was the desideratum of all opticians who wished to use a combination of lenses, or of lenses and solutions, to refract and focus light with a minimum of unwanted color dispersion. The

key, by Wollaston's analysis, was to match the dispersion of the refracting substances, not necessarily their refractive indexes.

3. When the dispersion of a substance is greater than that of the prism material, the violet rays are refracted through the prism at a larger angle to the normal than the red rays, resulting in a reversed order of refracted colors. Wollaston found that he could produce just such an inverted spectrum by placing sassafras oil on a flint glass prism, or many different natural oils on a prism of plate or crown glass.

The third section of this substantial paper, and a footnote in it, contained even more novel observations. One of them involved the spectrum of reflected sunlight. In Wollaston's words:

> I cannot conclude these observations on dispersion, without remarking that the colours into which a beam of white light [daylight] is separable by refraction, appear to me to be neither 7, as they usually are seen in the rainbow, nor reducible by any means (that I can find) to 3, as some persons have conceived; but that, by employing a very narrow pencil of light, 4 primary divisions of the prismatic spectrum may be seen, with a degree of distinctness that, I believe, has not been described nor observed before.
>
> If a beam of day-light be admitted into a dark room by a crevice 1/20 of an inch broad, and received by the eye at the distance of 10 or 12 feet, through a prism of flint-glass, *free from veins*, held near the eye, the beam is seen to be separated into the four following colours only, red, yellowish green, blue, and violet.[49]

It is important to recognize that Wollaston was looking at a very narrow beam of daylight in what must have been a very dark room. The light was of much less intensity than direct sunlight and had a different color profile, but Wollaston was so confident of his observations that he proposed a color spectrum different from those proposed by others, including Newton. The twentieth-century astronomer, H. C. King, has said of Wollaston's original four-color distinction, "Anyone who views the solar spectrum in this way, especially when the prism is held near the minimum deviation position, cannot but agree with this observation."[50]

Wollaston's acute observational powers also led him to perceive, for the first time, the dark lines that appear in the solar spectrum, as illustrated in Figure 3.6. His discovery has been the subject of some historical misinterpretation, so it is worthwhile quoting Wollaston directly.

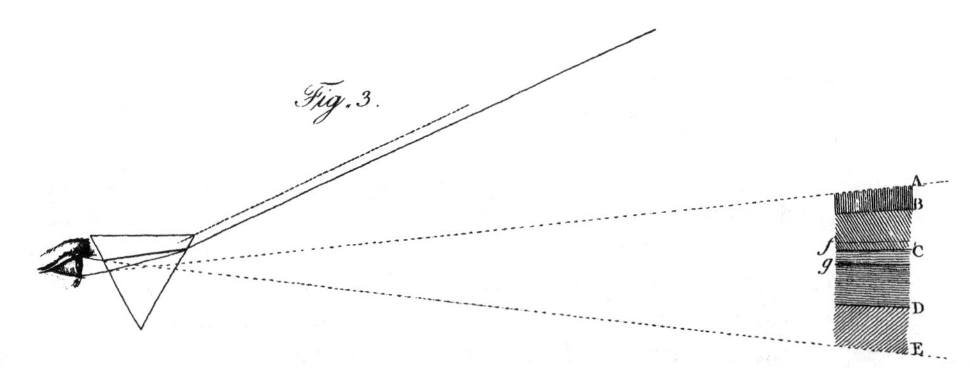

Fig. 3.6. Wollaston's Diagram of the Daylight Spectrum, Showing Its Dark Lines.

> [In the spectrum viewed as in Fig. 3.6] The line A that bounds the red side of the spectrum is somewhat confused, which seems in part owing to want of power in the eye to converge red light. The line B, between red and green, in a certain position of the prism, is perfectly distinct; so also are D and E, the two limits of violet. But C, the limit of green and blue, is not so clearly marked as the rest; and there are also, on each side of this limit, other distinct dark lines, *f* and *g*, either of which, in an imperfect experiment, might be mistaken for the boundary of these colours.[51]

This is the first mention of dark lines in the solar spectrum. Twelve years later, working with far superior prisms of his own making, the German optician Joseph von Fraunhofer independently rediscovered dark lines in the solar spectrum. He carried out a systematic study of hundreds of such lines with a spectroscope designed to reveal them more clearly, and designated the strongest ones with the letters A to K. The lines, now known to be wavelengths of light absorbed by chemical elements in the sun's outer layers or the earth's atmosphere (thus removing those wavelengths of light from the observed spectrum), are justly named Fraunhofer lines in his honor. There is no doubt, however, that Wollaston's lines are the same as some of those identified later by Fraunhofer. In 1832, David Brewster, himself a competent optician and astronomer, compared Wollaston's results with Fraunhofer's and concluded

> The correspondence of these [Wollaston's] lines with those of Fraunhofer, I have, with some difficulty, ascertained as follows:
>
> A, B, *f*, C, *g*, D, E, . . . Wollaston's lines
> B, D, *b*, F, G, H, . . . Fraunhofer's lines[52]

Brewster's comments suggest that the British, at least, recognized Wollaston's priority in the discovery of the lines, but some twentieth-century historians have suggested that he did not really identify the lines as narrow dark gaps in the spectrum, but saw them instead only as boundary markers for separate bands of color. It is true that Wollaston stated that the observed lines occupied positions that appeared to mark boundaries of the four colors (other than the faint lines f and g), but he unambiguously called them dark lines, and represented them as such in his diagram. Undoubtedly, Wollaston was the first to observe and identify some dark lines in the solar spectrum, a conclusion that does nothing to diminish the impressive and independent work of Fraunhofer.

What is generally overlooked as the most striking aspect of Wollaston's discovery, masked by the misleading clarity of his diagram, is the sighting of the lines with the aid only of a hand-held prism. It was often reported by his contemporaries that Wollaston was gifted with extraordinary senses of perception, but a more compelling explanation of his talents was offered by Charles Babbage.

> During the many opportunities I have enjoyed of seeing his minute experiments, I remember but one instance in which I noticed any remarkable difference in the acuteness of his bodily faculties, either of his hearing, his sight, or of his sense of smell, from those of other persons who possessed them in a good degree. . . .
>
> It was a much more valuable property on which the success of such inquiries depended. It arose from the perfect attention which he could command, and the minute precision with which he examined every object.[53]

And Babbage then used the observation of solar dark lines as an illustrative example.

> A striking illustration of the fact that an object is frequently not seen, *from not knowing how to see it*, rather than from any defect in the organ of vision, occurred to me some years since, when on a visit at Slough. Conversing with Mr. [John] Herschel on the dark lines seen in the solar spectrum by Fraunhofer, he inquired whether I had seen them; and on my replying in the negative, and expressing a great desire to see them, he mentioned the extreme difficulty he had had, even with Fraunhofer's description in his hand and the long time which it had cost him in de-

> tecting them. My friend then added, "I will prepare the apparatus, and put you in such a position that they shall be visible, and yet you shall look for them and not find them: after which, while you remain in the same position, I will instruct you *how to see them*, and you shall see them, and not merely wonder you did not see them before, but you shall find it impossible to look at the spectrum without seeing them." . . .
>
> It was this attention to minute phenomena which Dr. Wollaston applied with such powerful effect to chemistry.[54]

What Babbage finds in Wollaston is creativity, perhaps even genius, of a different sort. It consists of looking at things with an open, unbiased mind, and looking at every detail with every sense fully engaged, unclouded by preconception. Wollaston's detachment from his audience while inspiring nitrous oxide also comes to mind in this context. Such objective, careful observation is very difficult for most of us to do, but it is a route to discovery that we are all capable of, if sufficiently curious, motivated, and trained. There is genius in observation as well as in induction and deduction.

Wollaston made other interesting observations on dispersion spectra. When the blue flame at the lower part of candle light was used as a light source instead of daylight, he observed that the spectrum was divided into five bands, separated from each other. Those bands are colors emitted by the combustion of candle wax (hydrocarbon emission bands). They were studied more closely by William Swan in 1856 and have come to be known as the Swan bands.

Earlier in this chapter, I noted that Wollaston warned Young that some of his ideas might be endangered by the work of a "physician," and it is likely that the dispersion spectra discussed above were the basis for this cautionary statement. In a paper read to the Royal Society one week after the reading of Wollaston's dispersion paper, Young changed the color sensitivities of his proposed retina receptors from red, yellow, and blue to red, green, and violet "In consequence of Dr. WOLLASTON'S correction of the description of the prismatic spectrum."[55] Human retinal receptors are, it has been confirmed, optimized for red, green, and blue light.

The last discovery in Wollaston's impressive dispersion paper was introduced in a footnote near the end. In it, after acknowledging Herschel's study of heating rays less refractive than the red (now known as infrared rays), Wollaston mentioned his discovery of other invisible rays, "more refracted than the violet" that revealed their presence by their action on white muriate of silver (silver chloride).[56] He was led to this discovery by the eighteenth-century observation of Scheele that muriate of silver turned black when exposed to violet light rays. Wollaston repeated Scheele's experiment and found that "the discoloration may be made to fall almost entirely beyond the violet" (into the region we now know as the ultraviolet), and he noted that a similar conclusion had been reached independently about the same time by Johann Ritter of Germany. He concluded therefore that sunlight was divisible by refraction into "six species" of rays, four visible, and two invisible.[57]

Wollaston's paper on refractive and dispersive powers was well received and was soon translated and republished in the *Annales de chimie* in 1803. It revealed the breadth of his capabilities as a natural philosopher and contained examples of his talents for designing scientific instruments, employing simple experiments to obtain accurate results, grouping disparate observations into helpful reference tables, and observing the fine details of phenomena. Henry Brougham, reviewing Wollaston's paper in the *Edinburgh Review,* stated that the internal reflection refractometer was an ingenious contrivance infinitely superior to any other method for the measurement of refractive powers. In addition he claimed that the discovery of ultraviolet rays was the most important discovery in the whole volume of the journal and, in fact, " the most important discovery that has been made for many years in physical science."[58] Brougham ended by encouraging Wollaston to continue his fertile studies on refracted light.

Wollaston did fulfill one of his reviewer's wishes by publishing a paper two years later on the chemical effects of light. The short paper contained the results of experiments conducted to refute Ritter's description of the rays beyond the violet as de-oxidating.[59] Wollaston concluded that it was incorrect to label ultraviolet rays that way because their effect on gum guaiacum, a yellow substance that turns green on exposure to light, was demonstrably an oxidating one. Furthermore the rays beyond the red acted to de-oxidate guaiacum by their heating effect. In general he claimed that it was preferable to describe the effect of both red and violet invisible rays as "chemical," instead of attaching theory-laden oxidative labels to them.

DOUBLE REFRACTION IN ICELAND SPAR

Iceland spar, a transparent crystalline form of calcite (calcium carbonate), has the unusual property of splitting an incident ray of light into two, an ordinary ray and an extraordinary ray, as shown in Figure 3.7. This property, generally called double refraction (birefringence in modern terminology), had been discovered in the seventeenth century. The ordinary ray is

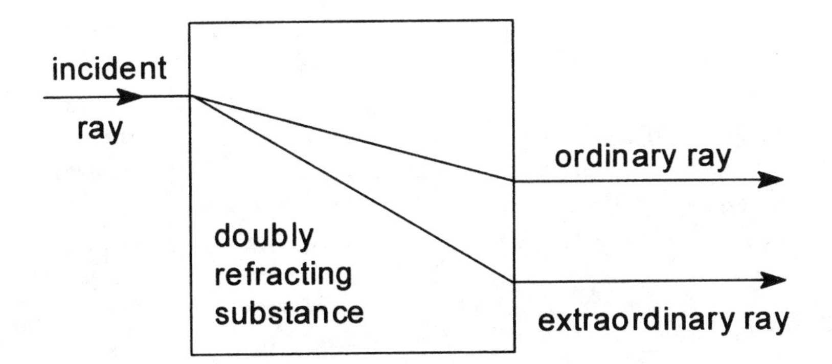

Fig. 3.7. Double Refraction in a Transparent Crystal.

refracted at the angle predicted by Snell's law, but the extraordinary ray is not. In 1690, Christian Huygens advanced an explanation based on the assumption that light in the crystal was propagated by two different pulsating wave fronts. Spherical waves moved through the crystal to give the normal refracted ray while elliptical waves moved through the crystal in a different direction to give the extraordinary ray.[60] Newton, who was skeptical of a wave theory of light, proposed in the second edition of his *Opticks* (1717) an alternative explanation on the assumption that light was a rectilinear propagation of corpuscles. His derived rules of refraction also gave results that accorded fairly well with observations. By the beginning of the nineteenth century, the corpuscular explanation had become dominant, due primarily to Newton's great prestige and a consequent widespread antipathy to a vibratory theory of light.[61]

In the prismatic reflection paper discussed earlier in this chapter, Wollaston reported values for two different refractive powers for Iceland spar; the larger angle of refraction had n = 1.657 and the weaker had n = 1.488. He must have communicated these results to Young, who later disclosed that he advised Wollaston to read Huygens's explanation of the phenomenon for the best relevant information.[62] This Wollaston did, and his results were subsequently read to the Royal Society on the same night as the refraction paper previously discussed, and published shortly thereafter.[63]

Wollaston realized that his refractometer could be used to test the predictions of Huygens's theory. This could be done by measuring the refractive power of Iceland spar at a number of places on selected crystal surfaces after cementing the crystal to the prism with balsam and orienting it in the desired way. By applying spherical trigonometry to the measured angles of the crystal, Wollaston was able to get good agreement between Huygens's theory and his results. He informed the reader

> The observations that I have made on this substance, accord throughout with this hypothesis of HUYGENS; the measures that I have taken, correspond more nearly than could well happen to a false theory, and are the more to be depended on, as all my experiments, excepting the last, were made prior to my acquaintance with the theory, and their agreement was deduced by subsequent computation.[64]

Wollaston transmitted two messages with this statement. The first is that he believed the number, precision, and accuracy of the results were sufficient to inductively support Huygens's theory, at least in its application to double refraction. It was, therefore, unlikely to be false (even though he could not bring himself to say he believed it to be true). The second is that he believed the results to be free of one type of experimenter bias; they were measured in the absence of a guiding theory and so could not have been bent toward more conforming numbers. Remember that Wollaston had warned Young about seeking data that he wished for. He had no wish to be censured for the same methodological error.

Claiming that his results were consistent with a wave interpretation of light was an unusually bold statement about a controversial theory. Wollaston recognized that his experimental measurements on Iceland spar were insufficient to establish the validity of Huygens's wave theory for a wider range of optical phenomena (such as the effect on the rays of two crystals of Iceland spar aligned transversely to each other). Yet, despite his theoretical leanings, Wollaston chose not to criticize Newton's corpuscular explanation in the paper. We know that he had read Newton's *Opticks* carefully, and was cognizant of his corpuscular interpretation, but he preferred to render his judgments in positive terms, not negative ones.

A careful modern analysis of Wollaston's results has demonstrated that, although the trigonometrical construction he used for some of his calculations is not strictly applicable to all of the angles measured, his results fit remarkably well with Huygens's predictions.[65] Its only weakness, when compared to later work, was its limitation to measurements of refraction at angles of total internal reflection, as dictated by the requirements of his refractometer. For many, Wollaston's paper provided sound observational evidence in support of the vibrational theory of light, and it became a reference point for later investigations in France on the topic. But not everyone was prepared to accept his results with equanimity. Henry Brougham, for example, in his anonymous review of Wollaston's paper in the *Edinburgh Review*, wrote:

> We were much disappointed to find, that so acute and ingenious an experimentalist had adopted the wild optical theory of vibrations. . . . [Nonetheless] it must be acknowledged, the near coincidence of the experiments, which are extremely well contrived, and appear to be accurately conducted, give this theory a plausibility which it did not before possess.[66]

Such remarks confirm that Wollaston's reputation as an influential natural philosopher had quickly spread beyond London, and his views had begun to carry such weight that they required censure by those who opposed them. Antipathy to vibratory interpretations of light by Newtonians such as Brougham made the theory controversial, and it is easy to see why Young, who also suffered editorial censure via the pen of Brougham, was so keen to recruit Wollaston to his cause. This could well explain why, a few years later, Young came to Wollaston's defense in a priority dispute on the matter.

In 1807, the Frenchman Étienne Malus undertook an extensive investigation of refraction in Iceland spar, using an instrument that allowed measurements of refractive index over a whole range of incident light angles. He was able to verify Huygens's predictions and confirm Wollaston's results with impressive accuracy.[67] In 1809, Pierre Simon Laplace succeeded in constructing a corpuscular interpretation of double refraction that accorded fully with the results obtained by Malus. In his publication on the topic, Laplace gave Malus full credit for verifying Huygens's theory of double refraction in Iceland spar.[68] Thomas Young objected to this (as well as to the larger conclusion that a particle theory of light was superior to a wave theory) and, in a review of Laplace's ideas, re-emphasized Wollaston's contributions.

> We know nothing of the extent of Mr. Malus's researches, but we know that Mr. Laplace sometimes reads the Philosophical Transactions, and he either must have seen, or ought to have seen, a paper published in them by Dr. Wollaston, as long ago as the year 1802, which completely establishes the truth of the law in question, on the most unexceptional evidence, and by the most accurate experiments.[69]

In all likelihood Laplace had not known of Wollaston's work when he wrote his treatise on double refraction, because he was quick to give credit when he shortly after became aware of it.[70]

OPTICAL INSTRUMENTS FOR NAVAL USE

Wollaston's researches on refraction were well received by the Royal Society, and he was chosen to give the Bakerian Lecture for 1802. His lecture, read November 11, 1802, contained some new observations on atmospheric refraction and a suggestion for a new approach to measuring the dip in the horizon due to refraction.[71] The lecture extended his earlier observations of refraction at sea to similar phenomena observed over much shorter stretches of the Thames. The added observations were made in 1800 and 1801 from a boat on the river or from the shore near his Cecil Street residence. By fitting a plane reflector to the object end of a small pocket telescope at an angle of 45°, Wollaston was able to measure refractions very near the water's surface by holding the telescope vertically within an inch or two of the water. He found that refraction varied greatly with height on different days; sometimes the refraction was greatest at the height of a foot or two. His conclusion was that atmospheric refraction, which resulted in inversion of portions of objects at the horizon (such as the oars of barges), was strongest when the air temperature was lower than the water temperature. Such conditions were common in the early morning when cool air from the land moved over warmer water under relatively still conditions.[72]

Drawing on these observations, he warned that nautical observations made with a sextant (which involved measuring the angles of celestial objects such as the sun relative to the horizon) near land masses could be subject to large errors due to increased refractive "dip" in the horizon. Navigators were aware of the general problems caused by normal atmospheric refraction on the open seas, and used tables to correct for average values of horizontal dip and height of the measurement above sea level. But there was no way of accounting for larger errors near land, where calculations of latitude were even more crucial. In early 1803 Wollaston took his idea for an improved instrument for measuring the dip of the horizon to the Commissioners of the Board of Longitude. After considering the proposal, the Board passed a resolution "That an Instrument upon Hadley's principles be constructed by one of the best Artists under the direction of Dr. Wollaston."[73] We learn of the evaluation of this, and two other instruments of Wollaston's design, from later accounts of voyages of discovery, such as the North Polar Expeditions.

Instructions for the use of all three, known as a dip micrometer, a dip sector, and a macrometer, were published under the auspices of the Royal Society in 1818.[74] The dip micrometer was a small telescope with lateral openings at the object end to sight opposing horizons at the same time. The instrument was first to be held in a vertical position with the sighting windows at the lower end and the position of the two horizons brought into coincidence by a micrometer adjustment. Then the instrument was inverted and a similar measurement of both horizons taken. The difference in the accumulated micrometer reading was easily converted to a value for the dip in the horizon. The macrometer was a device intended to measure the distance to a remote object. It consisted of a sextant-like device with two reflectors positioned about 1.5 yards apart. The images of the object are drawn into coincidence by an adjusting screw for one of the reflectors, and the adjustment required is converted to a virtual angle of triangulation. Reference to a table gives the distance of the object in yards, over a range from 70–30,000. There is little known about the use of either instrument on the voyages, although it is likely they were both tested.

In contrast, the dip sector (the instrument first suggested to the Board of Longitude) was employed, albeit with mixed success. It was described by a contemporary to be a variation of an instrument first used by Samuel Vince of Cambridge, although the earliest design was probably more similar to Hadley's reflecting quadrant.[75] Similarly to the dip micrometer, it had a pair of half-silvered reflecting glasses mounted at the object end of a small telescope such that they reflected the lines of opposing horizons to the eye-end of the telescope, which was placed at a 90° angle to the telescope body for easier viewing (Figure 3.8).[76] The instrument provided values for the dip of

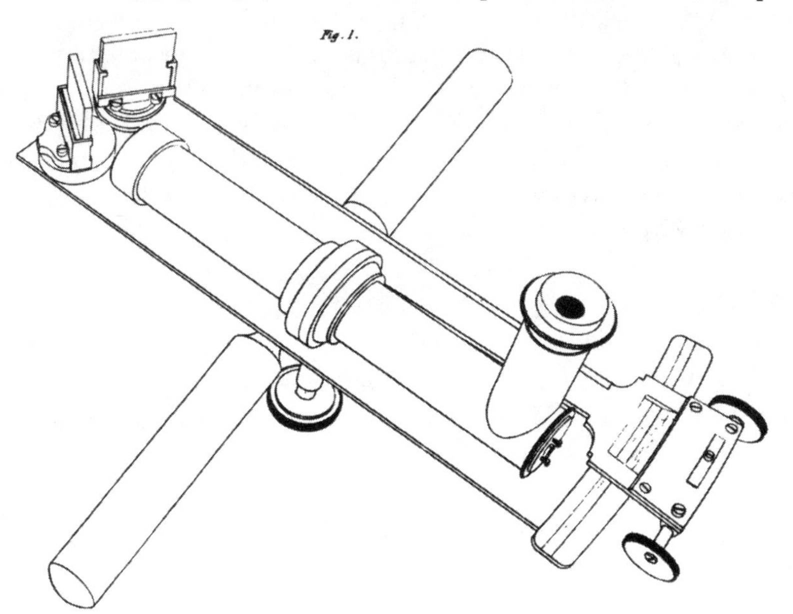

Fig. 3.8. Wollaston's Dip Sector.

the horizon accurate to fractions of a degree. Captain Basil Hall used the dip sector on his voyage of discovery to Korea and concluded, "It is much to be wished that this excellent instrument should be brought into general use in navigation."[77]

Captain John Ross was instructed to evaluate the dip sector on his first voyage in search of a northwest passage through the arctic in 1818. He found the instrument gave unreliable readings in the presence of ice. William Parry also used the dip sector during his exploration of the arctic and he, too, found it to be of limited utility in the presence of ice.[78] Although the dip sector remained in use well into the nineteenth century, its corrections to tabulated mean values of dip were only significant under special, and variable, conditions so it never became an essential navigational instrument.

Wollaston's contributions to natural philosophy in his first few papers were honored in November 1802 by the Royal Society, which awarded him its premier prize for research in the physical sciences, the Copley Medal.[79] Humphry Davy was also to receive the award in 1805 but, surprisingly, Thomas Young, the third of London's trio of brilliant natural philosophers, never did receive it.

It is evident that Wollaston had, by the end of 1802, made a successful transition from unhappy physician to notable natural philosopher, but without any improvement in his financial situation. His contributions to scientific knowledge and development of various instruments brought him some measure of intellectual acclaim, but not one shilling of income. But, as mentioned earlier, Wollaston was leading a double life. Unknown to his contemporaries he was working undercover during this time in his home laboratory to produce marketable goods from crude platina ore and raw natural products. The first results of those investigations were to appear in 1803, and we turn next to those studies.

CHAPTER 4

Malleable Platinum
1800–1801

> There can be no doubt, however, that [Wollaston] produced the best platinum so far made and that it was many years, perhaps as long as a century, before its equal was seen again.[1]

EARLY STUDIES OF SPANISH PLATINA

The mineral known as platina first became generally known to Europe in 1748 when the Spaniard Antonio de Ulloa described it in the account of his voyage to South America.[2] Silver-colored "white" platina was found intermixed with gold as an alluvial deposit in the beds of the San Juan and Atrato rivers, in the rainy lowlands of the Choco region of the Spanish viceroyalty of New Granada (now Colombia). During gold recovery from the river deposits, the silvery platina grains were tediously separated and routinely discarded, or transported to government repositories, because of the fear the dense metal could be used to debase gold. To prevent such a possibility, and to reserve whatever use it might find for the king of Spain, the Spanish Minister for the Indies ruled that platina could only be exported to Spain, from where it was distributed to interested parties. Not surprisingly, a vigorous smuggling trade in platina gradually developed, which moved illicit ore from the Choco region north to the port of Cartagena and then on to the English colony of Jamaica.[3] Traders there could legally sell any platina that came into their hands, and they did so at a price of 5–10 Spanish dollars per (Troy) pound.[4]

The first experimental investigation of platina was carried out by Charles Wood, an assayer who had brought a sample from Jamaica to England in 1741. He determined that the impure metal was extremely hard, very dense, and could not be melted in the hottest fire. It could, however, be combined with other metals such as lead or tin to form an alloy, which could be melted.[5] His work attracted interest throughout Europe, for the discovery of a new metal chemically similar to gold and silver was completely unexpected. Other key discoveries made in the last half of the eighteenth century were that platina could be dissolved in aqua regia, a combination of aqua fortis (nitric acid, HNO_3) and muriatic acid (hydrochloric acid, HCl), and precipitated as a yellowish-orange powder by sal ammoniac (ammonium chloride, NH_4Cl). The precipitate decomposed on strong heating to give a relatively pure powder of platinum.[6] Little could be done with platinum powder, as no reliable methods to consolidate it into a compact solid could be established. But a Parisian apothecary, Antoine Baumé, discovered in 1773 that he could get two very hot pieces of purified platinum to cohere by hammering them together.[7] Shortly after, in 1778, Macquer succeeded in heating platinum powder until the particles combined to form a solid mass, but one that only became malleable upon reheating to even higher temperatures.

Greater success was achieved by those who combined platina with arsenic to form a low-melting alloy. The alloy could then be strongly heated in air to drive off the arsenic, leaving behind solid platinum which was malleable enough to be hammered into useful shapes. Franz Karl Achard was able to make the first known platinum crucible by this method.[8] Working with arsenic was hazardous, however, for its toxicity was well known, and there was no way to use it without exposing oneself to quite dangerous levels.

The most successful consolidation of platinum powder was achieved later in the eighteenth century by Pierre Chabaneau in Vergara, Spain. He developed a technique by 1786 (which he kept secret) that involved the consolidation of purified platinum powder by heating and hammering. Nonetheless he, like everyone else who produced some form of malleable platinum at this time, struggled to get a consistent product. Even so, Chabaneau's sporadic successes caught the attention of the Spanish government, which reacted by paying miners a nominal amount for New Granada's platina and encouraging its accumulation and export to Spain. There never was a large amount of the ore available in New Granada (the only source known at the time), but about 2500 kilograms had reached Spain by 1800. Most of that was subsequently shipped to France, where there was some market for it.[9]

I mentioned in Chapter 1 that Tennant had worked with Crell in Germany in 1784 on methods to purify platina, and Wollaston had himself made an unsuccessful attempt to melt it in a blacksmith's forge during his undergraduate years at Cambridge. Furthermore, Wollaston's oldest brother, Francis John Hyde, the Jacksonian professor at Cambridge, included the basic knowledge of platina in his lectures of 1794.[10] So it is not surprising that both Tennant and Wollaston did some preliminary work on the chemistry of platinum in the late 1790s. In 1797, for example, Tennant published a method of making a soluble nitrate of platina.[11] Wollaston began his own studies of the metal by purchasing one ounce in June 1799, and a further four ounces in January 1800.[12] Their interests were then further stimulated by the appearance of a paper that may well have prompted them to establish their partnership.

On January 9, 1800, Richard Knight, proprietor of a London chemical supply business, read a paper to the British Mineralogical Society on a new process for making malleable platinum, which was shortly thereafter published in the *Philosophical Magazine*.[13] The brief paper reported that the platinum available from France, although as costly as gold, was of poor quality. In Knight's improved process initial treatment of the crude ore followed customary procedures. The crude platina was boiled in a glass vessel with a large excess of aqua regia until all the ore was drawn into solution. Sal ammoniac was then added until the precipitation of a yellow-colored solid was complete. The precipitate was washed with water until it was free of acid, and dried. The novel aspects of the process lay in Knight's subsequent treatment of the dried precipitate. He placed it in a hollow cone of crucible earth, covered it and heated the crucible to white heat in a chemist's furnace. The heating decomposed the yellow precipitate, expelled the volatile components and left behind a porous, grey metallic solid. While still in the furnace, the crucible cover was removed and a hot plug shaped to fit the crucible was placed on top of the metal and tapped carefully but forcefully to compress the spongy metal into a solid "button." The platinum button, Knight claimed, "may be drawn, by repeatedly heating and gently hammering, into a bar fit for flatting, drawing into wire, planishing (hammering sheet metal against a shaped surface), etc."[14] It appeared as if Knight had solved the problem of producing malleable platinum, and his process invites careful examination.

The purification of crude platina involves three steps: solution of the ore, precipitation of a purified platinum salt by sal ammoniac, and thermal decomposition of the precipitate to yield platinum as a porous, spongy solid, as shown in Equation 4.1. Solution of platinum in aqua regia is a surprisingly complex process, and I will look at it from a modern perspective to help understand Wollaston's crucial innovations. Nitric acid (HNO_3) first oxidizes the platinum and muriatic acid (HCl) provides the chloride ions to form chloroplatinic acid (H_2PtCl_6). The action of the nitric acid will vary depending on reaction conditions. If it is reduced to nitrogen dioxide (NO_2), the solution process can occur according to Equation 4.2. Alternatively, or

$$\text{platina ore} \xrightarrow[\textit{regia}]{\textit{aqua}} \text{dissolved platinum} \xrightarrow[\textit{ammoniac}]{\textit{sal}} \text{yellow precipitate} \xrightarrow{\textit{heat}} \text{platinum powder}$$

Equation 4.1. Three-Step Purification of Platinum.

$$Pt + 4\ HNO_3 + 6\ HCl \longrightarrow H_2PtCl_6 + 4\ NO_2 + 4\ H_2O$$

Equation 4.2. Oxidation with Nitrogen Dioxide Production.

$$3\ Pt + 4\ HNO_3 + 18\ HCl \longrightarrow 3\ H_2PtCl_6 + 4\ NO + 8\ H_2O$$

Equation 4.3. Oxidation with Nitric Oxide Production.

additionally, if reduced to the more stable nitric oxide (NO), Equation 4.3 applies. By Equation 4.2, 1 mol of platinum requires 4 mols of nitric acid to oxidize it to soluble chloroplatinic acid, whereas by Equation 4.3, each mol of platinum requires only 1.3 mols of nitric acid. Both reactions require 6 mols of muriatic acid to completely dissolve 1 mol platinum. It is important to realize that all chemical equations, like the two shown, represent the complete conversion of reagents to products when there are no other complicating factors. Such ideal circumstances are almost impossible to obtain for the reactions under consideration. For example, aqua regia also reacts with metallic impurities in crude platina, thus reducing the amount available for reaction with platinum. In addition aqua regia itself is an unstable solution as the two component acids react with each other to produce gaseous products that escape from solution, again reducing the amount of acid available for platinum oxidation. These and other competing processes make it necessary to add more than the stoichiometric amounts of the two acids to dissolve a desired weight of platina. Equations 4.2 and 4.3 predict that each mol of platinum requires a minimum of 1.3 mols of nitric acid for solution, and any amount in excess of 4 mols is unnecessary. Amounts between those two extremes will give the same chloroplatinic acid, but different nitrogen oxide products (which are unimportant to the solution of platinum). In either case, a minimum of 6 mols of muriatic acid are required for complete solution of 1 mol of platinum.

The limitations to Knight's purification process can now be better understood. He gave no weights of his reagents or products, and only stated that his aqua regia was composed of equal parts of nitric and muriatic ac-

$$H_2PtCl_6 + 2\,NH_4Cl \longrightarrow (NH_4)_2PtCl_6 + 2\,HCl$$

Equation 4.4. Ammonium Chloroplatinate Precipitation.

$$(NH_4)_2PtCl_6 \xrightarrow[\text{heat}]{\text{strong}} Pt + 2\,NH_3 + 2\,HCl + 2\,Cl_2$$

Equation 4.5. Formation of Powdered Platinum.

ids of unspecified strengths. Contemporary recipes for aqua regia generally specified some mixture of nitric and muriatic acids, without specifying the strength of either (and these varied widely in commercial acids). Thus, anyone repeating Knight's process would not, except by chance, be able to prepare aqua regia identical to his. Different formulations of the dissolving acid would give solutions containing different compounds and the type of erratic results that others had encountered.

The second, precipitation step of Knight's process is chemically simple, as shown in Equation 4.4. Chloroplatinic acid formed by the action of aqua regia reacts with added sal ammoniac to form ammonium chloroplatinate ($(NH_4)_2PtCl_6$) which is highly insoluble in aqueous solution and consequently precipitates as a pale yellow solid. Each mol of the acid requires 2 mols of sal ammoniac. As Knight reports, the sal ammoniac is added in small portions until precipitation ceases; the reaction signals its own endpoint. This is the key step in the separation of platinum from impurities in the crude ore, because most of the impurities remain in solution while the platinum-containing salt precipitates.

The third and final step of the process is the thermal decomposition of the solid precipitate, as shown in Equation 4.5. All of the decomposition products, except platinum, are gases, which dissipate into the air. The platinum is left in a finely powdered, very porous, sponge-like state, which Knight claimed he could compress into a compact, malleable mass by pressing it together in the heat of a furnace. If the purification and compaction process did indeed work as described, it would certainly have been "a new and expeditious process," but Knight never attempted to sell malleable platinum, and there is no evidence that anyone else using his process did either. Obviously, there were some complicating factors involved that were not addressed in his paper.

WOLLASTON'S PLATINUM PURIFICATION PROCESS

Shortly after Knight's paper was published, Wollaston purchased an additional twelve ounces of platina, perhaps from Knight himself, who had some crude ore for sale.[15] If Wollaston began to evaluate Knight's method, it was not done systematically until the fall because, as described in Chapter 3, Wollaston was busy with several other things at the time. But sometime before November 8, he carried out a series of small-scale platinum purification trials, which yielded results essential to the later success of the platinum business. The details of the trials were cryptically recorded in the pages of a small notebook containing miscellaneous observations on a variety of topics. The sequence of preliminary experiments appears on two pages, reproduced in Figure 4.1.[16] The individual trials have been labeled to aid discussion, although they may not necessarily have been performed in the sequence implied by the numbering.

In his research notebooks, Wollaston generally used the system of weights, volumes, and symbols commonly employed by apothecaries and physicians. English apothecary weights were identical to troy weights: 1 pound was made up of 12 ounces, 1 ounce of 8 drams, 1 dram of 60 grains. The apothecary/troy pound (373 g) was lighter than the commercially-common avoirdupois pound (454 g). The English gallon in use before 1824 (128 fluid ounces) was made up of 8 pints, 1 pint of 16 ounces, 1 ounce of 8 drams. One fluid ounce of water weighed one apothecary/troy ounce. The symbol for a dram is ʒ and the number of drams is given by lower case roman, or arabic, numerals; one-half is denoted by ſs. Chemical substances are denoted by their then-common symbols, or by alphabetical abbreviations. With such information at hand, trial 1 reads as follows:

> [crude] Platina 1 dram [60 grains], left 18 grains untouched by 4 fluid drams muriatic acid [and] 1 dram nitric acid, dried and washed. There was also a brown powder, undissolved by water, but dissolved in great measure by muriatic acid. The two solutions precipitated by 5/8 dram sal ammoniac, gave precipitate 1 dram 4 grains = 64 grains. Solution evaporated and heated left 8 grains.

Clearly, Wollaston is following the normal process for purifying platina: solution in aqua regia followed by precipitation with sal ammoniac. Characteristically, he elects to experiment with very small quantities, only 1 dram (3.9 g) of crude platina and a total of 5 fluid drams of aqua regia, just over one-half ounce, in trial 1. He could get all the information he needed

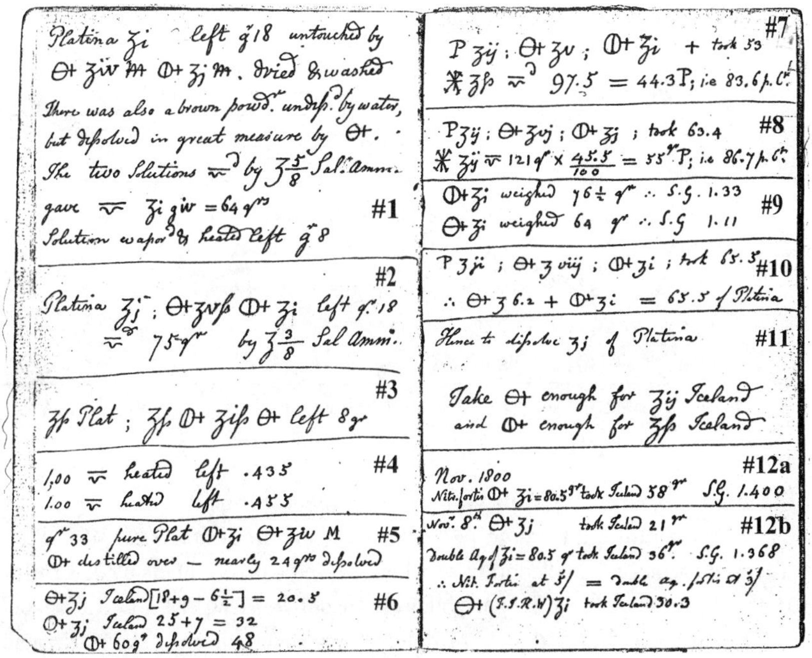

Fig. 4.1. Early Platina Solution Trials, Oct.–Nov. 1800. Reproduced by kind permission of the Syndics of Cambridge University Library.

working on this small scale, without consuming his small supply of platina too quickly. But there are several important innovations evident in these trials. The weights and volumes of all substances are recorded, and systematically varied, to optimize reaction conditions. Muriatic acid is present in greater volume than nitric acid, and the total amount of aqua regia used is less than required to dissolve all the platina. In trial 1, for example, only 42 gn. out of 60 (70 percent) is dissolved.[17] Unlike Knight and all of his predecessors, Wollaston opted to dissolve only part of the crude ore, likely on the suspicion that the crude ore contained poorly soluble material that could contaminate the final product if drawn into solution. To understand the full rationale behind Wollaston's purification trials, I will focus first on the information he provided on the acids used.

The entries labeled 6, 9, 12a and 12b are not records of solution trials. Instead they give information about the strength of the acids Wollaston used for the aqua regia solutions. Commercial muriatic acid was generally sold as a aqueous solution of specific gravity 1.11, which corresponds to a weight percentage of actual acid in water of about 23 percent. Nitric acid could be

TABLE 4.1. Platina Purification Trials, 1800

Trial #	Pt (gn.)	HCl (fl. ʒ)	HNO_3 (fl. ʒ)	Pt dissolved (gn.)	$HCl:HNO_3$	AqRegia:Pt (vol.:wt.)
1	60	4.0	1.0	42	4: 1	5: 1
2	60	5.5	1.0	42	5.5: 1	6.5: 1
3	60*	3.0*	1.0*	44*	3: 1	4: 1
7	60*	2.5*	0.5*	26.5*	5: 1	3: 1
8	60*	3.0*	0.5*	31.7*	6: 1	3.5: 1
10	60*	4.0*	0.5*	32.7*	8: 1	4.5: 1
11	[60]	[5.7]	[0.9]		[6.3: 1]	[6.6: 1]

* Numbers have been scaled to equal a starting weight of platina = 60 gn.

purchased in a variety of forms, ranging from nearly 100 percent acid to aqueous solutions ranging from 30 percent to 80 percent actual acid. Wollaston measured the actual acid content of all purchased aqueous acids by determining the amount of Iceland spar (calcium carbonate, $CaCO_3$) a dram of acid would neutralize. Entry 6, for example, shows that 1 dram of muriatic acid was neutralized by 20.5 gn. of Iceland spar (18 + 9 grains added, 6.5 remained unreacted). Entry 12b gives a similar result. By knowing the weight of muriatic acid that reacted completely with a given weight of Iceland spar, Wollaston could calibrate the strength of both acids employed in each trial. Entries 6 and 12b used an acid solution that was 23 percent muriatic acid by weight. Entry 9 corroborates this by recording a measured specific gravity of muriatic acid as 1.11, which also corresponds to an acid content of 23 percent.

The strength of nitric acid was determined similarly. Trial 5 gives the results for nitric acid of two strengths, one neutralized by 32 gn. of Iceland spar and a second by 48 gn. Appropriate equivalency values yield the weight percentage of nitric acid in these two solutions as 53 percent and 100 percent, respectively. Entry 9 gives the specific gravity of nitric acid as 1.33, which also corresponds to an acid strength of 53 percent. Therefore, on the assumption that Wollaston used 23 percent muriatic acid and 53 percent nitric acid for his solution trials, it becomes possible to better quantify the information in all the solution trials, such as those numbered 1, 2, 3, 7, 8, 10 and the optimized data given in 11. The results are summarized in Table 4.1.

In all the trials Wollaston uses a volume of muriatic acid three times greater, or more, than that of nitric acid. Some contemporary recipes for aqua regia recommended a volume of muriatic acid greater than that of nitric acid, but 50:50 mixtures were the norm. Nonetheless, trials 1, 2, and 3 show that, when the volume of nitric acid is held constant at 1 fluid dram, changing the volume of muriatic acid from 3.0 to 5.5 fluid drams had little effect on the weight of platina dissolved (about 42 gn., 70 percent). Trials 7, 8, and 10 showed that reducing the volume of nitric acid to 0.5 fluid drams diminished the weight of platina dissolved, but the decrease could be offset in part by increasing the volume of muriatic acid from 2.5 to 4.0 fluid drams. The maximum weight of platina dissolved using 0.5 fluid drams nitric acid was 32.7 gn. (55 percent).

From these results Wollaston reached the important conclusion recorded as entry 11, "Hence to dissolve 1 dram of Platina, take muriatic acid enough for 2 drams Iceland Spar and nitric acid enough for 0.5 Iceland Spar." Wollaston determined that the optimal volume of nitric acid for the solution of 1 dram platina was 0.9 fluid drams, nearly the volume used in trials 1, 2 and 3, and that of muriatic acid was 5.7 fluid drams, close to the volumes used in trials 2 and 8. We can compare Wollaston's optimal amounts with those predicted by modern stoichiometry, as given in Equation 4.3 (use of the smallest effective volume of nitric acid, as Wollaston did, makes that reaction equation more applicable than Equation 4.2). The comparative numbers are given in Table 4.2.

The fit between Wollaston's solution parameters and those given by the stoichiometric relationships for an ideal reaction of pure platinum is remarkable. Wollaston used only slightly greater amounts of muriatic and nitric acids together with somewhat more aqua regia than the minimum required amounts. Since not all of the aqua regia remains unchanged long enough to react with the platina, actual experiments do require more than the stoichiometric amounts. The excess of nitric acid over the stoichiometric amount exceeds the corresponding excess of muriatic acid, so Wollaston's ratio of marine to nitric acid ends up being slightly below the ideal ratio. But, because an excess amount of each acid was used, this difference has little impact on the results. What allowed him to achieve such good results?

TABLE 4.2. Wollaston Platina Solution Amounts vs. Modern Stoichiometry

	Platina	HCl	HNO_3	HCl: HNO_3 (by vol.)	AqRegia: Pt (vol.: wt.)
WHW, orig	60 gn.	5.7 fl. ʒ	0.93 fl. ʒ	6.3: 1	6.6: 1
WHW, mol	0.02	0.15	0.04		
Modern, mol	0.02	0.12	0.03	7.2: 1	5.2: 1

The most important thing he did was to measure carefully the strength of his acids by calibration with Iceland spar. In addition, by using a total amount of aqua regia that was insufficient to dissolve all the platina, he was able to determine the dissolving efficiencies of differing acid amounts, individually and collectively. If a large excess of aqua regia is used, such that all the platina is dissolved (as Knight and others customarily did), it is not possible to determine the minimum required amounts of muriatic and nitric acids, or their most effective proportion. Wollaston's quantitative and small-scale approach, so characteristic of his experimental style, led him to the most efficient formulation of aqua regia, one that dissolved the most platina with the least consumption of reagents. Such reaction efficiency makes both chemical and economic sense. The more efficient he could make the purification process, the lower would be the cost of the raw materials and the greater the resulting profit.

The purification trials in notebook J also contain information about the final two steps of the purification sequence, the precipitation and thermal decomposition steps. As shown in Equation 4.4, addition of sal ammoniac (ammonium chloride, NH_4Cl) to the solution containing the dissolved platinum causes the precipitation of the yellow compound ammonium chloroplatinate ($(NH_4)_2PtCl_6$), a reaction used from earliest times for the separation of platinum from other metals soluble in aqua regia. The precipitation data listed in trials 1, 2, 7, and 8 are given in Table 4.3, together with the stoichiometric values derived from Equation 4.4. To enable comparison among the four trials for which Wollaston gives data, the weights given for the added sal ammoniac and the solid precipitate (ppt) are scaled down proportionally to 1 gn. dissolved platina. These scaled values are given in square brackets. The stoichiometric entries are those expected for complete reaction with pure platinum.

There is no opportunity for innovation in the precipitation step. Sal ammoniac is added in small portions until precipitation ceases, although the

TABLE 4.3. Precipitation Trials, 1800

Trial #	Pt Dissolved (gn.)	NH4Cl Used (gn.)	ppt Formed (gn.)
1	42 [1]	37.5 [0.89]	64 [1.52]
2	42 [1]	22.5 [0.54]	75 [1.79]
7	53 [1]	30 [0.57]	97.5 [1.84]
8	63.4 [1]	120 [1.89]	121 [1.91]
Stoichiometry	1	0.54	2.26

Note: Numbers in square brackets are scaled to 1 gn. dissolved platina.

end point can be difficult to discern in a colored solution. A bit of excess sal ammoniac causes no problems, as it is very soluble in the aqueous solution and its only effect is to render ammonium chloroplatinate even less soluble. A weight close to the stoichiometric amount in trials 2 and 7 gave more precipitate than the greater amount used in trial 1, presumably because less aqua regia was employed. In trial 8, Wollaston added a large excess of sal ammoniac to see if the amount of precipitate could be increased, but no significant change resulted. The most significant information gained by comparing his results with stoichiometric values lies with the weight of precipitate formed. In every trial, less precipitate is deposited than expected if the weight of dissolved platina were entirely composed of platinum. It is obvious that the weight of crude platina dissolved included base metals such as iron, well known to be the major impurity in the crude ore. But this is why the precipitation reaction is one of the key features of the purification sequence—metals other than platinum do not form insoluble salts upon treatment with sal ammoniac. Consequently, a precipitate weight less than the stoichiometric value must result when anything other than pure platinum is dissolved.

The third and final step is the thermal decomposition of the dried precipitate, as shown in Equation 4.5. Wollaston's results from trials 4, 7, and 8 are shown in Table 4.4, together with the stoichiometric value. In trial 4, Wollaston records the weight of platinum powder remaining after two ignitions of 1 dram of precipitate (the weights have been converted to grains in the table to ease comparison with the other trials). He gives the weight of platinum obtained in trial 7 (44.3 gn.) from 97.5 gn. of precipitate and uses that percentage composition (44.3/97.5 = 45.4 percent) to calculate the weight of platinum expected from the 121 gn. precipitate in trial 8, which gives 55.0 gn. For each of the four trials, the percentage of platinum in the precipitate is very near the stoichiometric value of 44.2 percent. This close agreement suggests the purification process produces acceptably pure precipitate (ammonium chloroplatinate), and subsequently correspondingly pure platinum. In trials 7 and 8, Wollaston also compares the recovered weight of platinum powder to the weight of dissolved platina. Trial 7 isolates 83.6 percent of the dissolved platina as purified platinum; trial 8 isolates 86.8 percent. The missing weight, Wollaston knew, was mostly due to the base metals in the crude ore that were dissolved by aqua regia but not precipitated by sal ammoniac. This lost weight would have provided the first quantitative evidence that his purification process yielded a significantly purified platinum powder. This was the first goal of his research program. But whether or not his platinum powder could be successfully compacted and transformed

TABLE 4.4. Thermal Decomposition of Sal Ammoniac Precipitate, 1800

Trial #	Pt dissolved (gn.)	ppt (gn.)	Pt powder (gn.)	% Pt in ppt	Stoich % Pt	Percent Pt recovery
4		60	26.1	43.5	44.2	
4		60	27.3	45.5	"	
7	53	97.5	44.3	45.4	"	83.6
8	63.4	121	55.0*	45.5	"	86.8

* Value calculated by Wollaston from trial 7.

into malleable metal (as Knight claimed to have done) could only be determined by subsequent consolidation trials, and those required purification on a larger scale.

One disconcerting result of Wollaston's method, at least from an economic perspective, was the low yield of purified metal separated from the crude ore. Trial 7 yielded 44.3 gn. of purified platinum from a starting weight of 120 gn. of crude platina, an overall yield of only 37 percent. Trial 8 gave 55 gn. of platinum from 120 gn. of crude ore, a slightly better but still poor yield of 46 percent. Some of the lost material resulted from base metals in the ore that were left in solution after the precipitation step, but the greater part was in the undissolved platina. This was an intended result because Wollaston had elected to treat the crude ore with an amount of aqua regia insufficient to dissolve all the ore, on the assumption that fewer harmful impurities would appear in the final product. When he later moved to purify platinum in greater amounts, the attempts would lead him to new and totally unexpected discoveries.

The trials discussed above, conducted over a few days in early November 1800, and recorded in two pages of notebook J, laid the chemical foundation for Wollaston's pioneering work in platinum refining, the achievement for which he has become most renowned. They involve no new chemistry, differing from Knight's published process only by attention to detail and careful control of aqua regia composition and amount. This prompts the question: What difference did Wollaston's purification process make to the purity of the platinum he produced?

Crude platina closely similar to that used by Wollaston is still being extracted from alluvial deposits in the rivers of Colombia. Given the detail provided in Wollaston's early trials, it is possible to repeat his purification sequence on platina from the same source with the same reagents, and to measure the purity of the platinum produced. My colleagues and I have done this, and we found his purification process could yield platinum that was 98.57 percent pure.[18] Applying Knight's process to the same alluvial ore yielded platinum only 94.04 percent pure. These results bring clarity to the problems encountered by those working to produce malleable platinum in Wollaston's time. When sufficiently pure (greater than 99.5 percent, for example), platinum can be worked as easily as silver, but any impurities increase its hardness dramatically. As the impurity level increases, platinum powder becomes both harder to consolidate into a solid mass, and increasingly difficult to hammer into useful objects. The impurity level in Knight's platinum would have been too high to give consistently satisfactory results. Like Chabaneau, whose secret process also gave erratic results, frustration at the lack of reproducibility would inevitably have led anyone who tried to use Knight's process to abandon it. Wollaston's platinum was of much higher purity and nearly at the stage where a consistently malleable metal could be obtained.

THE NEED FOR SECRECY

By the time of Wollaston's purification trials in November 1800, he and Tennant must have already had a tentative agreement in place to produce and market chemical products. There is no written record of the beginning or terms of the agreement, likely because the two men did nothing more official than commit to a sharing of expenses, labor, and profit. Wollaston's November platinum investigations, for example, were carried out at his own expense. The sharing of costs only began in December when his and Tennant's account books open with payments to a supplier named Hutchinson for the purchase of nearly 6,000 ounces of crude platina, at a total cost of £795.[19] It is unlikely that such an amount of South American platina would have been sitting somewhere in London waiting for a purchaser. It is more plausible that Wollaston and Tennant had placed an order for platina via Kingston, Jamaica, some months in advance of its arrival. The placing of such an order prior to any intensive investigation into a possible purification process was a bold, possibly foolhardy, initiative. What each man's responsibility was to the overall conduct of the business is not revealed in any known documents, but there can be no doubt that Wollaston accepted the responsibility of preparing malleable platinum. Every ounce of platinum produced by the partnership over its fifteen-year lifetime was purified and consolidated by him. If, as seems likely, Wollaston was aware that the large shipment of platina was soon to arrive at the London docks, his purification experiments would have had some urgency to them in the last two months of 1800.

As soon as the early small-scale solution trials were completed, Wollaston began a series of purification trials on a larger, but still uncommercial, scale. For example, the first of these, dated November 9, exposed 900 gn. of platina to heated aqua regia for 3.5 days, after which 492 gn. (55 percent) of platina had been dissolved. This was fifteen times the scale of his first set of experiments. In a total of eleven similar trials continuing into January 1801, Wollaston dissolved nearly 5 oz. of platina, which suggests he consumed most of the small weight of platina he had purchased the previous February.[20] These larger trials were intended to increase the percentage of platinum recovered from the crude ore and to improve the aqua regia recipe, if possible. One of the first things Wollaston did in these trials was to identify and separate visibly different materials from the crude ore prior to solution in aqua regia. The first two trials mention manual separation of small amounts of eisenglimmer, an iron-containing mineral. A few years later, as we will see in Chapter 5, he was to report the discovery of a second distinct mineral in the crude ore, which contained two metals unknown to science in 1800.

In the third of these trials, Wollaston encountered the first of the problems presented by the scaled-up procedure. In the hope of increasing the amount of platinum dissolved by the aqua regia, Wollaston heated the solution for a longer time than the norm of three to four days. Over that time much of the solution evaporated, and, as it became more concentrated, the contents bumped in the glass vessel and splashed out of the reaction vessel. The obvious remedy to this undesired occurrence was to add more water to the solution and, after a few more trials, he began to do so routinely. He also tried to reduce the amount of nitric acid in the aqua regia because it was the most costly ingredient in the purification sequence. But in every trial in which the amount was below the optimal level noted in the first set of solution trials, the amount of dissolved platina decreased. He then began to increase the amount of nitric acid until, by the end of the series of scaled-up trials, he had satisfied himself that the optimal recipe for aqua regia was that given in trial 11 (Figure 4.1) of the first series of solution trials. In fact, because that entry is undated, it is possible that it was entered only after this second series of solution trials was completed.

Wollaston treated the solutions of this second series of trials with sal ammoniac dissolved in water, unlike the first trials in which he used solid sal ammoniac. This served two purposes: it allowed closer control over the amount added, and it also diluted the solution from which precipitation occurred, leading to a product with fewer impurities. Wollaston carried only one trial all the way through to powdered platinum, and in that he obtained

413 gn. of purified metal from 900 gn. of platina, for an overall yield of 46 percent, similar to the results from the first set of trials. The precipitates from the other trials were set aside until just prior to their conversion to spongy platinum, which must have been done before the end of January 1801, when Wollaston began his first compaction trials.

There was one other important innovation introduced in this second set of trial solutions, implemented to reduce the loss of platinum in the solution that remained after the precipitation step. Wollaston had determined in both sets of trial solutions that some dissolved platinum remained in solution even when an excess of sal ammoniac was used to precipitate it. In the last two trials of this second series, he suspended an iron bar in the solution after the precipitate had been filtered off. The iron, it was known, would preferentially combine with the acids and displace other metals, such as platinum, from solution. Wollaston combined the aqua regia solutions from the last two trials, which had contained 709 gn. of dissolved platinum before precipitation with sal ammoniac. After precipitation was complete and the solid was filtered off, Wollaston placed an iron bar in contact with the residual solution and, after a period of time, collected 26 gn. of what he came to call a "metallic precipitate."[21] He believed this residue to be largely, if not entirely, platinum. The use of iron to recover more platinum from the aqua regia solution served two purposes. First, it enabled Wollaston to move closer to a weight balance for the dissolved platina. He had found from the first series of trials that he could recover about 85 percent of the dissolved platina by precipitation with sal ammoniac and now another 4 percent by iron displacement. The missing 11 percent, he would have concluded, was comprised of metals in solution not displaced by iron, either iron initially present in the alluvial ore or other base metals such as copper or lead. The second purpose was to maximize the recovery of platinum from the ore—every increase in the percentage recovery would increase profitability if the metal could be rendered malleable.

The two series of purification trials carried out from November 1800 to January 1801 were all that Wollaston needed to set the chemical parameters for the larger-scale purifications he was to start in February 1801. Not surprisingly, he made a few minor adjustments to the fundamental solution-precipitation-ignition sequence during the purification of nearly 50,000 oz. of platina he was to process up to 1821. But the critical chemical features of the purification sequence were established by the first few trials conducted over a short period of time on only 10 oz. of crude ore. The work could have been done by any good, motivated chemist with a keen eye for detail, and Wollaston knew it.

The key components of Wollaston's chemical process were (1) an amount of aqua regia insufficient to dissolve all the crude platina; (2) a novel composition of aqua regia that contained standardized amounts of muriatic and nitric acids, with the former in much greater amount; (3) dilution with water of both the aqua regia and sal ammoniac; and (4) enhanced recovery of dissolved platinum by iron bars. None of these features were innovative enough to be patentable for, as we have seen, all (except for the proportions of acids in aqua regia) had appeared previously in the literature. So how could Wollaston protect the commercial potential of the proposed business? There was only one way—secrecy. Wollaston and Tennant kept their collaboration secret, and they did this so well that the partnership only became generally known long after their deaths. Even more crucially, Wollaston did not disclose the details of his purification process until the end of his life. Only in 1828, when death was imminent, did he dictate the specifics of his procedure to Henry Warburton, who arranged for them to be read to the Royal Society on November 20 as the Bakerian Lecture (Wollaston's fourth). The lecture was published posthumously in the *Philosophical Transactions*.[22] In that paper, now generally regarded to be the seminal one in platinum metallurgy, Wollaston finally revealed the optimal proportions of the acids in aqua regia used to dissolve the crude platina.

> With regard to the proportions in which the acids are to be used, I may say, in round numbers, that muriatic acid, equivalent to 150 marble, together with nitric acid equivalent to 40 marble, will take 100 of crude platina; but in order to avoid waste of acid, and also to render the solution purer, there should be in the menstruum a redundance of 20 per cent at least of the ore.[23]

It is surprising in retrospect how closely the details of Wollaston's final chemical purification process matched with those he established almost three decades earlier, and how little of chemical novelty they contained. The chemical process was, except for critical detail, almost the same as that published by Knight. So why was no one else able to produce large amounts of good-quality malleable platinum during Wollaston's lifetime? One reason was Wollaston's and Tennant's near monopoly on the purchase of ore smuggled out of New Granada to Jamaica, and Spanish control of the rest, but another was the great difficulty of compressing platinum powder into a compact, malleable solid. By January 1801 Wollaston had solved only the chemical half of the problem, and later he would have to turn his attention to the troublesome metallurgical half.

Wollaston's second series of trial purifications netted him about 2000 gn. of platinum powder, only about 4 ounces and insufficient for a meaningful consolidation investigation. But he did try to see if he could at least transform a small amount of the powder into a solid whose specific gravity (s.g.) matched that of Knight's (and others') product. He began by shaking 373 gn. of platinum powder with water and allowing it to settle. By comparing the weight of wet platinum with a similar volume of water he determined its s.g. to be 4.2, meaning it weighed 4.2 times more than the same volume of water.[24] By pressing the powder together he was able to increase its s.g. to 7. He then opted to carry out further compactions in hollow metal cylinders, generically referred to as gun barrels. In the first gun barrel trial, he compressed some wet platinum "mud" to a s.g. of 8.86. Heating this in a common fire converted it to a firm solid with s.g. 10.2. Stronger heating in the fire of a charcoal blast gave a final solid of s.g. 14.[25] A second, similar compression trial yielded platinum of s.g. 15.5. The third and final compression trial in the gunbarrel was carried out on January 31, 1801. After the first compression of the wet mud, a piece of the plug "cracked off" and reminded Wollaston that compression was always going to be a challenging process. The remainder of the plug weighing 416 gn. was heated as usual to a s.g. of 10.6. Then, in what appears to be Wollaston's first attempt at working metallic platinum, he cold hammered the metal on an anvil to a s.g. of 18, and then forged it further to s.g. 21.1.[26] This final specific gravity value was very close to that reported for pure platinum by Knight (22.3)[27] and Lavoisier (22.1);[28] the modern value is 21.5. This encouraging result was, of course, marred by the fact that a piece of the platinum solid had broken off after being compressed, which would ruin a commercial product. To bring the consolidation procedure to a point of reliability he would have to produce much more platinum powder. And that only became possible when the South American platinum arrived in London.

About half (3,067 oz.) of the platina purchased in December 1800 was delivered to Cecil Street in early 1801. This was supplemented by another purchase by Wollaston and Tennant of 800 oz. from Richard Knight on February 12, 1801. The partners paid a higher price for Knight's platina, at 4 shillings per ounce, than they had for the larger amount of the imported platina, at 2 shillings 8 pence per ounce.[29] Despite the higher price, the purchase made good commercial sense, for it prevented others from obtaining crude platina. No further purchases were made in London until 1806, which suggests that there were no significant amounts available until then. In fact, the only other purchase was made by Tennant from a French source during the short-lived Peace of Amiens in 1802, totaling 559 oz.[30] These three purchases provided Wollaston, who processed all platina, with 7,318 oz.

in total. The refining of all that platina was to occupy much of his time for several years. And no one, other than Tennant and perhaps his father and one or two brothers, had any idea what he was doing in his small home laboratory. That is just the way he wanted it to be.

Wollaston's first series of purification runs on a scale large enough to give sufficient platinum powder for consolidation trials began on February 21, 1801. In the first batch, labelled A, Wollaston placed 12 oz. of platina in a large glass vessel and added 4 oz. of 66 percent nitric acid, 32 oz. of 23 percent muriatic acid, and 20 oz. of water.[31] Batch A used the only large glass vessel Wollaston had at the time, so batch B, begun a day later, exposed 6 oz. of platina to the action of 24 oz. of aqua regia, divided equally among three small glass vessels. In his account book for keeping records of business expenses and revenue, entries begin at this time for improvements to the laboratory stove and chimney and for raw materials and laboratory apparatus, including several larger glass vessels. The transition to larger purification batches, and maintenance of high solution temperatures for several days at a time, also raised manpower issues. Someone had to keep the furnace stoked and the flasks held at the appropriate temperature. These tasks were assigned to the manservant Wollaston had hired in January 1800, John Dowse, a man then in his early twenties. Over time, John took on more responsibilities in the laboratory, and by the end of the year entries begin in the account book for "John's book," which reveal that he ordered coal and candles, lab supplies, and chemicals. By 1802 John's assistance with Wollaston's laboratory work had become sufficiently time-consuming that part of his wages began to be entered in the account book as an expense to be shared with Tennant. John must have been a capable and trusted employee, for his duties in the laboratory continued to grow. He remained with Wollaston as manservant and laboratory assistant until his employer's death.

The third large solution, batch C, was begun on March 1 on 12 oz. of platina, as in batch A. On March 6, the platinum in each of batches A–C was precipitated by sal ammoniac and converted to spongy platinum powder by the usual methods. From the combined weight of dissolved platina, 12.5 oz., Wollaston obtained 10.25 oz. (82 percent) of platinum powder, a yield consistent with earlier trials. Immersion of an iron bar in the combined solutions of A, B and C gave 0.8 oz. (6 percent) of metallic precipitate, again as expected. So far, the large-scale purification runs were proceeding normally. But batch D produced a surprise. Wollaston saved the undissolved platina from batches A–C and, hoping to extract even more platinum from the residues, he treated 9.5 oz. of this residual crude platina with aqua regia and precipitated the platinum from hot solution ("hurried," he noted) with sal ammoniac. The precipitation proceeded with much effervescence and gave "a very red precipitate," instead of the normal yellow one.[32] This meant that some red compound had contaminated the precipitate. This was the first time Wollaston had observed such a thing, and he, not surprisingly, believed it might have been caused by carrying out the precipitation while the solution was still hot. In the following batch E, he treated more residual platina in the usual way but added sal ammoniac only after the solution had cooled. The normal yellow salt precipitated out of solution, so his suspicion seemed to be correct, and thereafter precipitation was always carried out on cooled solutions. More importantly, from an economic perspective, secondary treatment of residual platina (that which was not dissolved by the first solution in aqua regia), usually gave platinum powder in similar amount and purity to that obtained by the first solution of platina. Consequently, by cycling residual platina through the purification process a second or third time, Wollaston was able to increase the total amount of platinum extracted from the crude ore to about 75 percent. The amount of the crude ore that remained undissolved after repeated exposure to aqua regia was collected and set aside for future study. He left nothing unexamined.

In another effort to increase the recovery of purified platinum, Wollaston collected the metallic precipitate deposited by bar iron from the spent aqua regia solution, after the addition of sal ammoniac. He believed it was platinum adulterated with some base metals which could be further purified in the normal way. So, in batch N on April 6, he combined 1.5 oz. of the metallic precipitate accumulated from previous batches with 6.5 oz. of residual platina and processed the mixture in the usual way. The sal ammoniac precipitate, however, had an unusual " brown hue."[33] Cautiously, he set the platinum obtained from this precipitate aside for later investigation. This was the first sign that there was something in the metallic precipitate, and thus in the spent aqua regia solution from which it came, which did not behave either like platinum or the usual base metals. From this point forward, Wollaston kept the metallic precipitate separate from the solution of crude and residual platina. Future investigation of similar metallic precipitates would result in the unanticipated discovery of two new chemical elements, as I will discuss in Chapter 5.

The scaled-up purification batches continued until June 1, at which time Wollaston had treated 228 oz. of crude platina in 28 batches to obtain 160.5 oz. of purified platinum powder and 14.5 oz. of metallic precipitate. This represented a total recovery of 80 percent (platinum and metallic precipitate) from crude platina, about the maximum percentage of platinum Wollaston was ever able to isolate from the crude ore. The long-term average was closer to 75 percent, as alluvial ores less rich in platinum were purchased and processed. In comparison to the amounts of platinum obtained in the first trial solutions, 160 oz. was a large amount. But, since platinum is 21 times heavier than water, 160 oz. of platinum occupies a volume nearly equal to 8 oz. of water, enough to fill only a regular-sized coffee cup. This cupful of platinum powder would require even more labor to consolidate it, as Wollaston was soon to learn.

THE PRODUCTION OF MALLEABLE PLATINUM

By April 11, 1801, Wollaston had accumulated enough platinum powder to begin attempts to consolidate portions of it into solid, malleable form. The successful compaction of one small 416 gn. platinum plug in January to a specific gravity of 21.1 had been encouraging, but commercial production of platinum would require compaction on a much larger scale. In preparation for the crucial trials, Wollaston purchased a twenty-six-lb. anvil, and Tennant bought two presses (one of iron and one of wood) and two gun barrels, one small and one large.[34] This was one of Tennant's few contributions to the platinum processing business, intimating that he played only a minor role in the development phase. Wollaston recorded no specific details on how the April consolidations were carried out (which is why the account of the trials I presented in the Prologue was based in part on information revealed later), but continuation of the small-scale, successful January procedure appears most likely. If so, he would have shaken weighed amounts of platinum powder with water, and transferred the wet platinum sediment into a cylindrical metal barrel. Then, using one of the newly purchased presses, he compacted the platinum to the maximum capability of the press, squeezing out water in the process. The resulting platinum plug, sufficiently compressed to be handled, was heated in a hot fire for several minutes, then placed on an anvil and hammered into an increasingly dense mass. The results of these first consolidation attempts, which Wollaston called "casts," were entered into notebook G (Figure 4.2).[35]

A small gun barrel was used for the first compaction, on 1 oz. 3 dwt. 0 gn. of platinum powder. (In these consolidation trials Wollaston used a balance capable of measuring heavier weights than he did for the earlier solution trials.) His weights are recorded in Troy amounts.[36] The first, undated, trial was unsuccessful. Upon hammering the platinum plug was "broken" In the succeeding trials, begun on April 15, larger amounts of platinum powder were compressed in the large barrel, capable of holding 6–7 oz. of

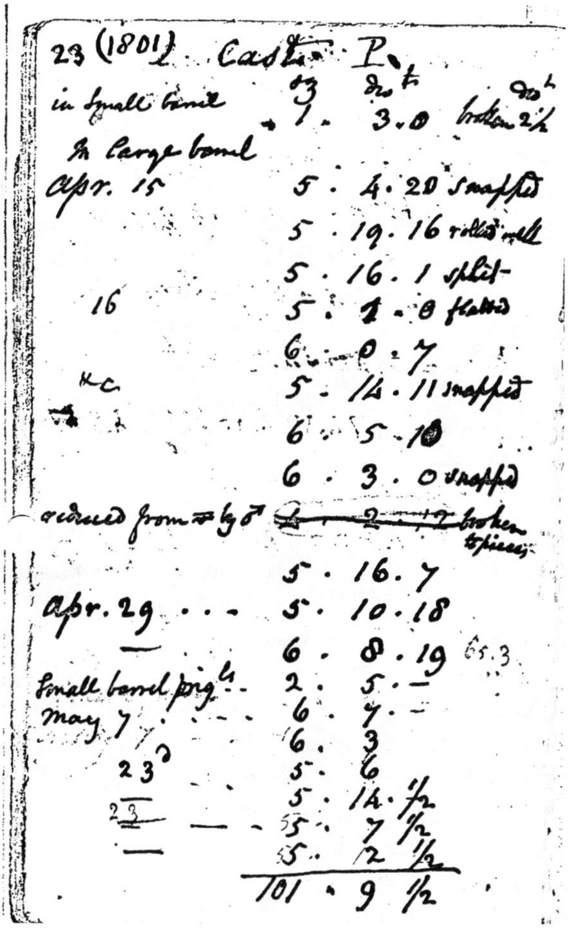
23 (1801) Casts P

oz dwts dwts

in small barrel 1 . 3.0 broken 2½

In large barrel

Apr. 15 5 . 4 . 20 snapped

5 . 19 . 16 rolled well

5 . 16 . 1 split

16 5 . 1 . 0 flatted

6 . 0 . 7

&c 5 . 14 . 11 snapped

6 . 5 . 10

6 . 3 . 0 snapped

reduced from ~~2 . 2 . 17 broken to pieces~~

5 . 16 . 7

Apr. 29 . . . 5 . 10 . 18

6 . 8 . 19 65.3

Small barrel pig 2 . 5 . —

May 7 6 . 7 . —

6 . 3

23 5 . 6

5 . 14 . ½

23 — 5 . 7 ½

5 . 12 ½

101 . 9 ½

Fig. 4.2. Wollaston's First Platinum Consolidation Trials, April 1801. Reproduced by kind permission of the Syndics of Cambridge University Library.

wet platinum powder. A separate entry in notebook **G** mentions that the first platinum plug in the "new" barrel, after heating to remove water, was a cylinder 1.875 inches long with a diameter of 0.76 inch and a specific gravity of 11.7.[37] It had a greater density than the small piece successfully compacted in January, so the newly purchased press worked satisfactorily. But the compressed, dried cylinder was only about half as dense as the pure metal; it still had to be hammered down to about half its original size. After setting the metallic cylinder upright on the anvil, Wollaston hammered it with increasing force to eliminate all porosities, but the plug then "snapped." This meant that, like the first broken ingot, there were fault lines in the compressed platinum, which did not meld together when hammered. Either the platinum powder still contained too many impurities to be fully consolidated without fracture, or the hammering process was not conducted with enough delicacy. We can imagine that Wollaston was beginning to become concerned about his ability to convert platinum powder into a malleable solid after the first two failures. But, happily, the third ingot was hammered successfully into a malleable ingot, one that was shortly thereafter taken to a metalsmith for further forging and flatting.[38] Wollaston noted, with customary brevity, only that it "rolled well." But he was very pleased with the result, because he converted a portion of the ingot into a souvenir blade. That blade, fitted with an ivory handle, is now in the custody of the Science Museum, London, England.[39] The inscription on this remarkable historical artifact reads

> This blade formed a part of one of the first masses of Platinum which was worked by W. H. Wollaston, M.D. and was by him given to his brother F. J. H. Wollaston, who gave it to his son F. H. Wollaston, by whom it was put into its present form.

The inscription, added long after the blade was formed, says that it was made from one of the first masses of Wollaston's platinum. Notebook entries indicate, in fact, that it was actually fashioned from the very first malleable ingot, as that was the only early ingot recorded as having been forged and flatted by a metalsmith. The third and last trial on April 15 was unsuccessful, and the ingot "split" during hammering. Thus, of the first four consolidation attempts, one in a small barrel and three in a large one, only one was successful. But that one gave promise of future success.

The following day, April 16, another six consolidations of platinum were attempted in the large barrel. Only two of the ingots snapped on hammering, suggesting that the quality of the platinum powder was not the primary cause of failure. As Wollaston's skill at hammering the platinum plugs into a compact, malleable mass improved, the number of successful forgings increased in tandem. All nine consolidations listed on the page for April 29, May 7, and May 23 were successful, as were ten more recorded on the following page. Overall, by the end of the consolidation trials on June 22, Wollaston had converted 151 oz. of the 160 oz. of available platinum powder into malleable ingots generally weighing 5–6 oz. He had indeed succeeded in achieving the first major goal of the partnership, producing consistently malleable platinum from crude alluvial ore. No one else had ever been able to do so.

Shortly after the completion of these experiments, some unsettling research results from the French analytical chemist, Louis Proust, appeared in an English scientific journal. Proust, who had replaced Chabaneau as the professor of chemistry at Madrid in 1799, carried out an extensive series of experiments on platina with the large amounts of ore made available to him by the Spanish government. His mandate was to discover a reliable method of extracting pure platinum from the crude ore.[40] He published the results of his investigation in a Spanish journal in 1799. The paper was subsequently translated into French and republished in the *Annales de chimie* in January 1801. It was then translated once more into English and published in two parts in the *Philosophical Magazine* in the summer of 1801, where it must have come to the attention of Wollaston.[41] The long paper, in a total of sixteen sections, reviewed the types of platina available from New Granada, its impurities and reaction with acids, and methods for precipitating its salts from aqua regia solution. Section XI was entitled "Solution of Crude Platina on a Large Scale" and, even more ominously for Wollaston, section XIV was headed "Of the Composition of the Nitro-Muriatic Acid for the Solution of Crude Platina." There, Proust reports his optimized recipe: each 3 oz. of crude platina required 4 fluid oz. of nitric acid mixed with 12 of muriatic to dissolve 2.25 oz. (75 percent) of crude platina.[42] Using acids similar in strength to those employed by Wollaston, Proust had independently discovered the composition of aqua regia that was most effective at dissolving platina. Moreover, he recommended using a total volume of aqua regia that was insufficient to dissolve all the crude platina. These results were published in the three countries where interest in producing malleable platinum was highest: Spain, France, and England.

Fortunately for Wollaston, Proust's paper had no discernible impact. For one thing, Proust made only passing reference to the precipitation of platinum salts by treatment with sal ammoniac and none at all to the production and consolidation of platinum powder. For another, the crucial aqua regia results were placed, without emphasis, near the end of the long paper. And finally, although Proust promised to continue his platinum researches, he never did, and platinum work ground to a halt in Spain, in part because of the French invasion of the country. Nonetheless, the publication of the paper in a widely read English journal must have made Wollaston aware that there was skilled competition in the race to produce malleable platinum. One way to stay ahead of the field was to buy up as much crude platina as possible to prevent anyone else from being able to launch a commercial business. A second way was to keep all details of the difficult purification and consolidation process secret. Wollaston and Tennant did both these things, and it is easy to understand why.

The successful platinum consolidation trials reproduced in Figure 4.2 included one unusual result, which, although not a complete surprise, added to Wollaston's interest in the composition of the metallic precipitate displaced from the spent aqua regia solution by iron. The sample of platinum powder, recorded as being "reduced from an iron precipitate," was especially unfit for hammering as it "broke to pieces" when Wollaston tried to consolidate it. This was the powder resulting from batch N. That batch, as mentioned earlier, started with a mixture of residual platina and 1.5 oz. of metallic precipitate and the sal ammoniac precipitate had an unusual brown colour. Of that failed consolidation, Wollaston observed "This [discolored platinum powder] was cast separate from the rest, gave white fumes when heated much & broke to pieces in forging."[43] Consequently, he decided to set aside all metallic precipitate collected from his purification process for future investigation. Those precipitates appeared to contain something unusual that was injurious to platinum consolidation and that something, drawn from crude platina into the aqua regia solution, piqued his curiosity. So, after significant amounts of the metallic precipitate had been accumulated over a period of many months, Wollaston turned his attention to its chemical analysis. The results, presented in Chapter 5, led to the discovery of two new metals, palladium and rhodium.

The critical features of Wollaston's compaction of platinum powder into malleable metal, a milestone in the evolution of powder metallurgy (the formation of metals from a powdered state), are not explicitly written out or explained in his notebooks. Correlation of dispersed entries gives a general picture of the process, but there is a great deal of tacit knowledge involved. Wollaston kept this acquired expertise secret. It is likely that only John Dowse, Smithson Tennant, and (maybe) one or two of his brothers ever got to witness his consolidation of platinum. The essential features of the process were only revealed in the posthumous publication of 1829. That paper gave many details accumulated over several years of hard work, but the fundamentals were little different from those established in the spring of 1801.[44] Wollaston's best platinum, compacted from the finest powder, was used for crucibles and very narrow wires. It had, he determined, a specific gravity of 21.5. The modern value for the specific gravity of pure platinum is 21.45. It is hard to quarrel with the conclusion of two knowledgeable platinum scientists that the quality of Wollaston's platinum was unsurpassed for nearly a century.[45]

The description of his consolidation process shows the careful attention to detail Wollaston paid to every step of the process. Such experiential knowledge was, in fact, critical to consistent success and probably erected a larger barrier to competitors than the chemistry of the purification sequence. We can also not fail to appreciate the hard work required to produce ingots of malleable platinum. By the time he ceased making malleable platinum in 1821, Wollaston would submit 53,500 oz. of crude and recycled platinum to the purification and consolidation process, in individual batches of (usually) 25 oz. of starting material. That means he performed over 2,100 purification and compaction sequences! Although there is evidence that John became knowledgeable enough to carry some of the purification steps through to the sal ammoniac precipitation step, there is no sign that he was ever involved in the consolidation process. It was Wollaston who hammered each plug of platinum into a malleable ingot. He was an incredibly hard worker.

No competitors to Wollaston in the platinum business arose in the two decades he marketed malleable metal (the topic of Chapter 7), so we might wonder how critical the details of his secret process were to making malleable platinum. One indication is the immediate success others achieved by employing his process after it was made public. Illustrative is the experience of Berzelius, who wrote to Friedrich Wöhler in May 1829.

> We are now re-casting all our old soldered platinum crucibles by Wollaston's method of making platinum pliable; it goes like a dance. I think Wollaston must have laughed inside over the many elaborate methods which have been used in vain for this purpose, when his is so simple.[46]

The value of Wollaston's platinum process to science was marked by the Royal Society's award to him, in 1828, of one of its two newly established Royal Medals. Wollaston's paper on platinum had been read to the Society as the Bakerian Lecture on November 20, 1828, just in time to make him eligible for the award presentation at the anniversary meeting of the Society a week later. Obviously, there was a movement to bestow the award on Wollaston before his impending death. The decision was applauded by most of Wollaston's contemporaries, but not all. Some believed that it was inappropriate to reward one who had kept key details of the successful process secret for so many years. One commentator, who was displeased with much about the Royal Society, wrote disapprovingly.

> the adjudication of a Royal medal to a philosopher who had already enriched himself by the very discovery which the council thought proper thus to reward, and of which death alone seems to have compelled the promulgation, was truely unjust. - By keeping that discovery to himself for many years - . . . Dr. Wollaston retarded the course of investigation and discovery respecting one of the most useful metals in nature.[47]

This criticism is not entirely unjustified. Wollaston's secrecy did retard progress in platinum refining; that was his clear intent. His and Tennant's chemical business was launched to become financially viable, and Wollaston's intellectual investment, together with monopolization of the supplies of crude platina, were the twin pillars of their success. The public revelation of processing details might have been good for science, but it would have been bad for business. The conflict between the scientific ethos of knowledge for the public good and that of entrepreneurial technology where commercial advantage trumps all permeates much of the Wollaston/Tennant partnership. The opposing modes of conduct were soon to draw Wollaston into public controversy, as I will discuss in Chapter 5.

After the Cecil Street purification and consolidation trials were completed in June 1801, Wollaston could do no more until September, when he moved into his larger house in Buckingham Street, with its well-designed laboratory space. Once settled there, Wollaston began in November to produce platinum in commercial quantities, while simultaneously investigating the unusual chemical properties of the metallic precipitates displaced from spent aqua regia by iron bars.

CHAPTER 5

Palladium and Rhodium
1801–1825

Wollaston's declaration of his being the inventor of Palladium has produced a very considerable sensation here; not at all in his favor.[1]

In the fall of 1801, as Wollaston began producing larger amounts of metallic precipitate, he began to search for the reasons why it behaved so differently from the usual platinum powder. To his great surprise, he ultimately discovered that the soluble portion of crude platina contained small amounts of two previously unknown metals not too dissimilar to platinum. As an added bonus, Smithson Tennant, who examined the insoluble portion of platina given him by Wollaston, discovered two more. The platinum business generated more than just financial rewards.

THE BATCH PROCESS FOR COMMERCIAL PLATINUM

Once settled into Buckingham Street, Wollaston began to produce platinum in commercially-viable amounts, and he had designed his new laboratory and furnace accordingly. But for these larger-scale batches, he changed the way he prepared the aqua regia. Wollaston always kept a close eye on the costs of his purification process, as any good entrepreneur would, and he introduced a more economical method of making the acidic solvent. During the earlier Cecil St. solution experiments, Wollaston calculated the average cost of the muriatic and nitric acids required to dissolve 1 oz. of platina to be 3.75 pence.[2] Since muriatic acid (HCl) could be made by combining the less costly ingredients marine salt (sodium chloride, NaCl) and vitriolic acid (sulfuric acid, H_2SO_4), he recognized that he could make muriatic acid more cheaply than buying it. In May and June 1801, he carried out a characteristically small-scale series of trial solutions to determine the amounts of vitriolic acid and marine salt needed to produce an amount of muriatic acid that, when formed in the presence of nitric acid, would give an aqua regia equivalent to that which he had previously found to be most effective at dissolving platina.[3] The new recipe for aqua regia, he calculated, reduced the cost of dissolving 1 oz. of platina to about 2 pence, thereby cutting the solution expenses almost in half.[4] There was another benefit to the cheaper way of making aqua regia solution. In the trials using the new recipe, Wollaston noted that a residue of gold remained with the residual platina when an excess of marine salt was used. This was the first time he had observed gold deposited during the solution process, and in subsequent years he found that, from some shipments of platina, he could recover enough gold to pay for half the aqua regia costs. But there were also two disadvantages to the newer solution process. Since the muriatic acid was produced during the course of the solution reactions, instead of being fully present at the outset, it took longer to dissolve the customary amount of platinum. In addition, the reaction to produce muriatic acid generated as a by-product a poorly soluble salt of vitriolic acid (hydrated sodium sulfate, $Na_2SO_4 \bullet 10H_2O$), which Wollaston generically called selenite.[5] The slower generation of muriatic acid and the coating of the undissolved platina by "selenite" meant that, instead of three to four days, the solution of platina took five to six days, and the resulting "selenite" had to be removed from the residual platina before it could be re-exposed to aqua regia. Despite these difficulties, Wollaston used the marine salt recipe for aqua regia in the commercial batches that began on November 30, 1801.

A brief description of Wollaston's first production runs, all recorded in a dedicated notebook, gives us a good idea of the basic process he used for the thousands of ounces of platina he purified over the next several years. Into each of eight large glass vessels he placed 16 oz. of crude platina, 17.3 fluid oz. (1 lb.) of 91 percent vitriolic acid, 26 oz. of marine salt, 5 fluid oz. of 65 percent nitric acid and 6 pints of water, for a total solution volume of nearly 1 gallon (128 fl. oz.), and a weight of about 9 lbs. All eight were placed on the laboratory furnace for three to six days, during which time they were slowly heated to an ever higher temperature. At the end of that time, they were allowed to cool to room temperature and the undissolved platina (and "selenite") were separated. The spent aqua regia with the dissolved platina from two or three runs was next combined in a very large glass vessel, and sal ammoniac was added to precipitate the yellow platinum-containing salt. The precipitate was well washed with water and set aside. Because the yellow platinum salt was very stable, Wollaston left it in that form and did not convert it to platinum powder until he wished to make malleable ingots. On many days, when there were other demands on his time, only two or four batches were prepared, but during the periods when he purified large amounts of platinum, there were generally several vessels being heated simultaneously, presumably under the superintendence of John.

In December Wollaston increased the starting weight of platina to 20 oz. and in January 1802 to 25 oz., which remained the norm for the following years. Of course, the volume and weight of the purification solutions increased accordingly, so the physical labor involved became greater. Working with strong acids in glass vessels was dangerous, as was functioning in what must have been a poorly ventilated workspace. There are only a few entries in the production records of broken vessels and lost platinum, and, in general, the work went on for two decades without major incident.

From November 1801 to September 1803, Wollaston treated all the crude platina he and Tennant had purchased in a series of four production sequences. The results are listed in Table 5.1, together with the totals from the Cecil St. sequence.[6] The most striking aspect is the large number of individual batches required to process all the platina. Wollaston chose to keep the production process at a scale that could be carried out by John and himself in his home laboratory. This decision was probably influenced by the desire for secrecy and independence. Of course, the total amount of platina available was small enough to permit its processing in 25 oz. quantities, if the personal tenacity existed to do so in 570 separate batches. To

TABLE 5.1. Platinum Production Sequences, 1801–1803

Sequence #	Dates	# Batches	Platina Ore Used (oz.)	Platinum Powder Produced (oz.)	Metallic Precipitate (oz.)
Cecil St.	Feb.21/01 to Jun.1	28	228	160	14
#1	Nov.30/01 to Jan.20/02	99	1129	852	38
#2	Mar.22/02 to Jun.15	102	1275	937	138
#3	Jul.20/02 to Dec.18	152	2000	1528	270
#4	Jan.5/03 to Sept.2	189	2395	1853	277
total		570	7027	5330	737

achieve the recovery of purified platinum shown in Table 5.1, (76 percent), Wollaston saved the undissolved residue from each batch and treated it repeatedly with aqua regia in subsequent batches. Finally, the very small amount that resisted repeated exposure to aqua regia, aptly dubbed "diabolite," was set aside for future study.[7]

Wollaston went on to convert the yellow precipitate produced by the purification sequences into platinum powder and consolidated it into malleable metal as time and amounts permitted, but no regular log of consolidations was kept during these early years. One entry does reveal that 725 oz. of malleable platinum had been made by April 30, 1802, but the finished product was not offered for sale until February, 1805. Why would Wollaston, so determined to establish a profitable platinum business, wait nearly three years to begin to market platinum, the valuable product he had first had ready for sale in 1802? The answer lies in the puzzling behavior of the metallic precipitate he had now begun to accumulate in reasonable quantity (see Table 5.1).

As mentioned previously, Wollaston had observed that the platinum salt obtained from aqua regia treatment of the first metallic precipitate was a darker yellow than normal and the spongy platinum produced from it could not be made to cohere. Not surprisingly, he wondered what component of the metallic precipitate did not reveal itself when first dissolved in aqua regia but became detrimental to the process when dissolved a second time in the same acid. He now had enough metallic precipitate to find out. He took the 38 oz. of metallic precipitate collected from the first sequence of purifications and redissolved it in aqua regia on March 23, 1802.[8] After standard treatment of the solution with sal ammoniac, iron bars were again immersed in the spent aqua regia to generate what Wollaston now termed a "second metallic precipitate."[9] Continuing his efforts to maximize the recovery of platinum, he treated the blacker, finer second metallic powder with yet another dose of aqua regia. To his great surprise he found that a portion of the second metallic precipitate, which consisted entirely of metals that had twice previously dissolved in aqua regia, was no longer soluble in that acid. Moreover, the portion that did dissolve gave an abnormally dark-colored solution, which, on treatment with sal ammoniac, gave a deep red precipitate instead of the normal yellow-colored one. Wollaston could only conclude that an unknown metallic impurity in crude platina had initially co-dissolved with platinum in aqua regia and had become increasingly concentrated in each successive metallic precipitate. Ultimately, the amount in the second metallic precipitate had become large enough that its chemical properties began to manifest themselves.

The challenge then became one of finding a way to isolate and characterize the suspected new substance.

THE DISCOVERY OF PALLADIUM

Using well-known chemical tests, Wollaston was able to detect lead, iron and copper in the second metallic precipitate, none of which could explain the anomalous observations. Treatment of the precipitate with nitric acid produced a dark brown solution, which could not have been formed by gold, silver, platinum, or mercury. He now was confident that the dark brown solution contained "some new body."[10] To extract the suspected metal from the nitric acid solution, he developed a technique that began with amalgamation of it with mercury. In his words

> I agitated a small quantity of mercury in the nitrous solution previously warmed, and observed the mercury to acquire the consistence of an amalgam. After this amalgam had been exposed to a red heat, there remained a white metal, which could not be fused before the blow pipe. It gave a red solution as before in nitrous acid; it was not precipitated by sal ammoniac, or by nitre; but by prussiate of potash [potassium ferrocyanide, $K_4[Fe(CN)_6]\bullet3H_2O$] it gave a yellow or orange precipitate; and in the order of its affinities it was precipitated by mercury but not by silver.
>
> These are the properties by which I originally distinguished palladium; and by the assistance of these properties I obtained a sufficient quantity for investigating its nature more fully.[11]

Wollaston's account of his discovery, written three years after the fact, is corroborated by numerous undated entries in his notebooks that appear to have been written in 1802, but the first unequivocal, dated entry on palladium is in July of that year (Figure 5.1).[12] It gives the specific gravity of the new metal as 11.8 (modern value 12.0), a value obtained by combining 4.35 gn. of the new metal "C" with lead and weighing the composite metal in a hydrostatic balance.[13] The retrospective note on the facing blotter says, "The upper part of the opposite page was written July 1802. I believe the C meant Ceresium a name which I once thought of giving to Palladium." Soon after, a notebook entry on August 3, 1802, refers to the new metal for the first time as "palladium."[14] The reason for the change can be tied to the discovery of the astronomical bodies that prompted the names. The asteroid Ceres, initially thought to be a small planet between Mars and Jupiter was discovered on January 1, 1801. A second asteroid in

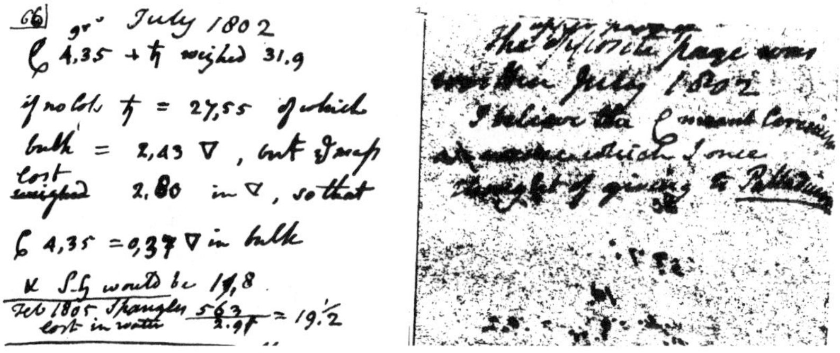

Fig. 5.1. Wollaston's First Dated Entry on the Discovery of Palladium, July 1802. Reproduced by kind permission of the Syndics of Cambridge University Library.

nearly the same orbit was sighted on March 28, 1802, and named Pallas. Wollaston then replaced the preliminary name for his newly discovered metal with palladium, perhaps because of the mythological connection of Pallas with Athena, goddess of wisdom and skills. Thus, all the available evidence suggests that palladium was isolated and characterized by Wollaston in July 1802, or somewhat earlier.

The discovery of palladium in crude platina set off a train of consequences for both Wollaston and Tennant. Its isolation did not resolve all of the chemical anomalies observed in the behavior of the second metallic precipitate. In the first place, as he sought to extract greater amounts of palladium in purer form, Wollaston began to suspect that the soluble portion of the second metallic precipitate contained a second unknown metal. Moreover, the insoluble portion of the same second metallic precipitate exhibited different chemical properties altogether. Because the "diabolite" appeared to have characteristics in common with the residue from crude platina that resisted repeated treatment with aqua regia, he gave both residues to Tennant for further investigation. This sharing of the discovery opportunities suggests that the partners agreed to split up the search for new metals in platina: Wollaston would seek those from the second metallic precipitate that were soluble in aqua regia, and Tennant would search for others in the insoluble portion. But Tennant, as we know, was not as assiduous as his partner, and his analysis of the residues was not completed until 1804. Because of these serendipitous, and time sensitive, opportunities for discovery of new elements, Wollaston and Tennant decided to delay the marketing of malleable platinum to avoid luring competitors into the field. Thus the thousands of ounces of valuable platinum that Wollaston was continuously making during 1802 and 1803 were left to sit in storage somewhere until the search for new substances in the crude ore was satisfactorily concluded.

The discovery of palladium, the first new chemical element to be found in crude platina, presented Wollaston with a dilemma. How could he make his discovery known without revealing any crucial details of his work with platinum? He ultimately made the regrettable decision to prepare a few samples of palladium and offer them for sale anonymously, so that priority in the discovery could be established without giving competitors any knowledge of his platinum researches. This plan of action was to damage Wollaston by drawing him into a controversy that was to tarnish his reputation for many years. But, before describing the circumstances of the impending dispute, I will briefly review Wollaston's production of small amounts of malleable palladium.

To produce the newly-discovered metal in amounts and quality good enough for sale, Wollaston had to develop an improved process for its isolation. The mercury amalgamation method that led to its discovery was unsatisfactory, for it did not extract all of the palladium from the second metallic precipitate. He found instead that he was able to selectively dissolve palladium from the second metallic precipitate with a solution of nitrate of potash (KNO_3) dissolved in muriatic acid. Evaporation of this solution produced crystals with an unusual property. As he explained

> I procured a solution from which by due evaporation were formed crystals of a triple salt, consisting of palladium combined with muriatic acid and potash. [These crystals exhibited] a very singular contrast of colours, being bright green when seen transversely, but red in the direction of their axis; the general aspect, however, of large crystals is dark brown.
>
> From the salt thus formed and purified by a second recrystallization, the metal may be precipitated nearly pure by iron or by zinc, or it may be rendered so by subsequent digestion in muriatic acid.[15]

Wollaston was now in familiar territory working with well-defined crystals, and he was confident that the palladium recovered from the recrystallized triple salt would contain fewer impurities than that produced by mercury amalgamation. The crystalline triple salt of palladium, potash and muriatic acid (K_2PdCl_4), which exhibited "a very singular contrast of colours," is an early example of a dichroic crystal, one that exhibits different colors when viewed from different directions.

The palladium powder produced by the action of iron on a solution of the triple salt could not be consolidated as formed, so Wollaston reacted it with sulfur to form the sulfuret of palladium (PdS) which was then heated and reduced anew to palladium powder.[16] Palladium formed this way could be successfully compacted. This process was in use by November 2, 1802, but was not made known until the same posthumous 1829 paper in which he published his platinum procedures. The compaction of palladium required much acquired skill, and would have presented a significant challenge to anyone who had not previously honed his technique consolidating hundreds of ounces of spongy platinum. It involved many alternating slow heatings and gentle hammerings, carried out with the "utmost patience and perseverance."[17]

Wollaston's production of malleable palladium in 1802 was a remarkable achievement, both chemically and technically. The barriers to success were, in fact, so large that one wonders if he really did have much to fear from potential competitors. Indeed, it was not until two decades later, in the early 1820s, before Bréant in Paris was able to obtain several hundred grams of palladium from 1000 kilograms of Spanish platina.[18]

By good fortune, some of Wollaston's original palladium still exists. Several years after producing it, Wollaston gave some samples to Michael Faraday. After Faraday's death the samples, inserted into a piece of paper labelled "Palladium from Dr. Woolaston" (a common phonetically-correct variant of the surname), found their way to the Science Museum, where they are now held.[19] My analysis of one of the slips of palladium showed that it contained significant amounts of impurities, which explains the difficulties Wollaston encountered in consolidating it. The metal was only 89.35 percent pure, and the major impurities were copper (6.29 percent), platinum (2.27 percent), lead, iron, and tin (combined 1.30 percent), together with other metals present in smaller amounts.[20] It is possible that the palladium in these samples was purified by an inferior process described by Wollaston in his 1805 paper, which differed from the two methods previously discussed. In this third method for isolating palladium, Wollaston did not need to extract it from the second metallic precipitate. Instead he added prussiate of mercury ($Hg_2[Fe(CN)_6]$) directly to the spent aqua regia solution, from which the platinum-containing salt had been previously precipitated by sal ammoniac, to give a pale yellowish-white precipitate of prussiate of palladium ($Pd[Fe(CN)_6]$). The precipitate from this simple process could be reduced to palladium powder by heating, with concomitant release of poisonous prussic acid (hydrogen cyanide, HCN).[21]

Wollaston interrupted his ongoing platina purification sequence #3 (Table 5.1) in August 1802, to redigest the 138 oz. of first metallic precipitate that he had accumulated from the previous purification sequence.

From this he obtained 24 oz. of a second metallic precipitate, and from that he isolated 4.5 oz. of palladium (which amounted to only 0.35 percent of the original 1275 oz. of platina ore).[22] Knowing now how best to separate palladium, Wollaston subjected 500 gn. of crude platina to the usual purification routine with the specific object of determining how much palladium it contained. From it he isolated a mere 2.5 gn. of palladium, only 0.5 percent of the crude ore.[23] A similar analysis for the palladium in a sample of "Paris platina" in November 1802, yielded the same result.[24] This showed that the small amounts of platinum produced in France had not been freed of palladium, which suggested that the new metal was still unknown there. Intent now on isolating enough palladium to offer some for sale, and to establish his priority for the discovery, Wollaston processed the 270 oz. of the first metallic precipitate from purification sequence #3 (Table 5.1) over the period December 1802 to February 1803 to obtain another 10 oz. of palladium.[25] Finally, in April 1803, he was ready to make the existence of palladium known, and he elected to do so in a most unconventional way.

THE PALLADIUM CONTROVERSY[26]

Wollaston printed 1000 copies of a small (4.5″ × 3.5″) leaflet and had them distributed anonymously throughout London and beyond.[27] The leaflet advertised the sale of a new noble metal named "Palladium, or New Silver."[28] The list of palladium's properties included most of those observed by Wollaston during the discovery process, but the leaflet gave no hints as to the metal's natural source, mode of discovery, or the identity of its discoverer. For someone who cherished his intellectual and ethical integrity as much as Wollaston, this was a bizarre way to announce his discovery. The notice stated that samples of palladium could be purchased from Mr. Forster of Gerrard St., Soho at a cost of five shillings, half a guinea (10.5 shillings), and one guinea (21 shillings). The owner of the shop, Jacob Forster, was a mineral collector and vendor who was away in Russia at the time of the notice. His shop was managed by his wife during his absence.[29]

We cannot be sure that Wollaston dealt personally with Mrs. Forster, for the palladium was delivered to her by a young man acting on his behalf. However, details of his account with her are recorded in a notebook devoted to palladium sales. The first entry records that she was given palladium samples, in the form of thin ribbons, on April 24. As vendor, she was allowed to keep 10 percent of the selling price as commission.[30] The largest samples measured about 3″ × 0.5″ and weighed about 25 gn.; the smaller were scaled in dimension and weight in proportion to their prices.[31] The samples were priced at about 1 shilling per grain, nearly £24 per ounce, or five times the price of gold. Considering the labor involved in making palladium, the price was not outrageous, but it would certainly have limited sales to those of ample means.

On April 29, the itinerant, abrasive Irish chemist Richard Chenevix, then resident in London, learned of the new metal and purchased a specimen for analysis. On May 12, a scant thirteen days later, he read the first installment of a long paper describing his researches on palladium to the Royal Society. He opened the paper with a statement of suspicion that many must have shared.

> The mode adopted to make known a discovery of so much importance, without the name of any credible person except the vender, appeared to me unusual in science, and was not calculated to inspire confidence. It was therefore with a view to detect what I conceived to be an imposition, that I procured a specimen, and undertook some experiments to learn its properties and nature.[32]

After working furiously, reportedly up to fourteen hours per day, Chenevix determined that palladium did, indeed, have all the properties attributed to it. But he concluded that the metal was not, as advertised, a new noble metal. He believed instead that it was merely platinum alloyed with some other element, even though he was unable to detect any known metals in palladium by the usual qualitative chemical tests. Undeterred by a succession of negative results, Chenevix claimed that he finally was able to synthesize an alloy that had all the properties of the purported palladium by a carefully-managed combination of platinum and mercury. He therefore announced that the palladium offered for sale was "a contemptible fraud directed against science."[33]

Chenevix foresaw two obvious criticisms (he preferred to call them prejudices) others would advance against his artificial palladium. The first was its anomalously low density, which was lower than that of either of the two metals it supposedly contained. The second was the so-called fixation of mercury, meaning it had been incorporated into a platinum alloy in such a way that it became undetectable by chemical means. Chenevix's discovery, as he claimed it to be, had no precedent in chemistry. Compound substances generally exhibited properties entirely different from their component bodies, but the individual components always revealed their presence on analysis (which was the reason they were identified as constituents of compound substances in the first place). Palladium, he argued, provided the first compelling example of a compound substance whose component bodies could not be detected by analysis. This revolutionary aspect of the paper had some appeal to those who shared Chenevix's desire to reduce the number of chemical elements to a much smaller store of primitive materials.

It is probable that Wollaston attended the meetings at which Chenevix's paper was read, for he had been elected in November 1802 to serve on the Council of the Royal Society for the customary (renewable) one-year term. While service on Council did not require attendance at meetings, it did lead to Wollaston becoming a member of the Committee of Papers, which met regularly to decide which of the papers read to the Society would be published in the *Philosophical Transactions*. It is reasonable then to assume he would have been present for the reading of all papers on which he would later be required to render a decision. He certainly did attend the meeting of the Committee of Papers on June 16, 1803, at which Chenevix's paper was approved for printing in the Society journal.[34] If Wollaston was in the audience during the reading of Chenevix's paper, he would have cringed to hear the author refer to his palladium as a shameful and contemptible fraud. But, if so, he remained silent. If he did not attend, it was by choice, for he must have been aware of Chenevix's investigations. For one thing, the Forster account notebook reveals that, on May 6, just after Chenevix had purchased nearly all of the palladium on hand, Wollaston's agent retrieved one unsold one-guinea sample and six half-guinea samples, and replaced them with newly prepared palladium samples: thirty of the one-guinea size, 30 of the half-guinea, and twelve of the five-shilling size.[35] Mrs. Forster would have had no reason to conceal the identity of the purchaser of most of the palladium.

Although he was supremely confident of his discovery, Wollaston did attempt to replicate Chenevix's findings, and recorded the results on an undated notebook page headed "Check for Chen[evi]x."[36] He was unable to make any alloy of platinum and mercury that bore the slightest resemblance to palladium, and metals, such as iron, which did form alloys with platinum, could always be easily detected. When he published these results about a year later, Wollaston said,

> [After repeating one of Chenevix's purported syntheses of palladium] I indeed obtained such a precipitate of metallic flakes as he describes; but, upon examination of these flakes, they yielded mercury by distillation; and the remainder consisted of platina combined with a portion of iron, but had not any properties which I could suppose owing to the presence of palladium.[37]

A colleague and I sought to shed light on the nature of Chenevix's artificial palladium by preparing it by one of his methods. We found, in agreement with Wollaston, the resultant metallic flakes to be composed of platinum, mercury, and iron, each of which was easily identifiable by characteristic nineteenth-century analytical tests.[38] But Wollaston, behind his shroud of secrecy, could do nothing to influence the public's reaction to Chenevix's paper. He must have been chagrined to observe the positive reception it initially received.

The announcement of palladium and Chenevix's denunciation of it, predictably at a time when chemistry was a very popular and accessible science, created a sensation. Word of the new metal spread quickly throughout Europe, aided no doubt by the mysterious circumstances of its origin. William Nicholson, for example, editor of a widely read science journal and author of a comprehensive chemical dictionary, received a copy of the notice and quickly reprinted the relevant details.[39] Nicholson had himself been sent a small sample by mail, and reported that Chenevix had proved "the pretended new metal" to be an alloy of two parts platinum to one of mercury. Chenevix also actively publicized his findings on the continent. On May 4, only a few days after he had first purchased palladium, he sent a letter with a sample of the metal to the skilled French analytical chemist Louis Vauquelin, with the added comment that in London "the learned world speaks of nothing but palladium."[40] The letter was quickly published in the *Annales de chimie*, the premier French chemistry journal of the time. In a note appended to the published letter, the editor reported that Vauquelin experimentally verified all the advertised properties of palladium, but could detect in it neither platinum nor mercury. As for Chenevix's claim that palladium was an alloy, the editor stated only that Vauquelin's results "must then give rise to doubt."[41] Obviously, Vauquelin and the journal editor were not as easily persuaded to jettison one of the cornerstones of chemistry as were many of their English counterparts.

As the controversy simmered, Wollaston went in August 1803 on a nine-day excursion with sisters Anna and Louisa to Dover and the surrounding area, his first summer vacation since the trip to the Lake District in 1800 with Hasted.[42] On his return, Wollaston learned that the Council of the Royal Society, of which he was a member, wished to recommend Chenevix for the Society's 1803 Copley Medal, largely because of his palladium paper. This placed Wollaston in a quandary. The Royal Society was about to bestow its top prize on a chemist who, Wollaston knew, had published an erroneous indictment of his own discovery. Stubbornly, he remained convinced that his own best interests depended on maintaining secrecy. So he approached Joseph Banks to tell him, in confidence, that he had discovered palladium and had announced its sale in such an unusual way. Banks was not happy to learn this (in fact he may well have been furious), and when he was able to reveal the secret to Chenevix two years later, he wrote, " [Wollaston] embarrassed me much by telling me his secret in November, 1803, under the strictest seals of secrecy: his reason was the fear of being blam'd should the Council have voted the Medal to you for your paper on Palladium."[43] Banks properly ignored Wollaston's intervention, and the Copley Medal for 1803 was given to Chenevix for "various Chemical communications printed in the Philos. Transactions," one year after it had been given to Wollaston himself.[44]

Wollaston was by now becoming discomfited by the international and institutional support given to Chenevix's palladium claims, and he reacted by supplementing his sequence of questionable actions with a decidedly bad one. He decided to offer a public, still anonymous, reward to anyone who could successfully synthesize palladium. The details were announced in an unsigned letter, dated December 16, to William Nicholson for publication in his journal.

> As I see it said in one of your Journals, that the new metal I have called palladium, is not a new noble metal, as I have said it is, but an imposition and a compound of platina and quicksilver, I hope you will do me justice in your next, and tell your readers I promise a reward of 20£, now in Mrs. Forster's hands, to any one that will make only 20 grains of real palladium, before any three gentlemen chymist's [*sic*] you please to name, yourself one if you like . . .[45]

The reward was to be in effect until midsummer 1804, and Nicholson announced in the following issue of his journal that he, Charles Hatchett, and Edward Howard had agreed to serve as judges of the competition.[46] However, their services were not needed as an entry in the Forster Account notebook records that the reward money was returned unclaimed to Wollaston on July 19, 1804.[47] If the reward announcement had any effect at all, it was to the reputation of its sponsor. In his summary of palladium's properties and Chenevix's researches, Thomas Thomson wrote in the 1804 edition of his comprehensive *A System of Chemistry* that the conduct of the person behind the reward was "unusual, at least in this country, and I think reprehensible."[48] These may seem like strong words from a modern perspective, when scientists are eager to commercialize their discoveries, but enlightenment values in the early nineteenth century placed the public good of natural knowledge far above personal gain. Wollaston's conduct was to be further criticized by others when he finally announced himself as the discoverer of palladium, and annoyance at his penchant for secrecy was to dog him to his death, and even beyond, as we shall see.

The palladium controversy continued in 1804, but related discoveries by French chemists also working on platina redirected attention to the constituents of the crude ore. Hippolyte-Victor Collet-Descotils had read a paper to the National Institute on September 26, 1803, on the cause of the different colors found in the salts produced by the aqua regia solution of crude platina. In that paper, translated into English and republished in 1804, Collet-Descotils suggested that the red salts obtained from the least soluble components of the ore contained a new metal, which he intended to name after he explored its nature more fully.[49] A second paper on the same topic was published immediately after Collet-Descotils's in the *Annales de chimie* by Fourcroy and Vauquelin.[50] They, too, in a lengthy study of platina, concluded that it contained a new metal, which they hoped soon to identify. If Wollaston did not learn of these researches in 1803, he certainly would have in early 1804 with the English version of Collet-Descotils's paper. He would have been relieved to learn that there was no mention of new metals in the soluble portion of platina on which his investigations were focused, but the work of the French chemists was an obvious threat to Tennant, who had taken on the task of characterizing the new metal(s) believed to be in the insoluble portion of platina. Clearly, the French chemists were well on their way to discovering new metals other than palladium in platina.

THE DISCOVERY OF RHODIUM

In the last half of 1803, Wollaston carried out numerous tests, mostly undated, on the components of the second metallic precipitate from which he had isolated palladium. He slowly closed in on reactions that appeared to be characteristic of another new metal he tentatively identified as "Novum," represented initially by the letter N, but he could not isolate it sufficiently free of contaminants to characterize it completely.[51] Frustrated by the confusion of results he had observed with conventional tests for metals, Wollaston went back to the technique that had served him so well in so many of his chemical studies—the formation of a crystalline salt of the suspected metal. To do this he dissolved a small amount of the second metallic precipitate in aqua regia, precipitated what platinum it contained with sal ammoniac and allowed the remaining liquid to slowly evaporate to dryness. From the mixture of crystalline salts deposited, Wollaston "selected for examination some that were of a deep red colour, partly in thin plates adhering to the sides of the vessel, and partly in the form of square prisms having a rectangular termination."[52] The very pure red crystals, being distinct from other crystals in the mixture, yielded the suspected new metal after strong heating drove off the non-metallic components. Because the crystals dissolved readily in water to give a rose-colored solution, Wollaston named the metal contained in them Rhodium.[53] Such hand-picking of distinctive crystals was unsuitable for isolating workable amounts of rhodium, but Wollaston quickly discovered a way to produce significant amounts of a triple salt containing it. This process is clearly described in notebook entries dated June 14 and 18, 1804 (Figure 5.2). The entries read:

> 14 June 1804. Metallic precipitate washed with weak nitric acid [to dissolve the base metals, which chemical tests] gave [to be] lead, copper & a little palladium.
>
> Aqua regia then took the remainder & with the addition of common salt was evaporated to dryness.
>
> Alcohol took the greater part [dissolving the triple salts of platinum and palladium] but left Rose coloured residuum = soda-muriate of a new metal call it Rhodium.
>
> [The metal was] precipitated [from an aqueous solution of the salt] by zinc & thoroughly dried [to give a] black [powder].
>
> June 18. [determination of rhodium's specific gravity by alloying with lead gave a value of 11.3, almost equal to the value of lead].
>
> This quantity & proportion [of rhodium with lead] with difficulty fused by blow pipe, very brittle, added [more] lead to weigh 50 gns now fusible, still brittle, but would not powder. Nitric acid [to remove the lead] left 17 gn. Aqua regia acted on rhodium as well as lead.[54]

These entries are the culmination of a long and complex series of chemical trials that extended from the fall of 1803 to the spring of 1804. They signal Wollaston's successful characterization of rhodium in the aqua regia soluble portion of platina. The rose-colored salt, Wollaston's soda-muriate of rhodium, is now known to be hydrated sodium chlororhodite, $Na_3RhCl_6 \cdot 18H_2O$. Wollaston's value of 11.3 for the specific gravity of rhodium is quite a bit less than the modern value of 12.4, suggesting the metal had not been fully compacted. Several years elapsed before he was able to melt rhodium (probably by the action of John Children's gigantic battery at Tunbridge Wells in the 1820s). However produced, small triangular pieces of solid rhodium, labeled "pure," are now in the possession of London's

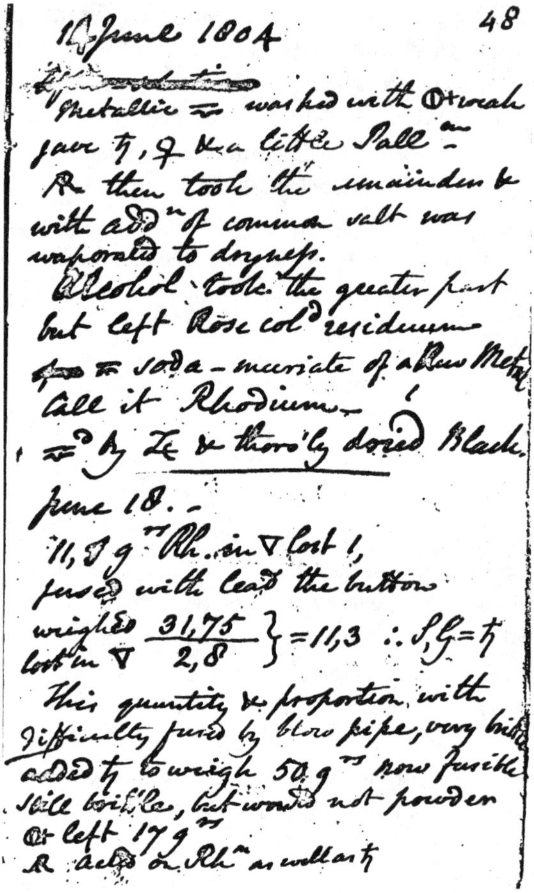

Fig. 5.2. Wollaston's Isolation of Rhodium, 1804. Reprinted by kind permission of the Syndics of Cambridge University Library.

Science Museum. My colleagues and I have analyzed one of these samples and found it to be 99.3 percent pure, with platinum and palladium being the principal impurities.[55] This remarkably pure sample was made by Wollaston several years after his first discovery of the metal, but it does indicate that the isolation of rhodium through formation of its triple salt was capable of yielding an excellent product.

With his search for new metals completed, and maybe chastened by the negative reaction to the manner in which he had announced palladium, Wollaston hurried to reveal his discovery of rhodium in the traditional way, by reading a paper announcing his results to the Royal Society on June 21, 1804.[56] That was a remarkable meeting at which three new chemical elements were announced, for Wollaston's paper was preceded by that of Tennant's describing the discovery of osmium and iridium, the two new metals he had finally isolated from the insoluble portion of platina.[57] Not surprisingly, there is no mention in either paper of the close collaboration between the two authors: the results were presented as if they had been made completely independently. That, we now know, was not the case. The discovery of osmium and iridium marked the apogee of Tennant's scientific career. He published nothing else of significance for the rest of his life. His lack of focus, a characteristic that grew even more pronounced as he aged, would soon begin to strain his business partnership with Wollaston, as I will describe in Chapters 7 and 8.

WOLLASTON REVEALS HIS SECRET

Although Wollaston did not reveal himself in his rhodium paper as the discoverer of palladium, no one at the meeting could have failed to suspect it, because he devoted the last third of the paper to an informed discussion of the controversial metal. After describing his rhodium work, Wollaston proceeded to give a detailed account of the recovery of palladium from the same metallic precipitate that had yielded rhodium. After delineating a number of characteristics by which palladium could be distinguished from platinum and all other known metals, Wollaston reported his own failed synthesis of Chenevix's artificial palladium, as discussed earlier in this chapter. After comments on the implausibility of any alloy having a specific gravity less than that of its lightest component (as Chenevix had argued to be the case for the mercury/platinum alloy that constituted his artificial palladium), Wollaston concluded by stating, "I think we must class it with those bodies which we have most reason to consider as simple metals."[58] The identity of the discoverer of palladium should no longer have been difficult to ascertain. The only thing that could have prevented anyone from making the connection was the belief that the Wollaston they knew would not have revealed such an important discovery in such a discredited way. Unfortunately Chenevix, ignoring the evidence placed before him, was one who did not make the connection.

Chenevix had departed for the continent in 1803 after his palladium paper was read to the Royal Society, and there he continued to work on the artificial synthesis of palladium, while simultaneously seeking support for his unorthodox claims. By June 1804 he had completed a second lengthy paper on the topic and sent it from Freyburg, Germany, to the Royal Society for publication. The paper, disarmingly titled "On the Action of Platina and Mercury upon Each Other" and composed shortly before Wollaston and Tennant had announced their discoveries, was read to the Society over three meetings in January 1805.[59] So this is where matters stood in the opening days of 1805. Chenevix and his supporters believed palladium to be a compound substance whose components could not be detected by normal chemical means and whose discoverer had perpetrated a fraud. Wollaston, and a few confidants, knew such was not the case but kept critical information from public view. And the majority of the learned public did not know where the truth lay. Fortunately, the reading of Chenevix's second paper on palladium helped bring an end to the controversy.

In a final twist to the palladium saga, Wollaston had been elected junior secretary of the Royal Society on November 30, 1804, a post he would hold until 1816. Therefore, he would have been aware of Chenevix's second paper before its reading and knew it contained no sign of a change of opinion. Surprisingly, Chenevix even admitted that Wollaston had in fact alerted him to the weaknesses in his first paper some time before he left England. One would have thought that Wollaston's hints and the added information on palladium added to the rhodium paper would have decided the issue, but such was not the case. Chenevix's 1805 paper reaffirmed his belief in the compound nature of palladium, but did not address the analytical anomaly in any detail and contained little new chemistry of relevance. However, it did prompt Wollaston to reveal his secret.

On January 31, one week after the final portion of Chenevix's paper was read, Wollaston read a paper to the Royal Society titled "On the Discovery of Palladium."[60] In it, he finally revealed himself as the discoverer of palladium and presented more details of its chemical behavior than he had stated in his earlier rhodium paper. For reasons that are not obvious, Wollaston withdrew this paper shortly after it was presented and replaced it with a fuller one read in July. But the secret was out, and Wollaston quickly gave it wider circulation by writing a letter on February 23 to Nicholson for printing in his journal. In it he gave the reasons for his actions.

> [I can now state] that a proportional quantity of platina, from which the whole [of palladium] was extracted, was purchased by me a few years since, with the design of rendering it malleable for the different purposes to which it is adapted. That object has now been attained, and during the solution of it, various unforeseen appearances occurred, some of which led me to the discovery of palladium; but there were other circumstances which could not be accounted for by the existence of that metal alone. On this, and other accounts, I endeavoured to re-

serve to myself a deliberate examination of those difficulties which the subsequent discovery of a second new metal, that I have called rhodium, has since enabled me to explain, without being anticipated even by those foreign chemists, whose attention has been particularly directed to this pursuit.[61]

The letter makes public for the first time that Wollaston was engaged in the production of malleable platinum. In fact, the first deliveries of malleable platinum to the London instrument maker William Cary had been made ten days previously, on February 13, as I will discuss more fully in Chapter 7. But Wollaston does not mention that the first ingots of malleable metal had been readied for sale a full three years earlier, nor does he mention the collaboration with Tennant. The existence of the chemical and business partnership remained hidden from view.

The palladium saga came to an end when the final version of Wollaston's palladium paper was read to the Royal Society on July 4, 1805, and published shortly thereafter.[62] The description of the chemistry employed differed, however, from that given in Wollaston's 1804 rhodium paper by inclusion of methods that involved mercury. Wollaston confessed that he had deliberately avoided any mention of techniques involving mercury in the 1804 paper because of their potential for misinterpretation by any (such as Chenevix) who might seek in them circumstantial evidence for the compound nature of palladium. Thus, Wollaston's first isolation of palladium by amalgamation with mercury, and his final method using prussiate of mercury (as described earlier in this chapter) are both described. What was also added to this paper from the January version was an introductory section on two unusual minerals found in crude platina, which had not previously been described. One, named "ore of iridium," consisted of extremely hard metallic particles that looked like platina but had a much higher specific gravity. Wollaston had found they contained osmium and iridium. The second was tiny red crystals of hyacinth, a mineral previously described by Haüy and today known as zircon (zirconium silicate, $ZrSiO_4$).

To ascertain whether or not there was anything else of interest in the newly-recognized ore of iridium, Wollaston "requested Mr. TENNANT to undertake a comparative examination, from whose well known skill in chemical inquiries, as well as peculiar knowledge of the subject, we have every reason [^] to expect a complete analysis of this ore."[63] In his author's copy of the palladium paper, at the point indicated in the quotation by [^], Wollaston wrote in the margin "barring indolence."[64] This is one of the very few disparaging comments on the work habits of his partner that exist in all of the Wollaston documents. In fact, Tennant never did carry out the analysis. Only in 1826, working with samples given him by Wollaston, did Thomas Thomson confirm that the mineral (now known as osmiridium) consisted almost entirely of osmium and iridium.[65]

At the end of the paper Wollaston described some properties of metallic palladium, which provide further illustrations of his inventiveness and ability to demonstrate interesting phenomena by the simplest means. To measure the heat-conducting properties of palladium and platinum in comparison to the other noble metals, silver and gold, he prepared thin sheets of the four metals and from them

> cut slips 4/10 of an inch in breadth, and four inches long; and having covered their surfaces with wax, I heated one extremity so as to be visibly red, and, observing the distance to which the wax was melted, I found that upon the silver it had melted as far as 3 ¼ inches: upon the copper 2 ½ inches: but upon the palladium and upon the platina only 1 inch each: a difference sufficient to establish the peculiarity of these metals, although the conducting power cannot be said to be simply in proportion to those distances.[66]

Wollaston often exploited the low thermal conductivity of platinum by using slips of the metal as beds on which to carry out micro-analyses, holding one end of the slip between his fingers and heating reagents at the other. To compare the thermal expansion properties of platinum and palladium, Wollaston riveted two thin strips of the metals together and heated them. The bimetallic plate became concave on the platinum side, indicating that palladium had the greater coefficient of expansion. Such heat-induced bending of coupled bimetallic plates was known to English watchmakers, but Wollaston's application of the principle to measure thermal expansion qualitatively is interesting. In 1807 he commissioned Charles Sylvester of Sheffield to plate platinum onto copper, and it appears that Charles Malacrida, a London barometer and thermometer maker, fashioned some sort of thermometer from the bimetallic strips.[67] In May 1807, Wollaston purchased two "platina thermometers" of unknown design from Malacrida for 19 shillings,[68] and Andrew Ure reports that Wollaston showed him in 1809 a bimetallic platinum/copper strip that curved when heated.[69]

The controversy that erupted with Wollaston's anonymous announcement of palladium is of broad historical interest both for its impact upon early-nineteenth-century notions of the simple bodies known as chemical elements, and for its role in illustrating the challenges confronting one who wished to establish priority for chemical discoveries without jeopardizing the success of an emerging chemical business. Wollaston's mode of action naturally had consequences for its two major participants. Chenevix published nothing further on palladium and moved on to literary pursuits, although he remained interested in science and attended the Royal Society Club several times as Wollaston's guest in later years. Happily there is no evidence that either man bore any enduring enmity toward each other as a result of the palladium incident.

It is possible, as some have suggested,[70] that Chenevix's chemical reputation was diminished by the incident, but it is certain that Wollaston's was. A few of his colleagues, some with very long memories, judged his conduct to be inappropriate for a man of science. For example, the president of the Royal Society, Joseph Banks, expressed the opinion of many in the scientific community in a letter he wrote to Chenevix shortly after Wollaston had revealed his discovery of palladium.

> The keeping of secrets among men of science is not the custom here; & those who enter into it cannot be considered as holding the same situation in the scientific world as those who are open and communicative; his reason for secrecy is, however, a justifiable one; tho' not such a one as either you or I should wish to avail ourselves of. . . .
>
> The surprise of the Society was very great when he read his Paper; & the opinion now, I think, generally goes to his being considered as holding a different rank in Science from those who are open & communicative, as every man but himself certainly is; for I do not think any Chemist in or out of the Society will ever converse with him in the same open & undisguis'd manner hereafter, as they used to do, & still will do with each other.[71]

Perhaps Banks, having been placed in a very awkward position by the actions of both men, overstated the negative consequences of Wollaston's conduct. He certainly believed Wollaston to be the architect of his own misfortune and, in this and other letters to Chenevix, apportions none of the blame to Chenevix. But the palladium announcement would have generated much less notoriety had Chenevix been a more accurate, and less opportunistic, chemical analyst.

The dispute that arose over palladium obscured the remarkable chemistry employed by Wollaston to identify, isolate and purify both it and rhodium from crude platina. To those who were more interested in chemistry than controversy, Wollaston's reputation as a superb chemist owed much to those two discoveries. A typical assessment is that of Vauquelin, who had been one of those pursuing new elements in platina. In 1813, in a paper on palladium and rhodium, he wrote

> Although M. Wollaston operated on only 1000 grains of native platinum, and had at the most 6 or 7 grains [half a gram] of each of the new metals at his disposal, he yet recognised their principal properties, which does infinite honour to his sagacity, for the thing appears at first to be incredible. For my part, although I employed about 60 marcs (15 kilograms) of crude platinum I found many difficulties in separating exactly the palladium and rhodium from the platinum and the other metals that are present in this mineral, and in obtaining them perfectly pure.[72]

Vauquelin was, of course, mistaken in his belief that Wollaston originally extracted palladium and rhodium from 1000 grains of crude platina. We know that he isolated both metals from several hundred ounces of metallic precipitate which was greatly enriched in the two metals. But, once he had discovered the best methods of separating each of the metals, he was able to isolate them from 1000 grains (sometimes even less) of dissolved platina, as he reported in his papers when he estimated their natural abundances as about 0.4–0.5 percent. Vauquelin was right to marvel at Wollaston's chemical skills, a judgment shared by his scientific contemporaries, but it is doubtful that Wollaston would have made his discoveries had he not sought to recover more platinum from the first metallic precipitate. The search for economic efficiency, in synergy with a keen natural curiosity and experimental talent, provided the opportunity for serendipitous discovery.

COMMERCIAL APPLICATIONS OF PALLADIUM AND RHODIUM

Palladium was the first product of the Wollaston/Tennant partnership to be offered for sale. Although Wollaston had accumulated enough second metallic precipitate by the fall of 1803 to generate about 45 oz. of palladium, he only produced 28 oz. of the metal in 1803.[73] Of this amount, 2.8 oz. (1350 gn.) were deposited with Mrs. Forster for sale in early 1803. She sold only 420 gn. from April 1803 to May 1805, most of it (332 gn.) to Chenevix.[74] Obviously there was little commercial interest in palladium. So Wollaston took back all the unsold palladium and closed the account with Mrs. Forster on May 23, 1805.[75] Prior to seeking new markets for the metal, Wollaston paid Tennant £6 as his share of palladium revenue,[76] and an additional

£20 for the sole rights to future sales of the palladium on hand.[77] Thereafter, from 1805 until 1814, Wollaston was able to sell only about 6 oz. of palladium for a total of £40.13.0, about double what he had paid Tennant earlier, even though he reduced the price during that time to 84 shillings per ounce, equal to the price of gold.[78] Most of the palladium sold was mixed with gold to produce a bright, silver-colored, hard and non-tarnishing alloy used for numerical scales in astronomical instruments. This application was developed by Wollaston in collaboration with Edward Troughton, one of England's leading instrument makers.

Makers of astronomical instruments, such as sextants, often used thin silver inlays for the numerical scales because its brightness made the engraved divisions more easily visible. But silver tarnishes over time and requires frequent cleaning to retain its luster. That cleaning ultimately wears away the surface and renders the engraved lines and numbers increasingly less visible. Gold was not a suitable alternative because, while it does not discolor over time, it is even softer than silver and more expensive.[79] Platinum imparts its silvery color to gold/platinum alloys, even when it constitutes only about one-quarter of the alloy. In 1807, Wollaston gave Troughton two free samples of platinum/gold alloys for trial as an inlay material.[80] The alloys were returned in November as being unsuitable (for reasons unnoted), and Wollaston, knowing that palladium also combined with gold to give a silver-colored alloy, went on to test the properties of a few palladium/gold alloys. The first auropalladium alloy made, a 3:1 ratio of gold to palladium, was found to be "brittle and bad," but a 4:1 ratio of the same metals turned out to be "white & free as need to be."[81] Wollaston sent a total of 9.5 oz. of the improved alloy to Troughton from May 21, 1808, to April 18, 1810.[82] Some of this was used in sextants, but the largest portion was used for one of the two engraved inlays on the newly-commissioned six-foot Greenwich mural circle, the other inlay being pure platinum.[83] It is not clear why two different metals were used for the two inlays. Perhaps it was to compare their performance over time, or, more likely, it was to showcase Troughton's ability to incorporate both into his instruments. The mural circle was installed at Greenwich in 1812 and was used until 1851. It is now on display at the National Maritime Museum, Greenwich.

Troughton's use of auropalladium led to its occasional use by other instrument makers. For example, in 1819 the skilled French instrument maker, Jean Fortin, elected to use auropalladium in the mural circle for the Paris Observatory. At Fortin's request, François Arago, a mutual acquaintance, wrote to Wollaston with a request to purchase the needed 10.2 oz. of auropalladium.[84] Wollaston prepared and sent the requested alloy, together with directions for shaping it into a suitable form for inlaying into the mural circle. Nonetheless, this and other entries in Wollaston's palladium notebook give information on the fate of a mere six oz. of the metal. From all the platina processed before 1820, Wollaston would have been able to isolate nearly 300 oz. of palladium. So 97 percent of all palladium in his possession went unused. Obviously, the metal had no market in the early nineteenth century and the one small application it found in astronomical instruments could be met equally well by much less costly platinum.

If the market for palladium was bleak, that for rhodium appeared to be even worse, for Wollaston could neither melt the metal nor render it malleable by powder consolidation methods. After announcing its discovery in 1804, he did little further work on rhodium until 1809, by which time he had separated from the second metallic precipitate a large quantity of the triple salt of rhodium. In August 1810, he found that he could obtain metallic rhodium by heating its salt in a crucible with borax, soda, and charcoal.[85] By 1819 he had worked out the most effective combination of ingredients and thereupon reduced 62 lbs. of rhodium salt to 8.5 lbs. of pure rhodium.[86]

The only use Wollaston could find for rhodium was as part of an alloy made into durable tips for the nibs of gold writing pens. This application depended upon the great hardness and resistance to acids in ink of an alloy consisting of 4 parts rhodium and 1 part tin.[87] Although goose quills continued to be the principal writing tool until well into the nineteenth century, Bryan Donkin patented a metal pen in 1808 whose nib was made of two flexible steel sections placed opposite each other to form a slit for the flow of ink onto a page. Acid residues in ink quickly corroded the steel nibs so gold was tried, but it proved to be too soft for long-term use. Some had tried to remedy this problem by fusing a hard, corrosion-resistant tip unto the gold, and Wollaston developed his rhodium/tin alloy for this purpose.

Most of the information on Wollaston's rhodium/tin alloys is contained in a notebook devoted to rhodium, now in the possession of the Science Museum, London. The first reference to the 4:1 alloy is dated June 26, 1822, and mentions that the alloy had been melted into a metallic button.[88] To make his pen tips Wollaston first poured the molten alloy into a small cylindrical mold to produce an ingot about 0.5 inch in diameter. The ingot was then given to a lapidary named J. Cuttell, who cut it into several circular discs, at a cost of 1 shilling per slice.[89] The discs were about 1/40 inch thick, similar to the width of the cutting wheel used by Cuttell, so half the rhodium/tin alloy was lost in the cutting process. The circular slices were then broken into small triangular fragments weighing about 0.5 grain each. From 1822 to 1825 Wollaston produced about 88 oz. of alloy for use in pen tips, of which 26.5 oz. are recorded as being sent to Cuttell for slicing. This was sufficient to make nearly 13,000 individual triangular tips and 6,500 pen nibs, since two tips were needed for each pen.[90]

Wollaston's unfinished tips were sold to T. C. Robinson, an optician and instrument maker, who attached them to gold nibs for fashionable gold writing pens with very hard, long-lasting writing points. Wollaston charged Robinson 6 pence per tip, which represented a return on rhodium of nearly £15 per ounce, over four times the price of gold.[91] Income from the sale of tips to Robinson grew from £10 in 1823 to £40 in 1824 and £80 in 1825, when Wollaston ceased keeping records of income.[92] These totals represent the purchase of 5,310 tips, sufficient for 2,655 pens. Wollaston purchased six of the pens from Robinson in April, 1822, one in May and a further twelve in June, the last at a cost of four shillings each.[93]

One of the first examples of Wollaston's rhodium/tin alloy employed as a hard tip for a metal pen is still in existence.[94] The tips on the extant pen are soldered unto silver nibs, probably a precursor to the later gold nibs. The pen was a gift from Wollaston in 1822 to William Codrington, then an eighteen-year-old naval ensign resident in Germany. It was sent to him by his father, the famed naval officer Sir Edward Codrington, who was a very close friend of Wollaston's at the time. Details of the gift, and of the status Wollaston was to acquire during his lifetime, are contained in an interesting letter from father to son.

> The accompanying pen is Dr. Wollaston's present to you, the nib being a bit of his own metal, Rhodium . . . I trust the knowledge you already have of Science & of the value of scientific men will have fixed in your mind the esteem & respect for so eminent a man as Dr. Wollaston. I consider his friendship for me as a much higher honour than that of any other person yet his rank be ever so great; and . . . the time will come when it will be a just pride for you to have to relate that he devoted some hours of his valuable life in teaching you the rudiments of that science which has borne his name all over the enlightened world. To whatever men of science you may meet with, his acquaintance is a passport to them, & his friendship a title to their regard.[95]

William Codrington took his father's advice to heart, for he took good care of the pen, which went through the Crimea campaign with him. He later bequeathed it to his own descendants.

Wollaston's discoveries of palladium and rhodium have given him a place of distinction among those who have discovered and characterized one or more of the eighty or so naturally-occurring chemical elements. Although his contemporaries were unaware of the large amount of platina he had at his disposal, they were greatly impressed by the quality of his analytical techniques and the reliability of his findings. Thus, by the end of 1805, when the marketing of platinum had first begun, and four years after he had abandoned medicine to begin life "entirely anew," Wollaston had established himself as one of Europe's leading natural philosophers in two disparate fields, chemistry and optics, with additional recognized expertise in galvanics. Serendipity, careful observation, and hard work had brought him scientific acclaim, but his finances would only begin to improve when sales of malleable platinum began to accelerate.

CHAPTER 6

Optical Devices and Social Networks 1804–1809

> In short, if Dr. Wollaston, by this invention, have not actually discovered a Royal Road to Drawing, he has at least succeeded in Macadamising the way already known.[1]

Although Wollaston intended to make a living by producing and marketing chemical products, for the first ten years of the nineteenth century he was forced to rely on the Hyde money received from brother George to sustain himself. The ongoing cost of producing malleable platinum was substantial, and although platinum sales had begun in 1805, the business did not become profitable until 1809. Unfortunately, Wollaston was unable to get much in return from the sale of palladium and rhodium either because the metals found no meaningful commercial applications. The organic chemicals business he initiated in 1802, to be discussed in Chapter 8, did not begin to return a small profit until 1807. So, while the chemical businesses slowly gained traction, Wollaston sought to supplement his own income in another way. He developed and patented two optical devices, which he offered for sale through London retailers. To maintain some chronological consistency I will now discuss Wollaston's invention of these and other optical devices, together with his entry into the higher echelons of London's scientific community, before moving on in Chapters 7 and 8 to discuss the commercial success of the platinum and organic chemicals initiatives.

PERISCOPIC SPECTACLES

In early 1804, Wollaston published in two of London's commercial scientific journals a novel design for the lenses of spectacles, which provided a wider

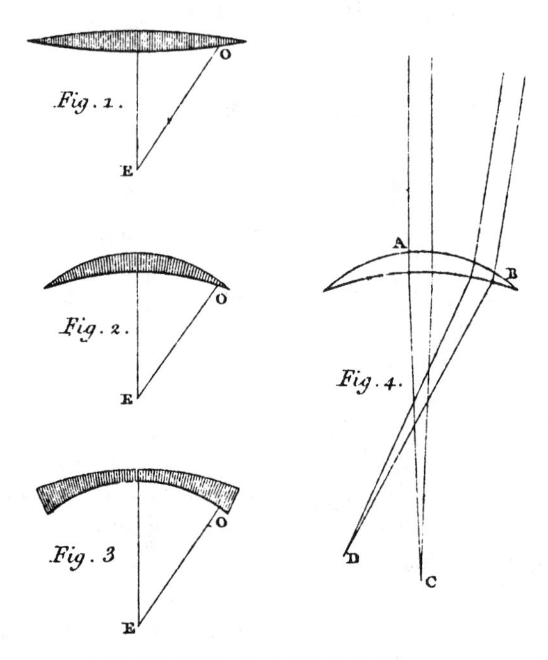

Fig. 6.1. Wollaston's Meniscus Lenses.

field of focused vision.[2] In the normal double convex eyeglass lens, labeled in Figure 6.1 as "Fig.1," the light rays that pass through the lens near its central axis are brought to a sharp focus at point *E*, and the curvatures of the lens's surfaces are fixed by the optician so that the focal point of the lens falls directly on the retina. Light rays that are more distant from the lens axis (shown in most extreme form as line *OE*) are refracted to a different point of focus. The difference in the point of focus resulting from the curved lens surfaces, known as longitudinal or spherical aberration, is shown more clearly in Figure 6.2, where the differing focal points of parallel light rays off the lens axis are shown. Since double convex spectacle lenses are properly crafted to provide sharpest vision for the bundle of light rays lying parallel and closest to the lens axis (the direct line of sight), rays lying further from, or at an angle to, the central axis become more indistinct as their point of focus moves increasingly farther away from the retina. Such spherical aberration is also generally accompanied by chromatic aberration because light rays of different colors are refracted at different angles at both lens surfaces. Thus light rays striking the eye from the periphery of double convex spectacle lenses can be both indistinct and oddly colored. Thicker

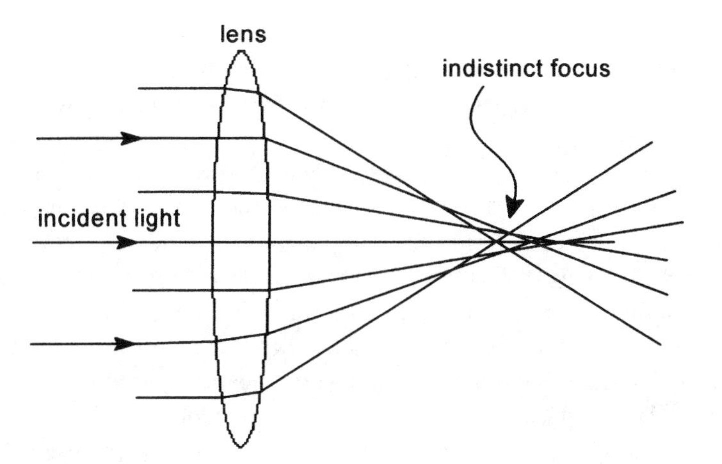

Fig. 6.2. Spherical Aberration.

lenses have shorter focal lengths, so the spherical and chromatic aberration are greater for those whose vision correction required thicker lenses.

Wollaston's proposed solution was to match the curvature of the lens to that of the cornea of the eye, so that oblique rays would enter and leave the so-called meniscus lens at angles that reduced refraction, simultaneously setting the radii of the outer and inner surfaces to values necessary for corrected vision. The ideal result is shown as "Fig. 4" in Figure 6.1, where the focal length, *BD*, of an oblique ray is the same as that of the central ray, *AC*. Because he believed his lens design made it easier for the wearer to look around at various objects with improved clarity, Wollaston named eyeglasses with meniscus lenses "periscopic spectacles." Other than the generalities of design mentioned above, he published no details as to the curvature of the surfaces of his lenses. Those he revealed only to the manufacturers and vendors of the spectacles whose design he had secured by patent, prior to publication of the design generalities.[3]

Wollaston selected the instrument company of J. and P. Dollond to produce and sell his spectacles. The Dollonds, John (who had died in 1761) and his son Peter, were leaders in the design and grinding of fine lenses, and their company produced some of Europe's finest telescopes and microscopes at the time. There are a few entries in one of Wollaston's notebooks on the radii of meniscus lens needed to achieve best results at a sighting distance of fourteen inches, but little more.[4] Although it is possible that he gave those numbers to Dollond, there is no doubt that the instrument maker had enough expertise to determine the best radii for lenses sold to each customer. The periscopic spectacles went on sale in February 1804, and Wollaston purchased a pair himself on February 24 at a cost of 10 shillings 6 pence, considerably more than the price of spectacles with double convex lenses.[5]

It did not take long for a London optician involved in the spectacle trade to dispute Wollaston's claims. William Jones, the chief optician at W. and S. Jones, published a rebuttal in the following issues of the same two London journals in which Wollaston's invention had been announced.[6] In short, Jones claimed that the meniscus lens was not novel, that opticians had avoided using it for sound reasons, and that the common double convex (and double concave) forms of lenses were "the best and most convenient that can be contrived, when clear glass, accurate tools, and good workmanship are used."[7] He added that opticians had abandoned meniscus lenses principally because of the greater image aberration caused by their greater spherical surface. To convince readers of his arguments (and undoubtedly to draw customers into his shop), Jones invited them to visit his Holborn Street location to assess the inferiority of meniscus lenses for themselves.

Despite his obvious desire to protect his own business, much of Jones's criticism of Wollaston's periscopic spectacles was valid. Meniscus lenses had indeed been previously known to opticians, and Jones expressed surprise that Wollaston was able to obtain patent protection.[8] Any focusing advantages they might impart were thought to be outweighed by distortions, both spherical and chromatic, introduced by the collection of peripheral rays. And, as Jones noted, they were heavier, more inconvenient to wear and care for, and cost more to make. But Jones did not directly address the chief advantage of periscopic spectacles: the larger field of focused image that included oblique rays. One can surmise, moreover, that he could not have been happy that a competitor had been given sole right to the sale of periscopic spectacles. Wollaston, not surprisingly, disagreed with Jones and discredited the optician's criticisms in the following issues of the same two journals via a short letter to the editors sent in late March 1804.[9] Wollaston did not think the comparative test of meniscus and double convex lenses recommended by Jones was relevant to spectacle wearers, and he proposed a different test of the two types of lens, which demonstrated the superiority of his design.

Of course, the tests recommended by Jones and Wollaston reflect their biases for the performance of spectacle lenses. Both considered only the optical properties of the lenses under study on the (unstated) assumption that the optical system of the eye itself is ideal. This simplifying assumption is understandable, for neither symmetrical double convex nor meniscus lenses can correct, without modification, individual eye irregularities such as astigmatism, identified a few years previously by Thomas Young. Furthermore, neither man mentioned the fact that the eye can rotate to focus on objects away from the lens axis, although the advantages of meniscus lenses become greater when this is taken into account. Moreover, even the image under view had an impact on the performance of spectacles, as the optimized tests of Jones and Wollaston demonstrated.

Showing much the same tenacity as had Chenevix in the cause of artificial palladium, Jones answered Wollaston's published letter with another one of his own written on April 10, 1804.[10] After reacting angrily to Wollaston's statement that he had been deceived by his own experiment, Jones reported a replication of his opponent's experiment using a pair of periscopic spectacles purchased from Dollond and a pair of double convex spectacles of his own manufacture, which sold for a third as much. He thus took the opportunity of injecting the comparative prices of the two types of spectacles into his rebuttal, a comparison much in favor of his own firm's spectacles. He concluded that Wollaston sought for clarity of a wider field of direct vision at the expense of increased aberration due to peripheral rays, while he himself placed greater emphasis on spherical and chromatic aberration than on a greater breadth of clear vision. Consequently, the comparative tests proved little, for each had been devised to deliver the desired results.

After matching each other paper for paper, claim for claim, and test for test, Wollaston and Jones published nothing more on spectacles until eight years later when another invention of Wollaston involving meniscus lenses reignited their feud. In the interim, Wollaston's periscopic spectacles achieved some commercial success, as judged from a few notes of his royalty income. One entry, dated November 1807, lists income from spectacles as £60 and another for 1810 records income of £37. The final entry occurs in 1818, the last year of patent protection, and lists an unspecified payment from "Dollond for 1409 pairs of periscopes, last account at 6d."[11] If Wollaston's share of the selling price of each pair of periscopic spectacles was 6 pence (5 percent of the selling price, a reasonable amount), it follows that his income in 1807 came from the sale of 2,400 spectacles and that in 1810 from 1,480 pair. Since these sales were made only by a London vendor, and may not be a complete record of sales, it is clear that periscopic spectacles were purchased in significant numbers, although with no great monetary return to Wollaston.

PERISCOPIC CAMERA OBSCURA AND MICROSCOPE

Wollaston recognized that the focusing properties of a meniscus lens could be exploited in optical devices other than spectacle lenses, and eight years

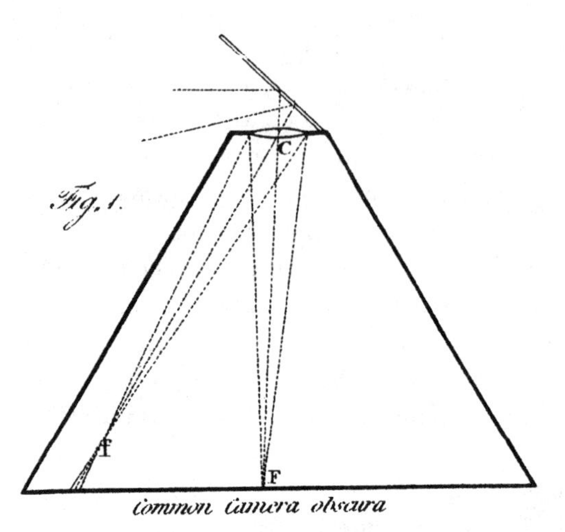

Fig. 6.3. Wollaston's Diagram of a Normal Camera Obscura, 1812.

after his periscopic spectacles were announced, he published a paper describing two additional applications.[12] Intending both to be for general use or because at this later time he was less concerned about supplementing his income, he sought patent protection for neither. The first was an adaptation of the well known image display device known as the camera obscura. In the traditional camera obscura, an image is admitted via a pinhole into a darkened room or an enclosed box, where it produces an inverted copy of itself on a surface in line with the pinhole. A common version of the camera obscura, much used at the time by those who wished to draw an accurate image on paper, contained a double convex lens to focus the image and a mirror mounted to reflect it through a 90° angle onto a piece of paper or a glass surface for tracing. Wollaston was quick to recognize that he could employ a meniscus lens to improve the image quality, although he did not publish a description of his improved device until several years after he first started work on the idea.

Wollaston included a diagram, shown here as Figure 6.3, of the traditional design to illustrate its optical weaknesses. The argument was similar to that used in his criticism of double convex lenses in spectacles. The pencil of incoming light rays, reflected by a mirror, that passed through the lens close to its axis were brought to a sharp focus at *F*, the plane of the paper (or glass) on which the image was projected. But those rays that passed through the lens at an oblique angle came to a focus above the image plane, such as position *f* shown for the rays on the left. Thus the projected image became less distinct the further away from *F* the refracted rays fell.

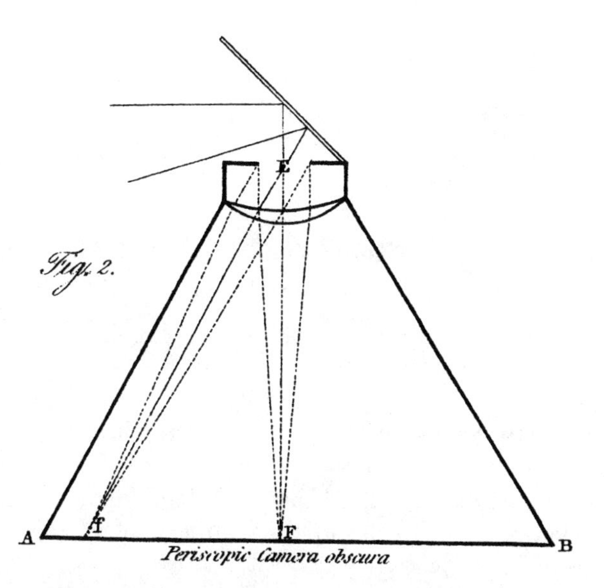

Fig. 6.4. Wollaston's Periscopic Camera Obscura, 1812.

Wollaston's improved instrument, shown in Figure 6.4, had two innovations. The first was replacement of the double convex lens with a meniscus lens four inches in diameter and a focal length of twenty-two inches and "with the curvatures of its surfaces about in the proportion of two to one."[13] The effect of the meniscus lens is to slightly lengthen the focus of the oblique rays, *Ef*, so they come into sharper focus on the image surface *AB*. However, because the field of view captured by the device, subtending an angle of about 60°, is so much wider than that of double convex lenses, even the introduction of a meniscus lens is insufficient to give a sharp image across the full width of the image plane. So Wollaston added a circular blocking device at position *E*. This restriction on the field of view reflected through the lens, known as a "stop," limited the obliquity of the rays that would pass through the lens and blocked what would be the most indistinct portion of the projected image. The two innovations incorporated into Wollaston's periscopic camera obscura, as he called it, combined to give a much sharper image, although one with a slightly reduced breadth of field and decreased image brightness. For anyone who wished to increase the sharpness of the image by decreasing the field of view, Wollaston published a geometric figure that related the desired focal length to both the larger and smaller radii of the meniscus lens. The diagram allowed one to find the desired lens parameters by sight, eliminating the need for calculation.

Fig. 6.5. Wollaston's Periscopic Microscope Lens, 1812.

Unlike the description he gave for his spectacles, which gave no essential design information, Wollaston gave everything needed for any instrument maker to construct periscopic camera obscuras of differing image widths. He was obviously trying to encourage production and use of his drawing instrument.

Perhaps to keep public awareness of periscopic spectacles high, or just to reveal the inspiration for his design improvements, Wollaston likened the improved camera obscura to a reverse application of the meniscus lenses used in his periscopic spectacles. This comparison provoked William Jones into action once again, even though the periscopic camera obscura provided no direct commercial threat. But before considering Jones's response, I will briefly discuss the second optical device introduced by Wollaston in the paper, the periscopic microscope lens, shown in Figure 6.5.

It was well known that a solid glass sphere acted as a powerful magnifying lens. In fact, a very small glass sphere formed the lens in van Leeuwenhoek's pioneering seventeenth-century microscopes. A lens system with a spherical outer surface has many advantages as a magnifying lens. The rays emanating from the object plane on the right (as in Figure 6.5) enter and exit the glass lens at nearly right angles to the lens surface, minimizing refractive distortions while magnifying images a few hundred times. Wollaston recognized that two plano-convex lenses placed back to back had nearly the same optical properties as a sphere, but the spherical aberration of peripheral rays could be greatly decreased by inserting a thin metal disk with a central circular opening between the two lenses, as shown in the Figure 6.5. Although the metal stop reduced the unwanted spherical and chromatic aberration of the magnifying lens, it also decreased the amount of light that could pass through the two lenses, making the magnified image less bright. Nonetheless, Wollaston's periscopic microscope lens was seen as a significant development by David Brewster, who ameliorated the dimming problem by filling the open space between the two lenses with Canada balsam, a transparent resin with a refractive index similar to that of the lens glass. A few years later an even better result was achieved by cutting an annulus-like ring into the equator of a spherical lens and leaving the central opening untouched, producing what is known today as a Coddington lens.[14]

OPPOSITION TO MENISCUS LENSES

When he learned of Wollaston's newest optical devices, William Jones renewed his criticism of the use of both meniscus and convex-plano lenses in place of the generally-used double convex ones.[15] He saw Wollaston's devices more as a threat to customary optical practice than as an opportunity for instrumental development. Intent on exposing "the error of [Wollaston's] reasoning, and the fallacy of his inference,"[16] Jones again criticized the meniscus lens and its greater spherical aberration, which he claimed Wollaston was forced to minimize by insertion of the circular stop. And finally, lest any reader fail to get the point of his paper, Jones concluded by denouncing Wollaston's innovations as if they were a deliberate attack on optical orthodoxy. He wrote, " From these remarks I presume there will be nothing to apprehend from the attempt of Dr. W[ollaston] to depreciate the excellence of the spectacles, Camera Obscuras, and Microscopes, as have been constructed by the most eminent Opticians of the day."[17] From such comments, one can see that Jones had taken on the role of defender of the interests of the London opticians' trade by opposing the recommendations of a threatening natural philosopher. Wollaston's optical devices certainly did have the potential to fragment the market for optical instruments, and it was this fiercer competitive environment that increasingly concerned Jones. His negative comments directed at Wollaston's devices had become both harsher and more personal in 1812 than they had been in 1804.

Wollaston waited a few months before replying to Jones's criticisms, but rather than responding at length to specific statements, he published his own translation of J. B. Biot's favorable assessment of periscopic spectacles, first published in French in the *Moniteur* on September 21, 1813.[18] Biot, an accomplished French natural philosopher who had learned of Wollaston's periscopic spectacles from their 1804 description, had the optician Robert-Aglaé Cauchoix make several pairs of spectacles of differing focal lengths to evaluate their performance. Cauchoix made his first periscopic spectacles with the exterior surface nearly concentric with the eye, as Wollaston had recommended. Biot reported on their effectiveness

> The field of view gained by this construction is really surprising, and it would require a person to be for some time trained to the use of the common defective glasses, to be fully sensible of all the superiority of these. For my own part, I have not been accustomed to wear spectacles commonly, and have only used them occasionally for seeing distant objects; but for the last three months I have regularly used the periscopic glasses, and I now shall never employ any others.[19]

Biot's account of his own use of periscopic spectacles is valuable as a first-person account of the utility of meniscus lenses, and his appreciation for the application of scientific thinking to the structure of lenses contrasts sharply with Jones's distrust of the same. It is obvious why Wollaston would want to bring Biot's unsolicited endorsement to the attention of an English audience.

Although Jones's continued opposition to meniscus lenses of any form in any optical application was biased by his commercial interests, the meager evidence we have on customer uptake suggests his claims about the negative effects of peripheral light had some validity. Both Dollond and Cauchoix acted to reduce the major design weakness of periscopic spectacles by reducing the curvature of the inner lens surface to a shallow curve, almost to the point of planarity. It is clear that the performance of periscopic spectacles depended greatly on the skills of the lens grinder and glass quality, and their appeal to purchasers was largely a matter of personal preference. Their use spread to the continent, where they were offered for sale in France and Germany from about 1813 on, and they continued to be sold in England after patent protection expired.[20] But, even if periscopic spectacles were not an immediate commercial or design success, Wollaston's innovation was one of the most significant advances in lens technology of the nineteenth century. A prominent historian of spectacles has even claimed that "Wollaston's was the only such influence in the spectacle industry between 1804 and 1898."[21] The meniscus shape of spectacle lenses has since become the norm in modern eyeglasses, although the curvatures of the lens surfaces have become much more complex to correct for a variety of individual visual defects.

THE CAMERA LUCIDA: A NEW DRAWING INSTRUMENT

The optical devices described thus far all involved the use of meniscus lenses. The first was introduced by Wollaston in 1804 and the last in 1812.

Between these dates, he also invented and patented a quite different optical instrument, which found widespread use as a drawing aid until well into the twentieth century. The genesis of his invention lay in his inability to draw well, a vexing problem for one who sought to record his visual observations accurately. A trip to the Lakes District with Hasted in 1800 led both men to acknowledge their poor sketching skills. A few months later, Wollaston had devised a solution, and Hasted mentioned that the first wire and sealing wax prototype of the drawing device was constructed in late 1800 or early 1801.[22] However, it took a few years before Wollaston had improved the design to his satisfaction. Finally, in late 1806 he secured a patent for his novel drawing instrument,[23] and in 1807 he published a general description.[24]

In his publication, Wollaston does not mention the prototype shown to Hasted, but he does say that the fundamental principle can be explained by viewing a sheet of paper through a piece of flat glass inclined at an angle of 45° to the paper. A distant image, whose light rays were parallel to the horizontal plane of the paper, impinging on the glass would be partially reflected to the eye, and that reflected (and inverted) virtual image could then be traced onto the paper seen through the glass. Even better, replacement of the single reflecting glass surface with two at successive angles of 22.5° to the incoming light rays, the first of which (upon which the distant image first falls) is silvered on one side to form a mirror, presents the virtual image in its correct vertical orientation. Such an arrangement also increases the fraction of reflected light, as shown in "Fig.1" of the diagram reproduced as Figure 6.6. The incoming light ray is reflected to the eye positioned directly above the clear glass reflector by following the course *fgh*. The eye then sees a virtual image of the object on the paper directly below the glass *ab*. To bring that virtual image to a focus on the paper so that both the image and the drawing pencil appear equally distinct, a convex lens *bd*, of twelve inches focal length, is placed beneath the glass *ab*. Consequently, the virtual image of the objects under view "will then appear to correspond with the paper in *distance* as well as *direction*, and may be drawn with facility, and with any required degree of precision."[25] Wollaston called the complete instrument, with all the optical components mounted in an appropriate wooden and metal frame, a "camera lucida."

Wollaston's patent for the drawing apparatus noted that the "see-through" optical system did not perform well for object images that were not bright, because the image reflected to the eye by the second, clear glass plate was not strong enough to permit tracing. To circumvent this problem he cleverly replaced the two glass elements with a four-sided prismatic reflector, as shown in "Fig. 2" in Figure 6.6. The prismatic reflector has sides *ab* and *bc* angled in the same way as the reflectors in "Fig.1." Thus a light ray entering the prism at right angles to its right-hand surface, as shown in the diagram, strikes both internal surfaces *bc* and *ab* at angles of 22.5°, which results in complete reflection (the angle is too shallow for any refraction out of the prism to occur). The light ray therefore exits the prism undiminished in intensity and enters the pupil of the eye at *e*. The eye cannot see directly through the prism to observe the drawing paper underneath it, so it must be placed so that a portion of the pupil sights the drawing paper beyond the edge of the prism. Thus, the four-sided prism ingeniously combines the functions of two reflecting surfaces and silvered mirror to form a simple and very effective "split pupil" optical device. To reduce the brightness of the reflected image to that of the simultaneously-viewed drawing surface, Wollaston added a brass image equalizer. In the optimized form of the camera lucida, as shown in "Fig. 3" of Figure 6.6, the four-sided prism was mounted on a pillar with an attached brass equalizer with a central circular aperture, marked by *c* in the Figure. The equalizer could be rotated to balance the light incident upon the eye from the prism and the paper.

A characteristic feature of Wollaston's instrumental ingenuity was the integration of scientific principles with practical ones, all combined with ease of use. Recognizing that the camera lucida could be used to trace images of both distant and near objects, including drawings and paintings, Wollaston attached two focusing lenses to the prism (shown in "Fig. 3" of Figure 6.6), one double convex and the other double concave. The convex lens could be rotated into a horizontal position beneath the prism to produce a sharp image for long-sighted persons. Alternatively, the concave lens could be rotated into place vertically in front of the prism to give a sharp image for those who were short-sighted. Use of one lens or the other also enabled the user to copy distant or near objects. In addition, the prism could be moved up and down the supporting pillar to allow the image size to be increased or diminished. The pillar had an engraved scale that indicated the relative distances at which the prism should be placed from object and drawing paper for clearest focus. In short, the camera lucida was a compact, portable drawing instrument of impressive novelty and versatility, and one that could produce a distortion-free image under the right circumstances.

The camera lucida was certainly far smaller and more portable than the camera obscura, but it was not entirely distortion-free when one of the corrective lenses was employed. Nor could its field of view be extended to the 80 degrees claimed by Wollaston without a small adjustment of the prism.[26] Interestingly, Wollaston himself improved the drawing capabilities of the rival camera obscura by adding to it in 1812 a stopped meniscus

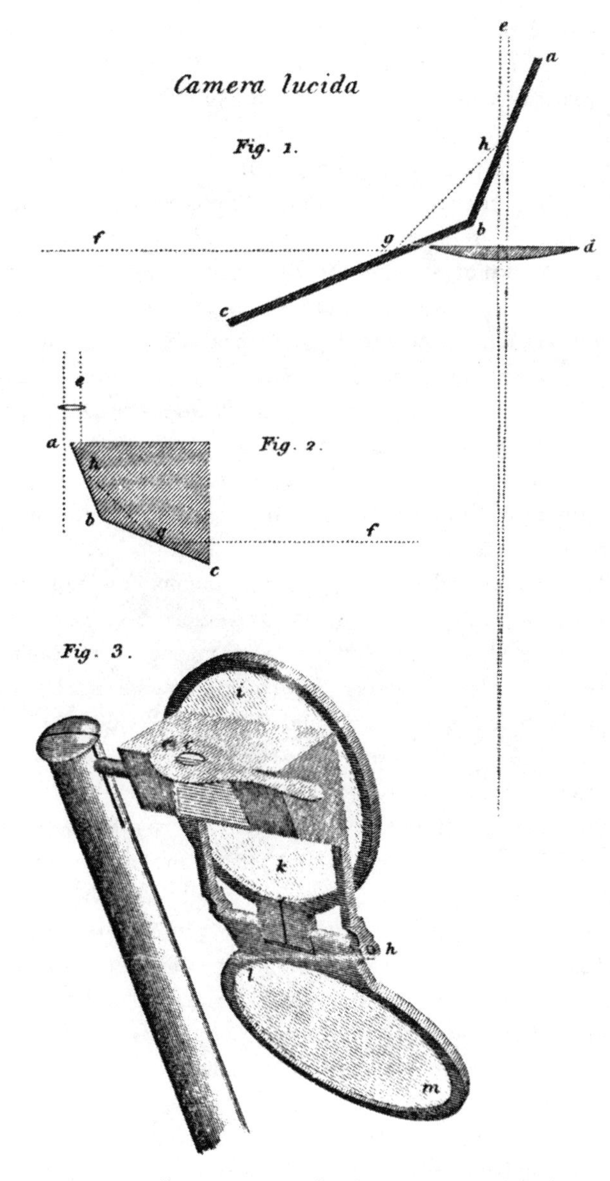

Fig. 6.6. Wollaston's Figures for the Camera Lucida, 1807.

lens, as previously discussed. But still, the camera lucida was a superior instrument in many ways and became quite popular for more than a century.

Wollaston gave no reason for his instrument's name in either the patent or the journal description, but the term was, unfortunately, an ambiguous and historically confusing one. Its predecessor, the camera obscura (Latin for *dark chamber/room*), was in general use in the eighteenth century as an image projection device. It was known from antiquity that an image could be admitted into a dark room through a small circular aperture and viewed in inverted form on an opposing wall, an imaging arrangement still on display at some tourist locations today. In box form, a version of which is the rudimentary pinhole camera, the image can be reflected off a mirror to a glass pane on top of a box for viewing and tracing. A lens can be used to admit and focus light, and it is in this construction as a portable viewer that the word "camera" slowly took on its meaning as an image-capturing device. But the camera obscura, as its name suggests, requires a dark enclosure to see the projected image. Wollaston's "camera" was quite different. It required no darkened space to function, so he obviously used the word "camera" in its newer and still-evolving meaning as an image-capture device. Furthermore, it did not project a direct image onto the drawing surface. Instead, the image to be traced was a virtual one seen by the eye after two reflections and merged with a simultaneous image of the drawing paper, such that the image could be seen and traced in bright daylight. For this latter reason it was christened "lucida," from the Latin word for light.[27] Unfortunately, Wollaston appears to have been unaware that the term camera lucida had already entered the English lexicon as a description of an imaging arrangement credited to Robert Hooke.

In the middle of the seventeenth century, Hooke had published a short article that described the projection of a bright image into a well-lit room.[28] Hooke's contrivance was nothing other than a room-sized camera obscura with a much larger, and brighter, image—so bright that it could be seen in an undarkened room. Hooke's viewing setup came to be known in later years as a "camera lucida," the Latin rendering of "light room," although he never used the Latinized name himself. Although Hooke's imaging display was not an instrument and was not intended, or suitable, for tracing images, its Latinized name led many writers on optical devices to confuse it with Wollaston's instrument. Some of them have even erroneously given priority for the drawing device to Hooke.[29] Although the etymology of the term "camera lucida" can, in one sense, be traced to Hooke, its name for a wholly different drawing device originates with Wollaston. As a consequence of this ambiguity in nomenclature, it is not uncommon even today to see Hooke's name inappropriately associated with the discovery and development of the camera lucida.

USE OF THE CAMERA LUCIDA

The camera lucida marketed by Wollaston was the four-sided prism type, and many examples still exist in museums around the world. The vendor's numbers scratched on them suggest that several thousand were made, although most were probably produced after expiry of the patent.[30] There are only two entries in Wollaston's daybook with information on his income from the sale of the instrument. The first, dated November 1807, lists revenue from camera sales as £140. This amount, accumulated only about nine months after sales began, represents the sale of about 1,000 instruments, on the assumption that Wollaston received 5 percent of the average selling price of £3. The second, in December 1810, records income of £100,[31] suggesting that royalty from camera lucida sales became meaningful after 1807. The popularity of the camera lucida is substantiated by the many drawings made with its aid by painters, engravers, artists, and travelers, in spite of the fact that it took a fair amount of practice to master.

Wollaston did not, in his published description of the instrument, give much direction on how to use it to best effect. Consequently, before vendors began to include printed instructions with it, several early purchasers found it difficult to get the desired results. After one user complained in print of his frustrations with the camera lucida,[32] a more-skilled user named Bate published a helpful guide for its effective use.

> In copying a landscape the instrument is to be fixed upon a steady table or board, on which a sheet of paper is stretched, and the prism brought over the middle of it: the open face of the prism is to be placed opposite the centre of the view; the black eye piece, or stop, being in a horizontal position, is to be moved till the lucid edge of the prism intersects the eye hole. The eye should now be brought close to this opening, and, upon looking through it vertically towards the paper, a perfect copy of the view will appear reflected upon it, and the reflected images will be large in proportion to the elevation of the prism. The eye hole should now be drawn farther off the prism, so as to leave a representation of the object barely distinct, for the more complete command of the pencil.[33]

Bate also reported that the camera lucida could be used to draw microscope and telescope images by placing the open end of the prism directly against the eye glass of each. Its use with a microscope, in fact, has continued well into modern times. Bate's instructions for use of the camera lucida are amplified elsewhere by illustrations depicting its correct use.

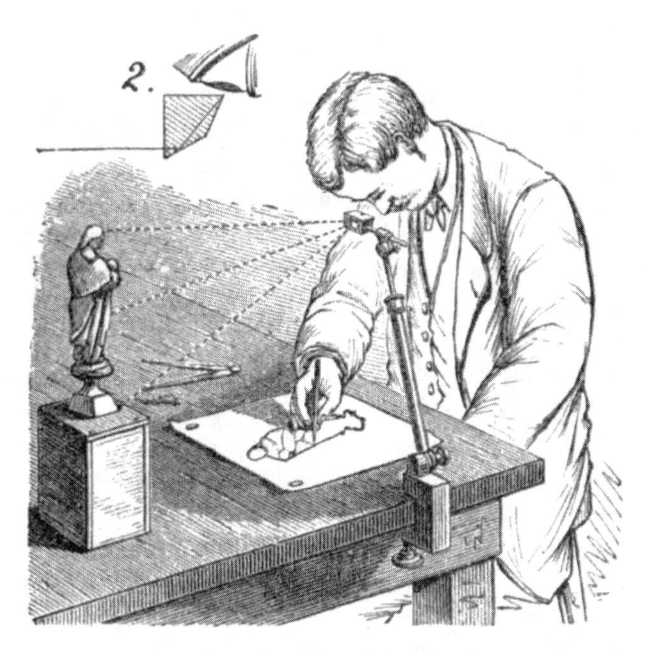

Fig. 6.7. Use of the Camera Lucida, 1875.

One example is shown in Figure 6.7.[34] It shows the user bent over his camera lucida and a closeup of his eye in the required position, but without inclusion of the image equalizer or corrective lenses. Similar diagrams are common in nineteenth-century technical and instructional manuals, suggesting widespread interest in, and use of, Wollaston's invention.

One of the first to obtain a camera lucida was the famed astronomer William Herschel, who purchased one from Wollaston for £2.12.6 in February 1807, likely for use in drawing images of his telescopic observations.[35] His son John, who would go on to become one of Britain's most accomplished scientists of the nineteenth century, used the camera lucida extensively in later years to make accurate drawings of scenes of interest.[36] One of Herschel's sketches, the Temple of Juno in Sicily, shown in Figure 6.8, illustrates the quality of drawing a skilled user could obtain with the camera lucida.[37]

Another noted user was the London sculptor Francis Chantrey, who became a good friend of Wollaston's in the 1820s. He used the instrument to make preliminary sketches of his subjects before sculpting their busts. His sketches of Wollaston, shown in Figure 6.9,[38] were made about 1827 and were the basis for the portrait shown as Figure 6.10[39] and the bust of

Fig. 6.8. John Herschel's Sketch of the Temple of Juno, Made with a Camera Lucida. © Science Museum/Science & Society Picture Library.

Wollaston that now stands in the entrance of the Royal Institution, London. Comparison of the two likenesses of Wollaston show dramatically the degree to which artistic embellishment is needed to bring life to the outlines made by use of the camera lucida.

Basil Hall, a naval captain and explorer, was another close friend of Wollaston's who used the camera lucida extensively during his travels throughout North America in 1828 and 1829.[40] In a preface to his book of etchings made during the tour, Hall writes glowingly of the drawing aid

> With his Sketch Book in one pocket, the Camera Lucida in the other, . . . the amateur may rove where he pleases, possessed of a magical secret for recording the features of Nature with ease and fidelity, however complex they may be, while he is happily exempted from the triple misery of Perspective, Proportion, and Form—all responsibility respecting these being thus taken off his hands.[41]

However, not all users of the camera lucida were as enthusiastic. One of the pioneers of photography, W. Henry Fox Talbot, for example, found that he was unable to produce acceptable drawings with the instrument. He wrote,

Fig. 6.9. Sir Francis Leggatt Chantrey's Sketches of Wollaston Using a Camera Lucida, 1827. © National Portrait Gallery, London.

> One of the first days of the month of October 1833, I was amusing myself on the lovely shores of the Lake of Como, in Italy, taking sketches with Wollaston's Camera Lucida, or rather I should say, attempting to take them: but with the smallest possible amount of success. For when the eye was removed from the prism—in which all looked beautiful—I found that the faithless pencil had only left traces on the paper melancholy to behold.
>
> After various fruitless attempts, I laid aside the instrument and came to the conclusion, that its use required a previous knowledge of drawing, which unfortunately I did not possess.[42]

Frustrated by his attempts to draw acceptable images with either the camera lucida or the camera obscura, Fox Talbot went on in future years to discover a way to fix an image projected onto a chemically-prepared piece of paper. Thus the camera lucida has a minor role in the history of photography as a catalyst for a better method of making and preserving images.

There are many references to the camera lucida and to drawings made with it throughout the nineteenth century, and it continued in use until being displaced by photography in the twentieth century. The nineteenth century also saw much innovation in the application of the four-sided prism ("split-pupil" image) and the two reflecting plates ("see-through"

Fig. 6.10. Sir Francis Leggatt Chantrey's Portrait of Wollaston, 1827. © Science Museum/Science & Society Picture Library.

image) to a variety of imaging applications.[43] Although the device is now largely a historical curiosity, it can still be purchased from art supply stores in a form corresponding to one of Wollaston's original designs.

ROYAL SOCIETY ACTIVITIES

As Wollaston was perfecting his light-focusing devices and establishing his chemical businesses, he was becoming more and more involved with

London's other scientific practitioners. At the Anniversary meeting of the Royal Society on November 30, 1804, as the palladium incident was nearing its conclusion, he was elected as one of the two general secretaries, replacing Joseph Planta, who had served in that position for twenty-eight years.[44] His selection must obviously have had the approval of the president Joseph Banks, suggesting that Banks's unhappiness with Wollaston's conduct in the palladium affair was not sufficient for him to oppose the election. As secretary, Wollaston joined with the multi-talented Thomas Young, who had become the Foreign Secretary in the preceding March. Just over two years later, in January 1807, Humphry Davy came aboard as the second general secretary. For the first time since the founding of the Society in 1660, three active and gifted natural philosophers served together as secretaries, a portent of the dominance of scientific members in the Society that was to take hold later in the century. The secretaries were also members of the Council of the Society, a governing group of twenty-one members who nominally superintended the governance of the Society, although the will of the president generally reigned supreme, especially during Banks's long tenure. All papers intended to be read to the Society passed first through the hands of the secretaries, and, if the author could not read the paper at the Thursday evening gatherings, one of the secretaries would read it in his place. In addition the secretaries formed part of the team of Society Councillors which acted as a Committee of Papers, deciding which papers read to the Society were also suitable for publication in the *Philosophical Transactions*.[45] In addition all three secretaries were ex officio members of the Royal Society Club, whose members met weekly for dinner and conversation at the Crown & Anchor Tavern.

Wollaston took his duties seriously and was a regular attendee at Royal Society scientific and Council meetings, and the Club dinners. The rewards of service to the Society were both personal and intellectual. He could meet and interact with nearly everyone who attended, or presented a paper to, the Society, as well as getting a first-hand look at the content of all submissions. Consequently it would have been nearly impossible for him not to become well informed about the latest scientific observations and ideas across the whole spectrum of natural, physical, and medical sciences. And, as an added bonus, he would have had the benefit of easy access to the opinions and judgments of Young and Davy, at least until Davy ceased being a secretary in 1812. Young remained Foreign Secretary until his death in 1829.

One of the duties of the Council members was to select a member of the Society to give the Bakerian Lecture, a prize lecture founded by Henry Baker in 1775 which has evolved to become today one of the most prestigious awards of the Royal Society. The annual lecture was to be on some aspect of natural history or experimental philosophy and was generally published subsequently in the *Philosophical Transactions*. The arrival in London of Young, Davy, and Wollaston over the short period from 1797 to 1801 brought a trio of excellent Bakerian candidates to the Society, and the Council turned frequently to one of its three secretaries in the early years of the nineteenth century. They were, after all, the best young natural philosophers in the Society. Save for one year, one of the three gave the Bakerian Lecture each year from 1800 to 1812. In 1805, the Council selected Wollaston, for the second time, to give the lecture. His lecture, a perceptive look at a century-old problem in dynamics, was on the force of colliding bodies.[46]

THE FORCES OF MOVING BODIES

The problem that Wollaston sought to clarify had its origins in the seventeenth century when natural philosophers, inspired by the work of Galileo, tried to discover the mathematical relationships that best described the collisions between two rigid moving bodies.[47] Descartes believed that the force transferable as motion when one body struck another was proportional to its bulk (what Newton was later to define as mass) multiplied by its speed. Descartes called this a body's "total quantity of motion." For clarity of argument (although anachronistically) this quantity can be represented by mv using the modern symbols for mass and velocity together with the simplifying assumption that the directions of motion of the colliding bodies lie in a straight line. This quantity is the antecedent of the force now known as momentum. In contrast, Leibniz believed the conserved quantity of motion, which he called *vis viva*, was equal to the bulk of a body multiplied by the square of its speed, mv^2, in modern symbolism, a forerunner of our expression for kinetic energy. Both Leibniz and Descartes argued that the quantities of the motion they described were conserved whenever two bodies collided. In his *Principia* of 1686, Newton argued that his laws of motion were consistent with conservation of mv, a quantity to which he gave the name momentum. He found, in addition, that the quantity mv^2 was not conserved in collisions between bodies that were deformed during impact. Therefore he considered momentum to be the key quantitative measure of colliding bodies.

Wollaston sought to achieve two things in his paper: to decide which of the two conserved quantities was more fitted to practical application, and to correct what he believed to be a misinterpretation of Newton's treatment of the problem. To convey the crux of his argument, he envisaged an idealized experiment that illustrated and resolved most of the relevant principles, and led him to two conclusions. The first was that momentum, mv, was conserved when two objects collided with each other. The second was that the percussive force (force exerted over a distance) of each object was proportional to its *vis viva*, mv^2. Therefore, if emphasis was placed solely on the motion of colliding bodies resulting from forces imparted at the instant of impact, one theoretical explanation was as good as the other. However, if one was interested in the amount of work (measured by the distance a mass could be moved by an impacting body) that could be effected, a consequence of much practical importance in the age of water wheels, then *vis viva* was the important quantity. In short, he suggested that the measurement of the percussive force, or work, that could be produced by a body in motion was a more practical and instructive concept than the mere transference of motion. He extended the results of his idealized experiment to the driving of piles into the ground, where practical experience had shown that the work performed by the driving ram correlated better with its *vis viva*, mv^2, than with its momentum, mv. The point had been made previously by the great eighteenth-century civil engineer John Smeaton in reference to the efficiency of water wheels, as Wollaston knew, but it had fallen on infertile ground in Britain because it was considered to be inconsistent with Newtonian principles.[48] Wollaston next sought to absolve Newton from the misleading "Newtonian" interpretation of his successors by suggesting that Newton's emphasis on the conservation of momentum was correct for the dynamical systems discussed in the *Principia*, where the work potential of motion was not a parameter under consideration. Only when motion was converted to work did *vis viva* become a more useful quantity than momentum.

Having thus established the relationship between work potential and *vis viva*, Wollaston was able to move on to the most prescient portion of his paper—the extension of the work concept to energy sources (as they are now called) other than moving bodies. Since the work potential, or mechanic force as he also called it, of a moving body is related to its *vis viva*, which in turn (as illustrated by the thought experiment) is proportional to the height to which a solid object can be raised, the work potential of energy sources such as gunpowder, coal, or horses, could be compared by determining the height to which each could raise a standard weight. For example, he stated that a weight of gunpowder with explosive power sufficient to raise a one-ton weight to a height of forty feet contained the same amount of mechanic force as a one-ton weight falling from the same height onto an object.[49] Similarly, the quantity of coal of equal mechanic force, acting through the intermediacy of a steam engine, could be determined. Such examples showed that the work potential of an energy source was independent of the time taken for the work to be performed. Only the percussive force and the space through which it was exerted are relevant, and Wollaston explicitly used the term "energy" for the quantity of work potential.

> In short, whether we are considering the sources of extended exertion or of accumulated energy, . . . the idea of mechanic force in practice is always the same, and is proportional to the *space* through which any moving force is exerted or overcome, or to the *square* of the velocity of a body in which such force is accumulated.[50]

Although it was not recognized at the time of Wollaston's lecture that heat was also a form of quantifiable energy, and consequently that the conversion of energy into work could not be complete in any process where heat was produced (as is the case in all real processes), it is clear that he recognized that work output was proportional to the energy content of the system that gave rise to it, whether resident in either its physical or chemical properties.[51] It is interesting to note that Wollaston uses the term "energy" to refer to the quantity of motive power in a moving body that was proportional to mv^2. It is likely no coincidence that Thomas Young used the term "energy" in exactly the same sense in his lectures on collision given about the same time at the Royal Institution. In the publication of those lectures in 1807, two years after Wollaston's paper, Young wrote, in what some historians have judged to be the first such definitive use of the term, "The term energy may be applied, with great propriety, to the product of the mass or weight of a body, into the square of the number expressing its velocity."[52] Neither Wollaston nor Young mentioned each other in their publications on motive forces, but their views and terminology for similar phenomena were intriguingly similar. It is impossible to say, from the available evidence, which of the two had the idea first. The most plausible explanation is that both secretaries of the Royal Society were trying to integrate ideas that were already under philosophical and practical discussion, and by conversation with each other had reached quite similar conclusions.

With the possible exception of Young, it does not appear that Wollaston's recasting of the *vis viva* debate had a significant impact on his contemporaries, and what little effect it had was limited to Britain. One commentary in the *Edinburgh Review* by John Playfair approved of Wollaston's distinction between work and energy. But Playfair missed Wollaston's key point—that it was possible to estimate the maximum work potential of an energy source by measuring the mechanical potential energy (to the extent possible by ignoring heat energy) it was capable of producing.[53] A more informed assessment of Wollaston's contribution was given several years later by Peter Ewart, a Manchester engineer, who wrote an influential paper that connected Wollaston's ideas on work and energy with the earlier pioneering studies of John Smeaton.[54]

Although Wollaston interacted extensively with civil engineers such as Marc Isambard Brunel (a frequent guest of Wollaston's at the Royal Society Club), he published nothing more on the work potential of energy sources. Consequently, his Bakerian Lecture marked both the beginning and the end of his thoughts on the subject and, without discounting the perceptiveness of its ideas, it did little to generate theoretical interest in a disinterested audience. So, despite the inventive genius of British engineers at the onset of the Industrial Revolution, and the discussions of energy transformation by Wollaston and Young, British theory continued to lag far behind that of the French, which culminated in the transformative work on heat engines by Sadi Carnot in 1824, and the foundation of modern thermodynamics.

THE LURE OF CAMBRIDGE

On February 18, 1807, Francis, William's eldest brother, was elected Master of Sidney Sussex College, Cambridge. Francis was the Jacksonian Professor of Natural Philosophy at the time, and he was prepared to resign that position if his brother William could garner enough support to succeed him. A few days after Francis's election to the Mastership William visited Cambridge for three days to assess his chances of winning a contest for the Jacksonian.[55] Finding that his chances of winning against strong local candidates were poor, Wollaston decided to not pursue the opportunity further, and Francis then decided to retain the Jacksonian Chair. It seems odd that Wollaston would even be interested in gaining a position at Cambridge at a time when his career in London was flourishing, and his days were full. The only clue one can find for his motives is found in a letter to Hasted, where Wollaston appears to confess to a certain degree of disappointment at his contributions to "the sum of human happiness."

> When I view our comparative progress it appears to me perfectly natural & perfectly right that he who contributes most directly to the sum of human happiness should be most directly rewarded. It can not be said that I have spent my time idly, but the secret enjoyment of my solitary labours is all that I am to expect; few know the good to which they tend,

> & still fewer care for any one who does not take proper pains to ingratiate himself with everyone.[56]

This is a puzzling, uncharacteristically morose and surprisingly inaccurate self-assessment, which makes sense only in the context of his disappointing polling for the Jacksonian chair. In any event, the despondency soon passed, and Wollaston's failed attempt to gain a Cambridge professorship in 1807, much like his unsuccessful candidacy for the post of physician to St. George's Hospital in 1800, worked out for the best. His chemical business was soon to generate substantial profit, and his scientific inventiveness and imagination continued to be displayed in a variety of ways. Several years later, when another opportunity at Cambridge arose and support for him had risen in tandem with his reputation, Wollaston was no longer interested.

NEW SOCIAL AND SCIENTIFIC NETWORKS

In the letter mentioned above, Wollaston admitted to Hasted that, even though he enjoyed much solitary labor, the rest of his time was not spent idly. We know both statements to be true. However, there is evidence that he had a much richer social environment than the letter to Hasted suggests, and that he was beginning to establish a London-based network of friends, most of whom were to interact extensively with him for the rest of his life. For example, Wollaston had become an active member of a small dining club by 1807, known as the Chemical Club. Although the Royal Society was the premier scientific society in the capital, membership was exclusive and expensive. Furthermore, papers presented at its meetings had to be authored, or transmitted, by one of the Fellows and publication in the *Philosophical Transactions* was vetted by the Committee of Papers. As a result, several less restrictive scientific discussion groups and societies appeared from time to time in London, some of which would evolve into enduring entities, although many of them had only a transient existence.[57] The Chemical Club was one of the more insignificant of these associations of like-minded men, as it had no official structure and was composed only of a self-selected group that liked to talk chemistry in an informal setting. Membership was small, and we know of its existence only through a few casual references to it made by members or their guests. In Wollaston's time, members met every second Tuesday to dine and discuss chemical topics.[58] Offhand remarks by a few of its members suggest the conversation was wide-ranging, free-flowing, and generally light-hearted. It is not known when Wollaston was elected to the Club, but he was present at a meeting mentioned by Dalton in 1809.[59] Wollaston was certainly a regular attendee at the Tuesday gatherings from the time of his joining until the 1820s, when references to the Club come to an end. Berzelius dined with Wollaston and four Club regulars in the summer of 1812, and he reported that the membership then totaled fifteen.[60] The other diners with Berzelius were Charles Hatchett, a wealthy chemist, and three others who were, or were about to become, some of Wollaston's closest London acquaintances—Alexander Marcet, Edward Howard, and Henry Warburton.

We know that Wollaston had interacted with Marcet since their simultaneous arrival in London in 1797. Marcet had fled Geneva as a young man and had gained an MD from Edinburgh before moving to London and becoming a Licentiate of the Royal College of Physicians (Wollaston had been one of his examiners). In 1804 Marcet was appointed physician to Guy's Hospital, and he also established a successful private practice while pursuing investigations into what was then referred to as animal chemistry, a precursor of what we would call physiological chemistry. Marcet was greatly influenced by Wollaston's characterization of different types of urinary calculi, and several letters between the two, dating from 1804, reveal their ongoing collaboration on a number of analyses.[61] His debt to Wollaston's analytical innovations had been made public in a footnote in an 1811 paper on analyses of animal fluids.

> The acuteness with which he [Wollaston] discriminates crystalline forms, however minute, the neatness of his chemical manipulations, and the dexterity with which he analyses the smallest quantity of matter, are known only to those who have seen him engaged in experimental researches. The chemistry of microscopical quantities is in great degree his own.[62]

Wollaston and Marcet visited each other's homes frequently for experiments and dinners and the two even traveled to Paris together in 1814.

The second diner at the Chemical Club in the summer of 1812 and a frequent guest of Wollaston's at the Royal Society Club was Edward Howard, whose eldest brother was to become the twelfth duke of Norfolk in 1815.[63] Howard had been educated in good preparatory schools in France and England, but could not matriculate at Oxford or Cambridge because of his Catholic faith. Undeterred, he applied himself to chemical studies and was elected as a Fellow of the Royal Society in 1799. In 1800 he discovered the highly explosive substance mercury fulminate ($Hg(OCN)_2$), which was later used in detonation caps, an application that was to render flintlock firearms obsolete (and, ironically, end the need for platinum touch-holes, the largest market for Wollaston's platinum). In 1802 he published analyses of a number of meteorites, establishing that they contained unusually high amounts of nickel, suggestive of a cosmic origin. He next turned his attention to a more economical process of refining crude sugar and, in the years 1812–1814 obtained three patents that protected an entirely novel way of purifying sugar that included low temperature evaporation at reduced pressure. Licensing of the patents brought him an annual income in the thousands of pounds in the few years before his death in 1816. The little that is known about Howard's personality suggests he shared many character traits with Wollaston—he was an independent chemist much interested in practical matters, and one with a reserved personality who lauded the accomplishments of others while deflecting praise for his own.[64]

The third diner with close connections to Wollaston (and Tennant) was the much younger Cambridge graduate, timber merchant, and science enthusiast Henry Warburton, who was elected FRS in 1809. He was elected to parliament in 1826, and his active political life undoubtedly deflected him from his later desire to write Wollaston's biography. These three men, Marcet, Howard, and Warburton, are representative of the new and diverse circle of chemical enthusiasts Wollaston was accumulating at the time of his unsuccessful Cambridge initiative—one was a Swiss émigré with interests in animal chemistry, one was a Catholic practical chemist, and one was a young intellectual recently graduated from Cambridge. Their common interest in chemistry had brought them to the Chemistry Club, where their friendship was nourished.

Another source of new acquaintances for Wollaston was the Geological Society, which came into existence in 1807.[65] Within a few months the members elected to transform their group into a full-fledged society dedicated to cooperative research in geology and to launch a dedicated publication for research results. That caused the president of the Royal Society, Joseph Banks, to oppose the formation of the Geological Society, believing that it was infringing on the interests of the older Society. Despite the opposition of Royal Society loyalists, interest in the rival society continued to grow and a year after its formation the Geological Society had fifty members (London residents) and eighty-eight honorary members (non–city residents).[66]

Wollaston's interests in mineralogy did not emerge until about 1808, but, perhaps because he too was unsure about the relationship of the Geological to the Royal Society, he did not become a member until late in 1812, at which time he also became a Council member.[67] Nonetheless, there is evidence he was an early supporter of the new society's objectives. In February 1811, Wollaston gave the first of two £10 donations to the Traveling Fund of the Geological Society in support of original geological research.[68] When the Geological Society signaled its independence by the publication of its first volume of geological papers in 1811, a complimentary copy was sent to Wollaston at the bidding of the President and Council "as a mark of their respect, for the liberal manner in which you have promoted the objects of the Society."[69] Although Wollaston was not a member in the formative years of the Geological Society, he would have known of the will of the membership through conversations with Marcet and Warburton, both of whom were elected to the Society in 1808. Smithson Tennant also joined in 1811, and the man John Herschel in 1820 described as Wollaston's most intimate friend, William Blake, joined in 1812. Blake was a wealthy landowner and sportsman with an interest in science. He was elected FRS in 1807 and often hosted Wollaston, first at his city residence and then at his Danesbury estate in Welwyn, Hertfordshire, after acquiring it in 1819. Wollaston served as godfather to Blake's youngest son, Henry Wollaston, born in 1815. William Blake was one of the first of the country estate owners and shooting enthusiasts that Wollaston was to befriend, and whose outdoor pursuits he was to take up in later years.

Wollaston's transition from a Cambridge and family-oriented man to one more widely connected to London scientific societies and a much more diverse circle of friends and associates occurred several years after his move to the city. This metamorphosis occurred slowly because he was at first concerned with building a new career as a chemical entrepreneur, and doing the foundational research needed to make his chemicals businesses viable. Once those initiatives began to turn a profit, Wollaston could begin to develop new interests and expand his social circle. Thus, there is no reason to believe that the short-lived disappointment he experienced from not finding support for his cause at Cambridge was anything other than a temporary setback.

CHAPTER 7

Commercial Platinum 1805–1820

> Wollaston succeeded in producing superior platinum ware through powder metallurgy because he was the first to realize *all* the intricacies connected with the manufacture of compact metal from platinum powder.[1]

THE FIRST SALES OF PLATINUM

After Wollaston had satisfied himself that there were no other new metals to be discovered in the soluble fraction of crude platina, he decided to begin selling the stock of malleable platinum produced from 1801 to 1803. He selected the instrument maker William Cary, with whom he had dealt on several occasions, to be the sole vendor of the metal.[2] Wollaston made his first delivery to Cary on February 13, 1805, ten days before he announced himself, in Nicholson's journal, both as the discoverer of palladium and the purchaser of large amounts of platina. That delivery was a small one, consisting only of five platinum crucibles and covers, together with a rod suitable for drawing into wires.[3] There is no evidence that the marketing of platinum was advertised by Wollaston or Cary, but it did not take long for news of its availability to spread. For example, in the letter to Chenevix in March criticizing Wollaston's actions in the palladium controversy, Joseph Banks reported that Wollaston "had purchased a large quantity of Platina . . . [and had] opened a Manufactory of Platina Crucibles, etc, which are sold in his name."[4]

The traditional laboratory crucible of the early nineteenth century was a small bowl-shaped vessel of heat-resistant material such as porcelain or charcoal, which was employed for carrying out high-temperature reactions, such as the conversion of minerals to metals and alloys. Such crucibles often cracked during heating and occasionally reacted with their contents.

Platinum crucibles were far superior to other substances in both these respects and had been used in the late eighteenth century, but their quality was poor, they were expensive, and supply was limited. We also know that William Allen, a London pharmacist and chemist, independently produced a few platinum crucibles in the early months of 1805 from a small amount of crude platina ore he was able to purchase, purify, and consolidate.[5] But only Wollaston had enough metal to supply platinum crucibles of consistent quality to all who wanted them.

In March 1805, James Hall, a Scottish chemist and geologist, wrote to his (and Wollaston's) London friend, Alexander Marcet, asking for six new platinum crucibles.[6] Soon afterward, Marcet sent the crucibles to Hall, who acknowledged safe arrival of the order in mid May, with the comment that "The crucibles are indeed beautiful—so much so indeed that I can hardly find in my heart to use them as in my experiments they run considerable risk of being destroyed by the fusible metal which I used."[7] Hall's assessment of the crucibles was a general one—nearly everyone who purchased a platinum crucible from Cary was impressed with its beauty, quality, and chemical utility. In fact, Wollaston's contemporary, Thomas Thomson, credited much of the era's progress in analytical chemistry to their availability.

> [Malleable platinum] furnished practical chemists with a most important utensil, to which chemistry is indebted for the great degree of perfection to which chemical analysis of minerals has reached. Every body now can analyse a mineral with tolerable accuracy; but before Dr. Wollaston supplied a platinum crucible, the analysis of the simplest mineral was a work attended with great labour, and a great waste of time.[8]

Although Wollaston had originally sent Cary five finished crucibles, they were probably meant to serve as examples of the desired form, for deliveries to Cary after the first were simply bar platinum ready to be fashioned by him into crucibles of different sizes. There are entries in one of Wollaston's notebooks that give the diameters of flattened platinum circles requisite for the making of crucibles ranging in weight from 1 to 6 ounces with a metal thickness of about 0.02 inches, and the size of the ingot needed for making a specified number of similarly-sized crucibles.[9] It is likely that Cary made Hall's crucibles, although a notebook entry dated March 26, 1805, for "6 covers" suggests Wollaston made the special lids for them.[10] Of all the applications to which platinum could be put, crucibles were the most demanding. The platinum in them had to be sufficiently pure that prolonged exposure to high temperatures would not cause volatile impurities to migrate through the metal to produce blisters on the surface, which could in turn lead to holes in the thin platinum. Unfortunately, Wollaston found that his platinum was not always immune to this fault. Within two months of the first sale of crucibles, two were returned to Cary in "blistered" condition, as was a platinum "arc" probably used by its purchaser as a high-temperature evaporator.[11] This meant that the compressed platinum powder still contained some of the volatile components of the sal ammoniac precipitate from which it had been obtained. To his regret, Wollaston could not connect the blistered platinum to any specific compaction process, for he had not kept individual consolidation information. Thus, he could not know at first if the problem was an isolated occurrence, or a systematic problem. Fortunately, most crucibles performed satisfactorily. Nonetheless, all future consolidations of platinum powder were individually recorded and linked to specific purification sequences, so any problem with the product platinum could be traced to its specific batch preparation. From this Wollaston learned to use only the finest, and purest, platinum powder for crucibles. Moreover, like any good businessman intent on establishing a reputation for the quality of his product, Wollaston refunded the full cost of all returned platinum. As in his other dealings with London craftsmen, Wollaston took a leading role in the design of the platinum items he wished to bring to market, often making the prototype himself. He also took an active role in the marketing and service aspects of his business. He was what we now might call a hands-on type of scientist, one who bridges the gap between idea-based science and object-based technology.

In 1805, Cary also began to sell bar and wire platinum, neither of which required finishing work, for 16 shillings an ounce, and crucibles, which he fashioned from flattened metal, for 17.5 shillings per ounce.[12] These prices remained unchanged for the twenty-year period he marketed platinum, even though there were wide swings in the prices of English goods during that period due to the economic consequences of war, then peace, with France. In the account books Wollaston kept for the shared chemical business, income from Cary is credited after a deduction of 10 percent for seller's commission. Thus, Wollaston received about 14s. 6d. for each ounce of platinum he sent to Cary.[13] For some direct sales of platinum he made personally, Wollaston charged 14s. per ounce, corroborating this amount as his customary base price for the metal. Since the platina ore had been purchased at an average cost of 3s. per ounce[14] and the average cost of chemicals required to purify the metal was about 2d. per ounce,[15] a reasonable estimate for the expenses involved in producing malleable platinum would be about 4s. per ounce, after sundry expenses and John's wages are included. This left Wollaston and Tennant with a profit of about 10 shillings per ounce of platinum sold, a justifiable amount considering the hours of labor invested in the processing. Complete sale of the 5,300 oz. of platinum Wollaston had prepared from 1801–1803 would thus net the partners a profit of about £2,650. Until overall revenue exceeded total expenses in 1809, income from Cary was shared between the two partners in rough proportion to their individual expenses.[16] Aside from payment for the shared purchase of the crude ore, Wollaston's expenses for the production of malleable platinum far exceeded those of Tennant (who played no part in the purification and consolidation of the metal), so for the first recorded, and typical, credit entry in the financial accounts on April 6, 1805, for platinum sales totaling £40, Cary retained £4, Tennant received £5 and Wollaston £31.[17]

Laboratory crucibles were an obvious commercial application of platinum, but Wollaston envisioned a variety of other uses for the metal. One notebook page, shown in Figure 7.1, on which the first entries appear to

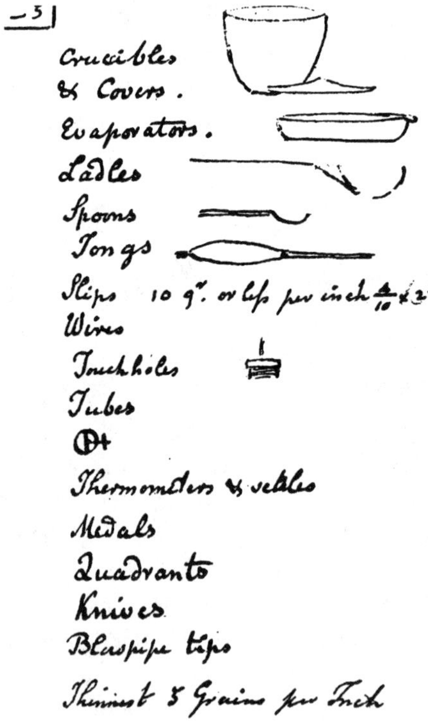

Fig. 7.1. Wollaston's Summary of Platinum Applications, ca. 1805. Reproduced by kind permission of the Syndics of Cambridge University Library.

have been made about 1805, lists the uses to which platinum was put over the years.[18] The first items on the list, with illustrative sketches, were simple laboratory utensils, such as crucibles and tongs. One would require those items to be made of platinum only for specific technical operations that involved highly reactive chemicals or that needed to be done at very high temperatures. Consequently, those items, together with some of the others on the list, such as wires, tubes, thermometers and scales, medals, quadrants, knives, and blowpipe tips were all niche applications, which, by themselves, would have slowly, if ever, exhausted the platinum prepared by the end of 1803. But two others, indicated on the list as touchholes and the symbol for vitriolic acid, were different. Both evolved to become sufficiently large markets to stimulate an expansion of the platinum business, and together they would ultimately account for nearly 85 percent of all platinum sold over the lifetime of the business.

FIREARM TOUCHHOLES

Flintlock firearms of the early nineteenth century had a small external pan in which the priming gunpowder was placed. When this powder was ignited by a spark from the triggered flintlock mechanism, its flame was transmitted to the main charge of gunpowder in the gun barrel through a small hole in the barrel called a touchhole. In most firearms, the touchhole was simply drilled through the steel gun barrel. But over time, because of the high temperature and corrosive nature of the gunpowder flame, the diameter of the hole gradually widened with the consequence that an increasing amount of the explosive force of the firing gunpowder escaped through the touchhole and the power of the firearm was reduced.[19] Makers of the finest firearms resolved this problem by lining the touchholes with gold, which was resistant to the sulfureous gases of exploding gunpowder. But gold was not ideal because of its relatively low (for a metal) melting point (1065° C). When gun makers, as one of the last steps in hardening the breech of a gun barrel, heated it to a high temperature, the gold touchhole melted if the temperature rose too high. Platinum, with corrosion resistance equivalent to gold but with a much higher melting point (1768° C), was a superior material, especially since it could be purchased at a price less than one-quarter that of gold. Consequently, London gun makers, especially the Manton brothers, John and Joseph, first began using platinum in touchholes about 1803 and 1804, and a trade label for Joseph Manton dating from 1805 refers to him as "Inventor of Platina Touch Holes."[20] It is possible that Wollaston supplied them with test samples of malleable platinum

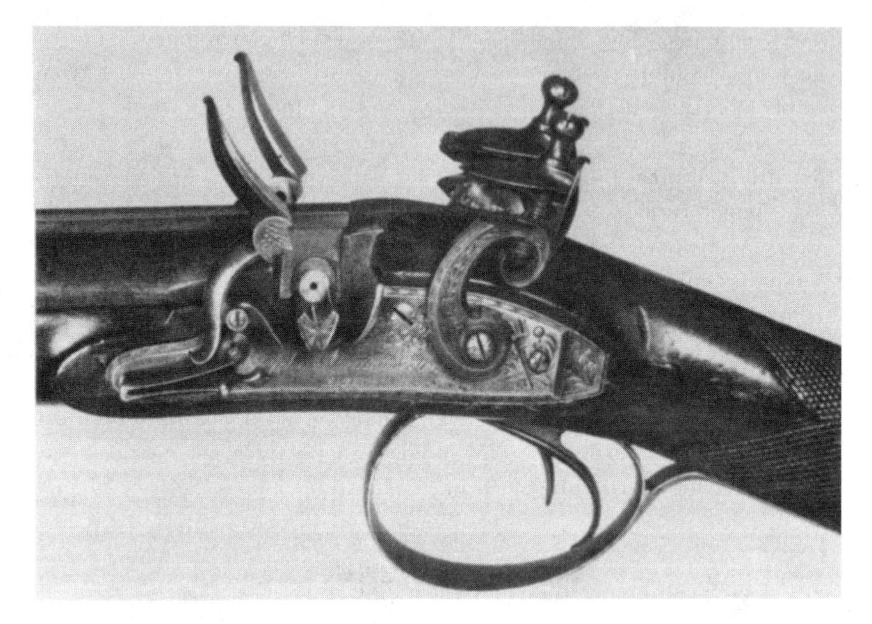

Fig. 7.2. A John Manton Gun of 1815 with a Platinum Touchhole. *Platinum Metals Review* 25(2), 1981, p. 76.

before general sale through Cary began in 1805, but there is no hard evidence that such was the case. The name of Manton appears frequently in the Cary account book from 1806 on in connection with platinum wires and ingots, both destined for touchholes. A picture of a platinum touchhole (seen as a bright-colored ring above the v-shaped priming pan) is present in the double-barreled flintlock gun made by Manton shown in Figure 7.2.[21]

In the early nineteenth century London was home to dozens of gun makers, who competed aggressively with each other for the lucrative trade in fine firearms. One directory for 1824, for example, lists seventy-four London gun makers, and most, if not all, of them quickly incorporated platinum into their products, mainly for touchholes but also for decorative and other purposes.[22] Although the weight of platinum used in each firearm would rarely exceed 1/8 ounce, its incorporation into tens of thousands of pistols, guns, and rifles created a huge market. By 1821, when bulk sales of platinum came to an end, gun makers had purchased about 25,000 ounces of the metal, approximately two-thirds of all the malleable platinum Wollaston produced.[23]

Touchholes were an ideal market for platinum, as that usage was a forgiving application. The platinum did not need to be of an exceedingly high purity; it had only to be malleable, hard, and corrosion resistant. Fortunately, the largest consumers of the metal, London's gun makers, were located in the same city as its sole English producer: there were no distribution difficulties to overcome. Finally, market demand correlated almost perfectly with product availability over the period 1805–1820. Thereafter flintlock firearms began to be replaced by percussion ones, in which the bullet was propelled by gunpowder ignited by a pressure-sensitive explosive set off by a firing pin. Such firearms did not require touchholes, and although some platinum continued to be employed in them for other uses, the principal application for platinum began to decline at the same time as Wollaston found himself unable to obtain more crude ore for processing. Although Wollaston may well have anticipated platinum's use in firearms, it is unlikely that he could have expected that application to become so extensive, or so lucrative.

SULFURIC ACID BOILERS

Sulfuric acid was an important industrial chemical widely used in the chemical and textile industries. It was produced in very large amounts in Britain by the lead chamber process, in which a combination of sulfur and niter was repeatedly burned over a period of weeks in a large lead-lined chamber flooded with water. The oxidized sulfur dissolved in the water to produce a weak solution of sulfuric acid, which was drained from the chamber into lead boilers and concentrated by boiling off much of the water. Then, in an increasingly dangerous part of the operation, the stronger acid was transferred to glass distillation vessels nested in sand-filled iron pots. Finally the acid was boiled at ever higher temperatures to drive off still more water to form the highly concentrated acid sold commercially in wicker-protected glass carboys as "oil of vitriol."[24] Not surprisingly, handling and heating of the glass vessels used in the final concentration step had to be done with great care to avoid breakage of the boilers, which would result in loss of product and acid damage to the surroundings and, possibly, the workmen. One report suggests that glass retorts normally survived only about five boilings before breaking.[25] Replacing glass with platinum for the boiler operation had obvious practical advantages: platinum does not react with concentrated sulfuric acid; it is a hard, durable metal that does not break or corrode at high temperatures; and it can be flattened and shaped into boiler form. But it was costly, and its use for a boiler would only be economically feasible for a large-volume producer. One of these was the London chemist and acid manufacturer, Philip Sandman, who ordered

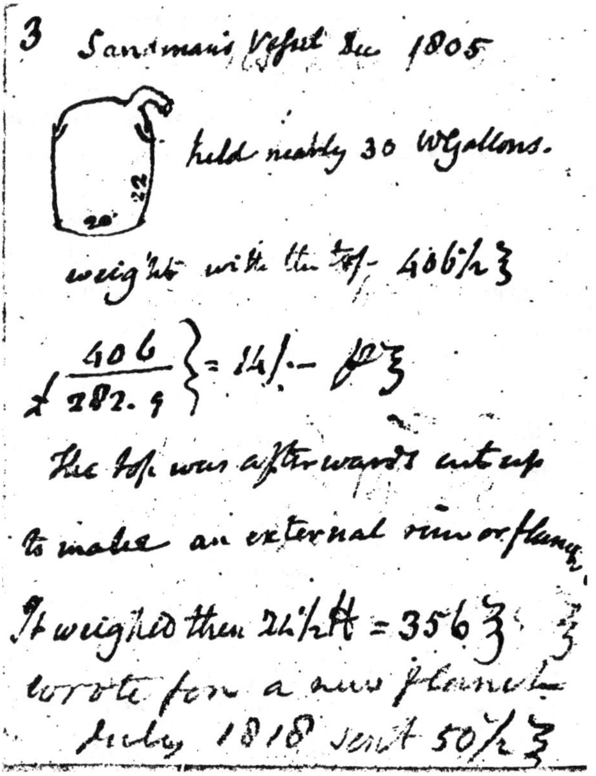

Fig. 7.3. Wollaston's Notebook Entry on Sandman's Platinum Boiler, 1805. Reproduced by kind permission of the Syndics of Cambridge University Library.

a platinum boiler from Wollaston in 1805. It was to be the first in England, probably the world, made for sulfuric acid producers.[26]

Sandman's vessel was fabricated in 1805 and delivered in December. Details on the boiler are recorded in one of Wollaston's notebooks devoted specifically to platinum boilers, and a sketch is reproduced in Figure 7.3.[27] The boiler was about 20 in. wide and 22 in. high, with a rim around its upper circumference into which a lid with an exit tube could be placed prior to boiling down the acid. It held nearly 30 (wine) gallons and was made with 406.5 oz. of platinum priced at about 14s. per ounce for a total cost of £282.9. The vessel was still serviceable in 1815 when it was sold by Sandman to the Glasgow chemical producer Charles Tennant,[28] and it was the latter who ordered extra platinum from Wollaston for a new platinum rim ("flange"), as mentioned at the bottom of Wollaston's notebook entry.

There are some interesting aspects of Sandman's vessel that shed light on the Wollaston/Tennant business partnership. There are entries in Tennant's partnership records for July 1805 that suggest he was involved in the construction of the boiler, for entries under the heading "Sundries for Sand[ma]n" include costs for "flatting," "blacksmith large platinum," and "forging."[29] Recall that Wollaston's process for purifying platinum yielded only enough metal in each batch for consolidation into ingots weighing 20 oz. or less. Thus a vessel of the size intended for Sandman required several sheets to be soldered together in leak-proof joints, a task that required considerable skill and expertise. Wollaston and Tennant selected Charles Sylvester, a metal worker based in Sheffield, to construct the vessel. An entry on one of the sheets of paper present in the Cambridge collection states that "Sylvester of Sheffield," in December 1805, was "soldering platina with gold by Blowpipe, while surrounded inside & out with hot charcoal."[30] Sandman's vessel contained several lengths of such carefully soldered joints and would have required many days of exacting work. Tennant packaged the platinum sheeting and sent it to Sylvester, and paid him the considerable sum of £50 for the fabrication of the vessel.[31]

Tennant's involvement in the construction of Sandman's boiler suggests that he might have been the one who first thought of using platinum for that application, a possibility made more likely by the employment of a Sheffield metalworker to construct the vessel. Wollaston, as previously noted, had very close relationships with London craftsmen and would have been less likely to engage someone so distant to make a boiler for a London merchant, even though Sheffield was a leading English center for silver working and steelmaking. It seems likely, therefore, that Tennant had initially taken on the task of finding and developing markets for the products of the chemical business, and the Sandman boiler is evidence for this hypothesis. Although Wollaston was certainly capable of developing applications for platinum on his own (as he did for palladium and rhodium), he did not have the outgoing personality of Tennant, so he may have been content to have his partner take the lead in finding markets for their chemical products. Unfortunately, but characteristically, Tennant's enthusiasm for platinum boiler fabrication quickly waned, and all boilers after the first were made under Wollaston's sole superintendence.

The financing arrangements for Sandman's vessel are also of interest. Although the cost of fabricating the vessel was substantial, Wollaston and Tennant billed Sandman only for the platinum used. Moreover, to encourage Sandman to be the first to use a platinum boiler, the partners allowed him to pay for it, free of interest, a year after he took delivery. And this he did, remitting payment in three installments in 1807. It is clear that Wollaston and Tennant encouraged Sandman to pioneer the introduction of platinum boilers into the sulfuric acid manufacturing process, both by selling the boiler to him at a platinum price below what Cary was charging, and by giving him generous terms for payment. Later purchasers were not given the same inducements. As the British economy began to improve in the early years of the nineteenth century, production of sulfuric acid increased dramatically, and the market for platinum stills expanded. But the second sale of a platinum boiler did not occur until December 1809. Perhaps because its construction was entirely superintended by Wollaston, there is much more information on its fabrication in his notebooks.

In late 1809, Thomas Farmer, a sulfuric acid producer with his father's firm of Richard Farmer and Son in south London, asked Wollaston for a platinum boiler similar to Sandman's. Letters to him about the boiler design, fabrication, pricing, and delivery reveal the close attention Wollaston paid to all aspects of the order, and illustrate his considerable entrepreneurial bent. The first we learn of the order is in a letter from Wollaston to Farmer in late November.

> But in truth the difficulties [making the boiler] have been greater than I had supposed & the trouble I have had is such as I should not have undertaken if I had foreseen it.
>
> I believe the bottom to be by this time firmly soldered into its place. The vessel will next require to be hammered into form; & then the rim to be fastened on. . . .
>
> Whenever it is finished I will trouble you again with a line in hopes that you will be able to make it convenient to see it here before you receive it.[32]

For this boiler, Wollaston engaged a young London silversmith named George Miles, who was a cousin of the London assayer and Wollaston colleague John Johnson (who was importing platina for Wollaston at this time). Miles was paid £14.10.0 in December for the work,[33] which turned out to be much more challenging than he had envisioned. The most difficult operation was the soldering of the separate platinum sheets together with gold, as Sylvester had done years before (at a cost of £50) for Sandman's vessel. A sketch of the vessel in a notebook, shown in Figure 7.4,[34] shows what needed to be done. The unnecessary lid had been eliminated, and the boiler now took on the shape of a large, open pot. Wollaston by this time had become

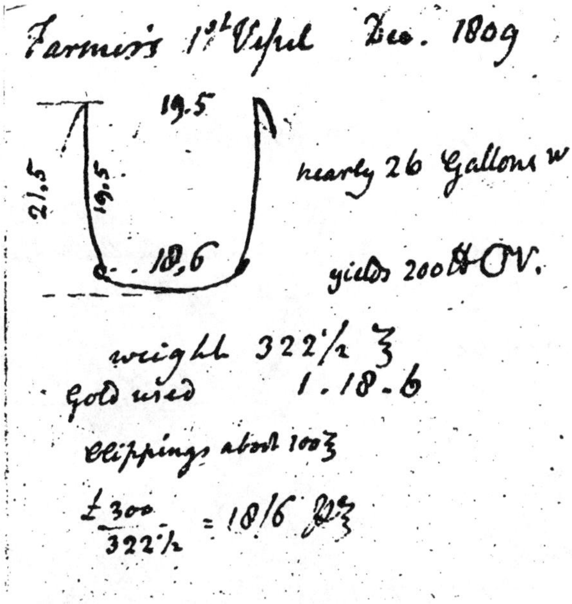

Fig. 7.4. Wollaston's Sketch of Farmer's First Platinum Boiler, 1809. Reprinted by kind permission of the Syndics of Cambridge University Library.

good enough at consolidating platinum that he was able to make large ingots weighing about 100 oz. and only four of these were needed for Farmer's vessel.[35] Three of the flattened sheets were soldered together to make the walls of the boiler and the fourth was soldered onto the bottom to form the base. To join the sheets together each edge was folded into a small "u" shape, such that one slotted into another, as roughly shown for the base union in Figure 7.4. The juncture was then sealed tight with gold solder. Considerable skill and perseverance were needed to make the seals leak free. The upper rim of the vessel was reinforced by a copper flange soldered, or riveted, around its circumference, which acted to hold the vessel in place in the furnace. It was to get Farmer's preference for the means of attachment of the rim that Wollaston wrote to him, but Wollaston was forthright in mentioning his own preference for riveting on the basis of strength and cost.

Farmer also asked for, or Wollaston suggested, a platinum siphon, so that the concentrated sulfuric acid could be removed without lifting the boiler out of its furnace mount. The diagram of the proposed siphon is very similar to that of a second siphon made for Farmer in 1812, shown in Figure 7.5.[36] The siphon fitted over the rim of the boiler and reached to the bottom with a slightly curved end. The longer delivery side, which carried the concentrated

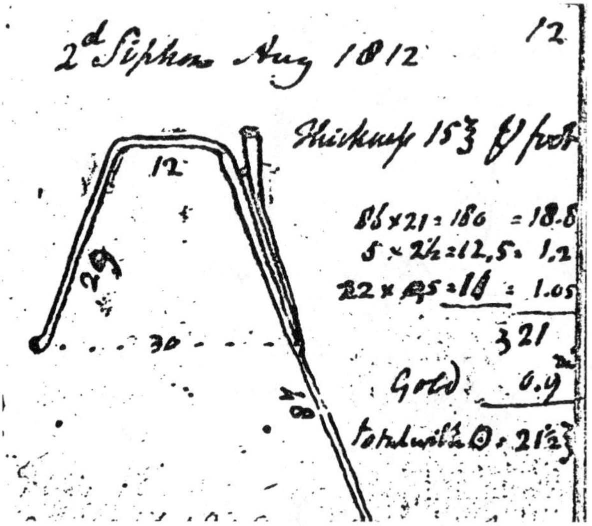

Fig. 7.5. Wollaston's Sketch of Farmer's Second Platinum Siphon, 1809. Reproduced by kind permission of the Syndics of Cambridge University Library.

acid into a collecting vessel, had an attached (gold soldered) tube which, by applied mouth suction (very cautiously applied, one expects), initiated the siphoning action. In mid December, Wollaston wrote a second letter to Farmer informing him that the vessel had been completed and was ready at his home for a final inspection. He also warned Farmer that the fabrication expense "has much exceeded any expectation I could have formed; for notwithstanding my own personal exertions it amounts to upwards of 30£."[37] By mid February the platinum siphon was finished, and Wollaston billed Farmer for the boiler and siphon.

The total cost of the vessel and siphon were calculated in a more sensible way than had been done for the Sandman boiler. This suggests both that Wollaston was a sounder businessman than Tennant and that incentives for the purchase of platinum boilers were no longer necessary. The platinum used in the vessel (322.5 oz.) and siphon (8.25 oz.) was priced at 16s. per oz., the same price at which Cary sold bars of the metal. To this was added all fabrication costs, including the gold used for soldering, to a total of £300 for both boiler and siphon.[38] Wollaston continued this pricing method for all future boilers. He supplied the platinum at a price equal to that charged by Cary (to protect Cary's retail sales), superintended the construction of the boiler, contracted out the fabrication work, and billed the purchaser for the total costs, charging nothing for his own (substantial) contributions. Furthermore, payment was due upon delivery of the finished vessel. Entries in the account books record payment by Farmer for the full amount on February 16, 1810.[39]

Wollaston's interest in, and responsibility for, his platinum boilers and siphons did not end with their fabrication. Frequently, he elected to fix construction flaws himself, and at his own expense. Wollaston even carried out performance tests on the boilers, and ensured they were leak-free before delivery. For another boiler sold in January 1811 to Samuel Parkes, Wollaston was eager to learn of its effectiveness and enquired one year later,

> As it is now more than twelve months that you have been in possession of your large vessel it may reasonably be supposed that you have full trial of its merits and I hope you will not think it unreasonable in me to request a few lines upon that subject, as it would gratify me to hear that it has fully answered your expectations . . . if you would take the trouble of adding any further information on points of economy which I presume you have carefully estimated (saving of time, saving of fuel, saving of breakage, saving of labour) you would confer an additional favour on [me].[40]

Unfortunately, there is no record of Parkes's reply, although in a later book he wrote that his vessel "holds 32 gallons, and costs a few hundred pounds; but the advantages which result from its employment are fully adequate to the expense."[41] Wollaston also received a favorable report for the Sandman boiler, which had been repurchased and put into service in 1815 by Charles Tennant of Glasgow. Tennant informed him that, by using the vessel, "they boil off 3 times per day and turn out 50 bottles of 150 lb per week & reckon to save the prime cost in 2 years."[42]

Wollaston's interaction with manufacturers and tradesmen, well illustrated by the preparation and delivery of acid boilers, was a consistent feature of the implementation of his scientific discoveries. He was no ivory tower intellectual, isolated from the commercial applications of science. On the contrary, he expended much time and effort in understanding the uses to which his inventions and innovations were to be put and to gathering feedback on their effectiveness, in the process contributing to the success of his customers and his products. This characteristic was emphasized in a nineteenth-century biographical entry, written by someone who was obviously well-informed.

> When he [Wollaston] proposed to manufacturers or tradesmen improvements in chemical processes, or in the construction of instruments or

> apparatus, he contracted to receive nothing, if they should prove unsuccessful, but to be paid a certain proportion of the savings or profits, in the event of their succeeding. In making a profitable business of practical science, he thus never abandoned the character of a professional man and a master-manufacturer, but always maintained the position of a gentleman.[43]

This assessment of Wollaston as philosopher and businessman was shared by many who dealt with him in the practical trades, but it contrasts sharply with the earlier views expressed by Joseph Banks, for example, who viewed his entrepreneurial and for-profit initiatives (which led to the palladium incident) as unbecoming for a man of science. Such differing opinions of Wollaston's contributions to early nineteenth-century British science were to resurface in the obituaries that followed his death, likely because so few details of his actual financial, technical, and entrepreneurial activities were generally known.

Wollaston ultimately supplied a total of sixteen platinum boilers to purchasers throughout the United Kingdom, as shown in Table 7.1.[44] The

TABLE 7.1. Platinum Boilers Sold by Wollaston, 1805–1818

#	Purchaser	Date	Weight (oz.)	Cost (£.s.d.)
1	Sandman, London	Dec. 1805	406.5	282.9.0
2	Farmer #1, London	Dec. 1809	322.5	300.0.0
3	Pepper, London	Feb. 1810	377	282.15.0
4	Parkes, London	Jan. 1811	377	330.0.0
5	Farmer #2	Jan. 1812	405	345.0.0
6	Farmer #3	May 1815	420.5	376.9.0
7	Tennant #1, Glasgow	Feb. 1816	445	378.0.0
8	Norris, Halifax	Apr. 1816	452	384.4.0
9	Tennant #2	Aug. 1816	440	374.0.0
10	Mackenzie	Feb. 1817	471.5	400.15.0
11	Tennant #3	May 1817	442	375.14.0
12	Tennant #4	Mar. 1818	445	378.5.0
13	Farmer #4	Apr. 1818	419	355.0.0
14	John Smith, London	July. 1818	443.5	358.10.6
15	Gregg & Boyd, Belfast	Nov. 1818	440	374.0.0
16	McKenny, Dublin	Dec. 1818	448.5	381.4.4
Total			6755	5676.15.10

last ten boilers were made by a London coppersmith named John Kepp, who proved to be skilled enough to relieve Wollaston of many of the fabrication problems Miles had encountered. In addition to the boilers, Wollaston designed and delivered eight siphons and one cover with delivery spout with a total weight of 412.7 oz. So the production of platinum boilers and related hardware, the major application of the malleable metal handled entirely by Wollaston himself, consumed over 7,100 oz. of platinum and brought in revenue of a little over £6,000, shared of course with Tennant until his death in 1815. This important market was, like that of touchholes for firearms, an ideal one for platinum, as the boilers proved to be very durable and cost-effective. In fact, demand for the boilers continued after Wollaston could no longer obtain crude platina, and his notebook lists seven unfilled orders after 1818.[45] The first platinum boilers in France, made with platina from Spain, appeared in 1819. English-made ones only became available once again after a new Russian source of platina arose later in the nineteenth century.

TOTAL PLATINUM PURCHASES AND SALES

As shown in Chart 7.1, sales of platinum to gun makers and sulfuric acid producers were principally responsible for making Wollaston's and Tennant's business a commercial success.[46] Of the nearly 37,800 oz. sold before the business came to an end, 28,300 oz. (75 percent) were designated either

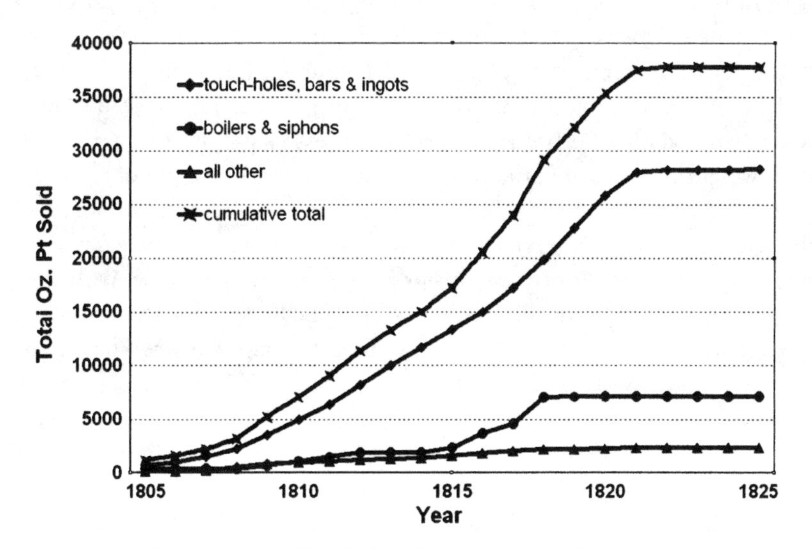

Chart 7.1. Sales of Malleable Platinum (Troy oz.), 1805–1825.

as for touchholes or as bars and ingots. All of these sales are grouped together because almost all of the bars and ingots were also purchased by gun makers. In some years, the bulk of the sales by Cary were listed as for touchholes, and in other years only as bars and ingots. The inverse relationship between these two sales categories, together with an assumed continuity of firearm manufacture, suggests that Cary did not use a consistent descriptive label for the platinum he sold. Nor did he need to, since the platinum sold for touchholes was similar in quality and form to that in bars and ingots. Certainly some of the bar platinum was purchased for uses not related to firearms, but that amount was not large enough to alter the fact that the gunnery business was the principal market for malleable platinum.[47] Acid boilers and siphons accounted for another 7,200 oz. (19 percent) of marketed platinum, and the remaining 2,300 oz. (6 percent) was nearly equally split between the highest-quality platinum used for crucibles, and all other miscellaneous applications.

I mentioned in Chapter 4 that 7,300 oz. of crude platina had been purchased by the end of 1802, and from this Wollaston had extracted 5300 oz. of purified platinum powder by September 1803. As Chart 7.1 shows, all the platinum from these early purchases was sold by 1809, and the potential for the sale of much more metal would have become obvious at least a year or two earlier. To meet demand, Wollaston first purchased back all the platinum returned to Cary, including that from blistered crucibles and other chips, cuttings, and remnants from purchasers. There are several entries in the notebooks that indicate he generally paid 5s. to 7s. per ounce for returned, or waste, platinum. By late 1807, he had accumulated 1200 oz. of platinum scraps, which amounted to about one-fifth of the metal he had refined and sold to that time. He continued to reprocess platinum scraps throughout the lifetime of the business, with the result that the total weight of platinum processed exceeds the overall weight of crude ore purchased because some of the metal was processed more than once. Although the processing of repurchased platinum scraps made good business sense, it meant little for the expansion of the business. For that, Wollaston needed to get more crude platina, and to that end he enlisted the assistance of London's only commercial assayer, John Johnson, who was knowledgeable in the buying and selling of bullion.[48]

Entries on platinum purchases in Wollaston's notebooks and payments to Johnson drawn on his Coutts's bank account reveal that Johnson acted on Wollaston's behalf to import over 38,000 oz. of platina from Kingston, Jamaica, from 1806 to 1819.[49] The importance of Johnson's intermediacy in the procurement of such large quantities of ore is illustrated in Chart 7.2,

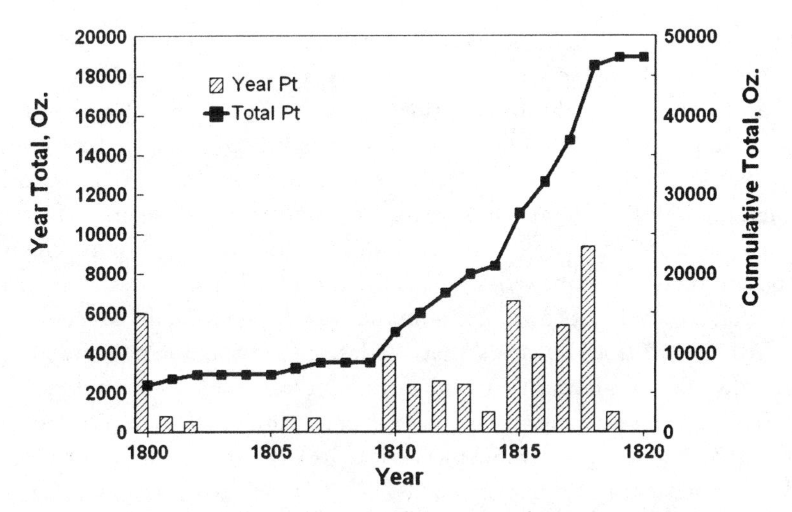

Chart 7.2. Wollaston/Tennant Platinum Purchases, 1800–1820.

which shows the quantities of platina purchased over the years by Wollaston and Tennant.[50] All platinum obtained after 1805 was purchased from Johnson, except for a total of 300 oz. in 1816 (from Fletcher) and 1,300 oz. in 1819 (684 oz. from Bollmann and 600 oz. from Hodgson).

The weights of ore that reached London are quite astonishing in light of the fact that export of alluvial platina from New Granada was tightly controlled by the Spanish government. The amounts imported from English traders in Kingston had all been smuggled out of the Choco region over narrow, mountainous trails to Cartagena and from there shipped to Jamaica, from where it could be legitimately purchased. A thorough study of the production and outflow of crude ore from New Granada has shown that legal exports to Spain fell off precipitously in the early 1800s as the contraband trade with Jamaica (destined for Wollaston) increased. In fact, shipments of platina to England amounted to about half of all the crude ore available from New Granada from 1805 to 1820.[51] Wollaston could not obtain platina from Spain, or its political and trading partner, France, during the Napoleonic era because of the imposed trade embargo with England, but importation from Jamaica was aided by Britain's naval control of the Atlantic shipping trade. In such a political environment, the contraband trade flourished. There is one entry in Wollaston's notebooks that gives the price of platina available in Kingston about 1820 as 1s. 6d. per troy ounce, which was sufficient to invigorate the contraband trade, especially since the workers were not paid anything at all for the alluvial platina, which was supposed to be turned over to the Spanish authorities in New Granada.

If one assumes the 1820 price was about the same during the years Johnson was bringing the ore to London, then his average selling price to Wollaston of 3s. per ounce brought him a return of about 1s. per ounce, for a total of about £1,900, before shipping and sundry costs are deducted. And at 3s. per oz. of crude platina, Wollaston was able to realize a profit of about 10s. per ounce of malleable metal. The miners, the smugglers, the traders, Johnson, Wollaston and Tennant, gun makers, sulfuric acid manufacturers, chemists, and instrument makers were all the ultimate beneficiaries of Spain's inability to develop its own platinum industry from its colonial resources.

All the evidence suggests that Wollaston had some sort of standing order for all the platina that traders in Kingston could acquire. Once it arrived in London, Johnson held it in storage until needed for processing, at which time it was delivered by carriage to Wollaston. Financial records of the purchases from Johnson include costs for delivery and for interest charges at 5 percent for the period of storage under Johnson's care.[52] The importing agreement with Johnson began slowly, with the delivery to Wollaston of 776 oz. of platina in late 1806 and another 700 oz. in early 1807 (see Chart 7.2). Then, presumably after Johnson informed the Kingston traders of the need for even more platina, transport of the ore to London resumed in 1810, reaching a peak of over 9,000 oz. in 1818. The importation of platina via this contraband route finally came to an end in 1819 with the declaration of independence of Colombia, after which the new country held the valuable metal for its own use.[53]

As Charts 7.1 and 7.2 illustrate, the year 1809 was an important one for the Wollaston-Tennant partnership. All the platinum from their joint purchases before 1803 had been purified, consolidated, and sold. Large-volume markets had been established, the production process had been regularized, and ample supplies of the requisite ore had been secured. The business venture had begun to turn a profit, and all that was needed for continued success was hours and hours of labor by Wollaston and his assistant John. Under these new circumstances, the 50/50 profit-sharing agreement between Wollaston and Tennant was no longer fair to the harder-working (egregiously so) partner, and a new agreement was negotiated. On February 1, 1809, details of the new arrangements were put on paper.

- WHW shall be paid before division
- for every lb of Bricks converted into SS beyond what is sold & paid for
- for every oz of Pl [platinum] sold by Cary & also for every oz prepared to be forged for other purposes

- & may reserve for his own use all Palladium which he may separate at his own expense from all Platina hereafter to be dissolved
- This agreement bearing the date February 1, 1809.[54]

The first item in the agreement refers to Bricks[alt] and SS (salt of sorrel), products of the partners' organic chemicals business, which I will describe in Chapter 8. The second item, the most important one from a financial perspective, indicates that Wollaston was to be paid (an unspecified amount) for platinum sold by Cary and himself, such as for boilers, before the remaining revenues were divided with Tennant. The third entry gave Wollaston sole rights to all revenue from palladium recovered from the new purchases of platina, rights that he had previously secured by payment to Tennant of £20 in 1805 for palladium isolated from the platina purchased before 1803. As previously discussed, sales of palladium never amounted to more than a few dozen pounds, so this provision never proved to be financially rewarding. The much more meaningful second item, however, fails to indicate how Wollaston was to benefit from future sales of platinum. But the details can be gleaned from entries in the business account books. Beginning in 1811, more than a year after the date on the new revenue-sharing agreement, Wollaston began to receive 10 percent of all revenue from platinum sales before equal division of the remainder with Tennant. For example, an entry recording payment of £330 to Wollaston from Parkes in January 1811 for a platinum boiler is reduced by one-tenth to £297 before being credited to the joint account of evenly-shared revenue. Other credits from Cary, who usually made remittances after each £100 of platinum sales, appear as shared revenue to the amount of £81, after Cary's 10 percent (£10) and Wollaston's 10 percent of the remaining £90 (£9), were deducted.[55] The new profit-sharing agreement was negotiated in 1809 presumably because that was the year that the chemical business (which encompassed both the platinum and the organic chemicals endeavors) began to show an overall profit.[56]

As Chart 7.3 shows, initial outlays for crude platina (and raw organic materials) in 1800–1802 placed the business in deficit by about £1700 in the early years. Revenue from the sale of organic chemicals, which began in 1803, and platinum products, starting in 1805, gradually brought the business to the break-even point in 1809. Up to then, all income from the chemical business was shared between the two partners in proportion to the amounts they had spent individually, so the question of how to fairly proportion profits was not a pressing issue. Although there is no indication

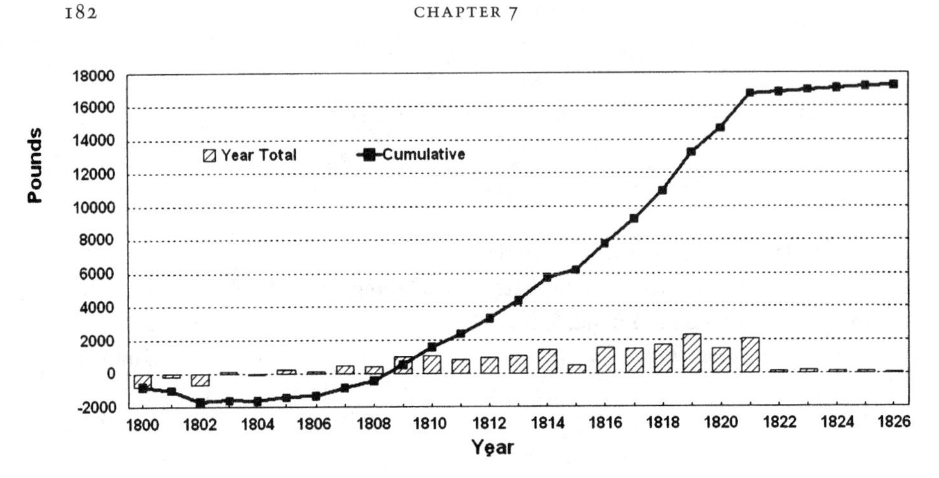

Chart 7.3. Financial Totals for the Wollaston/Tennant Chemical Business, 1800–1826.

that Wollaston was excessively unhappy with the business partnership, or with Tennant himself (other than his previously noted displeasure with Tennant's haphazard work habits), he cannot have failed to recognize the injustice of profits shared equally between one partner who did all of the development and production work and the other who did almost none. So, before Wollaston committed himself to processing the platina that began to arrive from Johnson in 1810, he obtained Tennant's agreement for the 10 percent personal share of profits mentioned above. By any measure this was still a revenue-sharing agreement much in Tennant's favor. It is hard not to judge the new agreement as a generous concession on Wollaston's part, and a strong measure of the continuing loyalty he maintained for his older colleague. Even as the labor required to process the large quantities of Johnson-brokered platina increased after 1809, Wollaston made no attempt to re-adjust the terms of revenue sharing, which remained unchanged right up to Tennant's death in 1815. It was unfortunate that Tennant did not live a few years longer, for he missed out on the most prosperous period of the chemical business. In the six years after Tennant's death, annual profits soared to amounts between £1400 and £2300. It was only then that Wollaston became a relatively wealthy man.

IMPROVEMENTS TO PLATINUM REFINING

Although Wollaston had large amounts of crude platina to process from 1809 on, he did not increase the scale on which he and John operated. They continued to dissolve the platina in batches of 25 oz. each, usually beginning two batches each day (often including Sundays), to a total of eight, which

TABLE 7.2. Platinum Purification Sequences, 1801–1821

Series #	Date	Material	Oz.	# Batches
Cecil St.	Feb.–Jun./01	ore	215	28
1	Nov./01–Jan./02	ore	1129	99
2	Mar.–Jun./02	ore	1275	102
3	Aug.–Dec./02	ore	2000	152
4	Jan.–Aug./03	ore	2395	152
	Dec./07–May/08	scraps	1272	113
5	Apr.–Jun./09	ore	1572	100
6	Jan.–Mar./10	ore	2006	108
	May–Jun./10	scraps	818	47
7	Nov./10–Jan./11	ore	1840	102
8	Sep.–Nov./11	ore	2400	124
9	Jul.–Sep./12	ore	3016	145
10	Sep.–Nov./13	ore	3000	142
11	Mar.–Jun./15	ore	3600	175
12	Sep.–Dec./15	ore	3000	146
	Mar.–Jun./16	scraps	1400	78
13	Jun./16–May/17	ore	5000	247
14	Aug./17–Jan./18	ore	3950	194
15	Mar.–Sep./18	ore	5000	247
	Nov./18–Feb./19	scraps	1466	77
16	Mar./19–Jul./20	ore	5200	264
	Nov./20–Jul./21	scraps	2040	108
Totals			53594	2950

likely was the working capacity of the furnace needed to heat the solutions. When solution was complete after a few days, the platinum was precipitated as a triple salt by sal ammoniac, washed, dried, and stored until more malleable metal was needed. The labor involved can be appreciated by looking at the total amounts of platina purified by the batch process over all years of production, as shown in Table 7.2.[57] From the first batch purifications of 1801 carried out in Cecil St. to the final batch processes in the summer of 1821, Wollaston and John together purified over 53,000 oz. of platina (including the secondary treatment of platinum scraps). The production batches, totaling nearly 3,000 over all years of production, were grouped together into a few consecutive months of each year. In addition, after the platinum precipitate from each of the batches had been collected, it was reduced to platinum powder by intense heating, and then washed, dried, sieved, compressed, and hammered into ingots. One notebook summary records that, to prepare 180 oz. of finished platinum, Wollaston spent 3 hrs. in washing and grinding the platinum powder, 2 hrs. in compressing it into compact plugs of metal, and 2 more hrs. in forging it into ingots.[58] This works out to 1 hour of labor for each 26 oz. of metal, or over 2000 hrs. over all the production years. Despite all the chemical and technical innovations that were essential to the successful production of malleable platinum, the commercial metal owed as much to hard, physical labor as to intellect and invention.

Beginning with production series 5, the first to use platina purchased from Johnson, Wollaston abandoned the in-flask preparation of muriatic acid by the reaction of sulfuric acid and common salt, and reverted to simple use of muriatic acid itself. Although this increased the per ounce cost of chemicals for purification from 4s. to 5s., the time required for digestion of the platina was reduced from five days to two. This resulted in a small saving in the costs of coal but more than doubled the weight of platinum that could be purified each day. Wollaston calculated that the change in aqua regia formulation enabled him to increase the rate of platinum purification from an average of 12.3 oz. per day in 1802 (series 2) to 31.5 oz. per day in 1809 (series 5).[59] Even so, this means that, for example, ten days or more were required to purify enough platinum for Farmer's boiler (which weighed 322 oz.), completed in December 1809. In addition, that weight of platinum would have required over twelve hours of extra labor to forge it into ingots. It is obvious that an equal sharing of the profits between the two partners was due for a change in Wollaston's favor.

As previously mentioned, when some of his first crucibles were returned because of surface blistering, Wollaston regretted that he had not kept records of individual consolidations, which would have enabled him to seek out causes of the problem. This oversight was remedied beginning with production series 5, when he began to keep specific records of each ingot he forged. These records of consolidations, "castings" as Wollaston called them, provide a wealth of information on the different qualities of platinum produced, its destination, occasional flaws, and the evolution of the refining process toward that described in the 1829 *Phil. Trans.* paper. These notebook records were retroactively summarized, with an explanatory comment, which indicated that Wollaston started to separate platinum powder of different grades by passing it through a sieve. The finest powder was reserved for crucibles.

> A regular journal was now kept of the casting. . . . The softer part [of platinum powder] which 1st passed thro a lawn sieve was cast separate & marked L. That which required more grinding [was] marked G [and] formed ingots much harder than the former, & consequently both were afterwards mixed together [and labeled] LG, some of the softest cream [of the platinum powder] alone being reserved for small ingots [for crucibles].[60]

Entries such as these in the records of the many production sequences indicate that Wollaston continued to encounter problems dealing with the variable composition of the crude platina obtained from Johnson. Consequently, he was forced to make continual improvements to the refining process.

In series 5, for example, the grinding mentioned above (by hand or with a wooden mortar and pestle) did not always yield platinum particles that could be consolidated into an easily-forged mass, and Wollaston initially thought that the larger platinum particles were contaminated with iron. But an analysis of coarse powdered platinum revealed that iron was present in amounts less than one part per thousand.[61] He concluded, therefore, that the quality of the malleable platinum obtained by compaction of the powdered metal depended on both chemical purity and particle size. The finest powder yielded the best platinum. Wollaston recognized that, as the particles became smaller, their surface area (relative to their weight) increased and they could be made to cohere with greater ease to yield ingots of greater malleability. Consequently he began to separate the powdered platinum into grades of different size by thorough and repeated shaking in water. As he was to report in 1829:

> By repeated washing, shaking, and decanting, the finer parts of the gray powder of platina may be obtained as pure as other metals are rendered by the various processes of ordinary metallurgy; and if now poured over, and allowed to subside in a clean basin, a uniform mud or pulp will be obtained, ready for the further process of casting.[62]

This was experimental ingenuity, and powder metallurgy, at its best. And by reserving the best platinum for crucibles, the most demanding application, Wollaston eliminated the blistering problems that had plagued the earliest crucibles.

Another problem arose in series 5 from heating the platinum precipitate too strongly when reducing it to a powder. Working in an era when good high-temperature thermometers did not exist, Wollaston had no means of reliably measuring the heat of his furnace, so that the ideal reducing temperature may either have been surpassed or maintained at a suitable level for too long a time. Either way, "over-burned" platinum powder lost its cohesive properties, and nothing he could do to it with a variety of chemical treatments (short of a complete re-purification) could restore a "weldable surface," as he called it.[63] Over-burning continued to be a problem throughout 1810 (series 6 and 7), and all that ruined platinum had to be redissolved in aqua regia to give a double-refined product. Wollaston does not record how he finally solved the problem, but over-burning of platinum powder is no longer mentioned as a recurring difficulty after 1810.

Not surprisingly, Wollaston also devoted much effort after 1808 to improving the purity of his refined platinum. He found that digesting the crude ore for a longer time gave a better product, so beginning with series 6 he allowed the hot aqua regia to act on the platina for four days instead of two. In series 7 he stopped combining the undissolved residuum with fresh platina (he used only enough aqua regia to initially dissolve only about half of the crude platina) and instead combined all residual platina for subsequent, and separate, purification and consolidation. He also paid careful attention to improving the removal and recovery of palladium and rhodium by the processes he had earlier developed for each metal, because he had found evidence of both metals in the platinum scraps he processed in 1810. Thereafter, the platinum obtained from the first and second metallic precipitates, which were treated to recover palladium and rhodium, began to be processed independently. But even as late as 1818, during production series 15, Wollaston found that all the platinum he recovered from the first metallic precipitate gave "very brittle & bad" metal. He was, however, able to obtain good product from the bad ingots in 1820 by redissolving them and adding extra muriatic acid before the sal ammoniac precipitation step. This final adaptation to the production process, included in the 1829 paper, "seemed to succeed, for none was returned by Cary."[64] This last modification to the purification process illustrates the daunting problem that confronted Wollaston through all of his platinum work: he had no way to determine easily or reliably the purity of the platinum powder produced other than by observing the color of the sal ammoniac precipitate and assessing the ease with which the platinum powder made from it cohered. Thus the notebook records of the batch processes have several comments on irregularities in the color of the sal ammoniac precipitate,

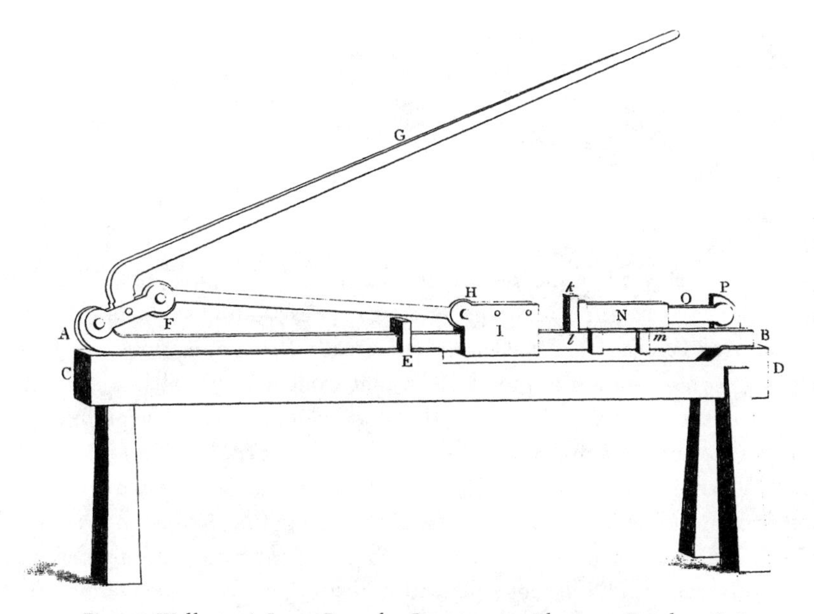

Fig. 7.6. Wollaston's Lever Press for Compressing Platinum Powder, 1816.

as well as on the varying degrees of hardness of some of the consolidated ingots. The poorest platinum, whether too contaminated with impurities or over-burnt, was redissolved in aqua regia, and that suspected to be of dubious, but usable, quality was made into ingots for touchholes. But, on the whole, remedies were found for each irregularity, and each successive production process yielded less and less unusable platinum.

One important modification to platinum consolidation was a great improvement in the power of the press used to compress the wet powder into a compact mass prior to heating and forging. In July 1816, Wollaston purchased a powerful lever press that had been made to his design.[65] The new press performed so well that Wollaston included a diagram of it in his 1829 paper, reproduced in Figure 7.6. Pressure on the handle *G* communicated a force to the sliding iron cradle *N*, which held the wet platinum mud in a cylindrical metal barrel. The platinum was compressed as *N* forced the barrel against the immoveable piston *O*. Wollaston calculated that the press multiplied the force applied to the lever by 60 times at a lever angle of 5°, by 300 times at an angle of 1°, and by a nearly infinite amount as the angle approached zero (when the lever was fully depressed).[66] One notebook entry mentions the time, in minutes, needed to consolidate a mass of platinum using the lever press as:

- apparatus made ready 8
- greased 4
- charged 14
- pressed & delivered 22
- all away 7
- [total] 55.[67]

If one subtracts the time needed for setup and cleanup, the consolidation of each platinum plug required 40 minutes, adding even more time-consuming labor to that needed for chemical purification and subsequent forging of the compressed metal. We might wonder why Wollaston did not hire an extra assistant to help with the more menial tasks of the production process, but his personal attention to every facet of the production of malleable platinum was probably the main reason for its commercial success. One twentieth-century metallurgist, in fact, credits Wollaston's accomplishment to his mastery of all facets of the powder-consolidation process.[68]

Wollaston's and Tennant's platinum business exemplifies characteristics of technological innovation that are immutable. Good ideas and good science can give rise to good products, but commercial success depends on the fit of goods with market demand. Palladium and rhodium (and osmium and iridium) were all impressive scientific discoveries that became valuable commodities only a century or more after their discovery because there were no meaningful markets for those metals in the early nineteenth century. Platinum, on the other hand, found a lucrative market as soon as it was offered for sale. With the benefit of short-sighted Spanish colonial politics, supply of the crude ore from the Republic of New Granada became available just when Wollaston and Tennant were in a position to utilize it. The timing couldn't have been better.

FINANCIAL SECURITY

As the chemical business prospered after 1809, Wollaston's concern for his financial well-being (as revealed in daybook entries) diminished as his share of the profits began to exceed his needs. There was, however, no outward sign of his greatly improved financial position and few outside of his immediate family circle (and possibly only some of them) could have known of the rewards of the platinum business. The letters to Hasted, which provide the most candid information on Wollaston's personal life and interests over the course of his full adult life, make no mention of the platinum venture or of its success after the early mention of the renunciation of medicine for a business in chemistry. Nor did Wollaston's lifestyle change in any appreciable way, although he did hire another cook/housekeeper about 1813 to join the one that had been in his employ since the early 1800s. Wollaston's three long-time domestic servants, the manservant and laboratory assistant John Dowse, and the cooks/housekeepers Mary Ann Couch and Martha Roblou, remained with Wollaston until his death and bore witness to his will. But three domestic servants were no great extravagance for a man of his eminence.

The success of the business did, however, allow Wollaston to reward the faith his father and brothers had shown in him. I described earlier how, in addition to the £8,000 in annuities given him by brother George from the West Hyde legacy, William and his younger brother Henry were given control of over £8,000 in Bank of England stock (with a cash value much greater than the nominal amount). In 1808, when it had become obvious that the chemical business would soon move into a period of continuing profitability, the stocks were dispersed equally to all of Wollaston's sisters, with each of the nine (married and single) receiving £600.[69] Payments of £800 to each of the brothers other than George and William had previously been paid out of this account, suggesting that the stock holdings had been intended from the start for dispersal among the Wollaston children, but only after William had successfully established himself in business. On October 27, 1808, each of the six unmarried sisters also received an additional £450 of 4 percent Annuities from a fund in William and Henry's name, which had been built up by contributions from the eldest son Francis John, who had also received some of the West Hyde legacy.[70] These transfers of dividend- and interest-bearing stocks among all the Wollaston children from money initially willed to George and entrusted to the care of Francis John, William, and Henry is a strong testament to the cohesion of the Wollaston family under the guidance of its patriarch.

It seems likely that much of Wollaston's drive to profit from his various scientific pursuits—chemical and optical—was motivated by his desire to return money to his siblings that had been loaned to him to launch his chemical business. If this assumption is correct, we can empathize with Wollaston's anxiety over quitting medicine in 1800 to launch a risky business venture. As I will relate in subsequent chapters, Wollaston continued to support family members and other good causes without manifesting any great interest in amassing personal wealth, although financial independence remained a top priority. Malleable platinum made all of this possible.

CHAPTER 8

Organic Chemicals and Multiple Combining Proportions 1802–1815

> [T]he inquiry which I had designed appears to be superfluous, as all the facts that I had observed are but particular instances of the more general observation of Mr. DALTON, that in all cases the simple elements of bodies are disposed to unite atom to atom.[1]

Following his happy departure from medicine in 1800, Wollaston's many discoveries and inventions moved him into the upper echelon of English natural philosophers, and earned him respect from both intellectuals and artisans. But there is one other large component of the Wollaston-Tennant partnership still to describe, which was also unknown to their contemporaries. That was the production of value-added "organic" chemicals from the dregs of wine fermentation.[2] This undertaking began in 1802 and, like the platinum enterprise, underwent a slow period of growth before becoming profitable several years later. Characteristically, Wollaston's close monitoring of the production process led him to some remarkable insights into the quantitative aspects of chemical combination. Those insights drew him inexorably into the debates surrounding the nascent atomistic ideas of John Dalton.

CHEMICALS FROM WINE DREGS

During the fermentation step of winemaking, a thick, solid crust of tartar (commonly called argol) is deposited on the sides of wine casks as a byproduct of the action of yeast on grape sugar. Argol consists mostly of a water-insoluble salt of tartaric acid ($H_2C_4H_4O_6$, hereafter H_2Tar) partially neutralized with potash. In its pure form, the salt is known as cream of tartar ($KHC_4H_4O_6$, hereafter KHTar).[3] Cream of tartar is a mildly acidic salt that was used widely in the textile industry to make dyes adhere more strongly to fabrics, functioning as a mordant. Most of the cream of tartar used in England was imported from the wine-making regions of the continent, but war with Napoleon in the opening years of the nineteenth century and the resulting economic blockade of France and its allies made the salt more difficult to obtain, and more expensive. Cream of tartar could also be converted by a sequence of reactions to oxalic acid ($H_2C_2O_4$, hereafter H_2Ox), a vegetable acid that, together with its salts, was commonly employed in the textile industry to remove stains from whitened areas of patterned fabrics, such as calicos. The most commercially valuable salt of oxalic acid, known as salt of sorrel (KHC_2O_4, hereafter KHOx), was traditionally obtained from the juice of the sorrel plant. But the yield was very low. One report from the early nineteenth century, for example, said that 50 lbs. of the sorrel plant could be pressed to give 25 lbs. of juice, from which only 2.5 ozs. of salt of sorrel could be extracted.[4] So, with a business plan that had an operational structure similar to the platinum enterprise, Wollaston and Tennant decided to purchase large amounts of argol for conversion first into tartaric acid, then into oxalic acid and its salts for sale to manufacturers in Manchester, the center of the massive British textile industry. The chemistry required to produce the desired materials promised to be less challenging than that needed for platinum purification because the fundamental reactions had previously been discovered by the Swedish chemist, Carl Scheele, whom Tennant had visited in 1784. All the evidence in the Wollaston manuscripts suggests that Wollaston agreed to carry out all the chemical operations and Tennant was to look after marketing and sales. Although notebook records suggest that Tennant played a somewhat more active role in the argol business, his contributions once again paled in comparison to Wollaston's, at least until a new revenue-sharing agreement was implemented in 1809.

Although the day-to-day details of the argol business were not as fully documented as those for the platinum business, Wollaston did keep good purchasing, production, and sales totals. His laboratory notebooks also contain a few pages of summary notes on the batch processes employed to make the desired products, so that the major features of the production sequence can be understood. Wollaston's financial notebook reveals that he and Tennant made their first purchase of argol in January 1802, buying about 200 lbs. at a total cost of £16.14.6.[5] To recover as much tartaric acid as possible from the argol, Wollaston treated the latter with an inexpensive form of powdered chalk termed "whiting" (calcium carbonate, $CaCO_3$). This converted the principal component of argol, KHTar, to an insoluble tartrate of lime (calcium tartrate, $CaC_4H_4O_6$, hereafter CaTar), with the co-production of soluble potash (K_2CO_3), carbon dioxide and water, as shown

$$\underset{\text{argol}}{2\,KHC_4H_4O_6} + \underset{\text{whiting}}{2\,CaCO_3} \xrightarrow{\text{water}} \underset{\substack{\text{tartrate of lime}\\ \text{(insoluble)}}}{2\,CaC_4H_4O_6} + \underset{\substack{\text{carbonate}\\ \text{of potash}\\ \text{(soluble)}}}{K_2CO_3} + CO_2 + H_2O$$

Equation 8.1. Primary Argol Treatment.

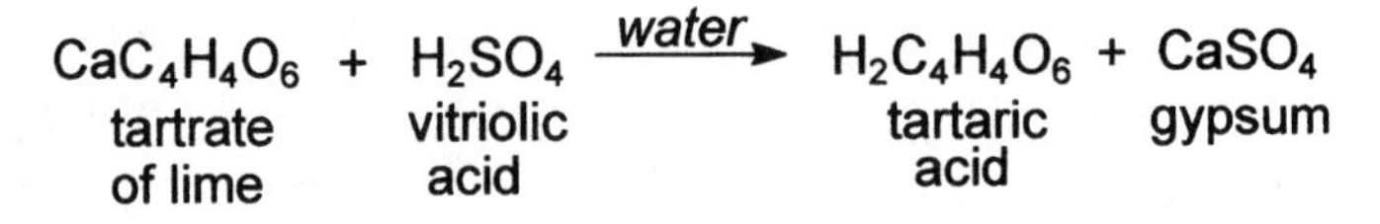

Equation 8.2. Tartaric Acid Formation.

in Equation 8.1. After collecting the tartrate of lime, Wollaston reacted it with vitriolic acid to produce tartaric acid and gypsum, as shown in Equation 8.2. Although much of the gypsum produced by the reaction precipitates out of solution, it is sufficiently soluble in water that Wollaston needed to separate the dissolved portion from the co-dissolved tartaric acid by sequential precipitation. To do this, he placed the aqueous solution of the reaction products into an evaporating vessel and heated the solution until its specific gravity reached 1.33, at which point the precipitation of the less soluble gypsum became nearly complete. He then transferred the remaining liquid to a smaller evaporator and concentrated it further to a specific gravity of 1.52, at which point most of the tartaric acid precipitated.[6] All of the reactions used to generate tartaric acid from argol were previously known, so Wollaston was able to increase the scale of the procedure to a commercially viable level after only a few weeks of experimentation. But the scale of the operations, described in a notebook entry dated February 22, 1802, was much larger than that used in the platinum work, and the physical labor required was much greater.[7]

To prepare tartrate of lime, Wollaston combined twenty-four lbs. (avoirdupois weights) of argol with six lbs. of whiting and eighteen gallons of water in a large vessel and brought the mixture to a boil. As the reaction proceeded, the cream of tartar present in the argol at the bottom of the vessel was converted to tartrate of lime, which replaced it as a solid sediment mixed with the insoluble and unreactive components of the original argol. The mixture was allowed to stand for a while before the liquid was poured off and the remaining solid was put into a large press and squeezed dry. This gave about eighteen to nineteen lbs. of impure tartrate of lime. This solid "cake" of product, as Wollaston referred to it, was placed in a second vessel with five gallons of water, mixed well, and left to stand until the following day, allowing the water to remove more soluble impurities. The water used to wash away the impurities was then poured off, and the vessel was refilled with fresh water to a total of 25 gallons. This process of washing the tartrate cake with large amounts of water was repeated five more times over the next two days. After the last of the six washings, the tartrate of lime was sufficiently pure for conversion to tartaric acid by treatment with vitriolic acid, as shown by Equation 8.2.

In the acidification step, Wollaston added twelve lbs. of vitriolic acid diluted with eight gallons of water to eighteen lbs. of tartrate of lime and left the mixture to react for one day. Then the aqueous liquid containing the desired tartaric acid and some by-product gypsum was poured off into a large tub. The solid left behind was treated again with six gallons of water for one day to recover more tartaric acid, and then a third and final time with five gallons of water for yet another day. The water extractions from all three washings (about twenty-two gallons) were combined and placed in a large lead vessel for concentration to the specific gravities required for crystallization of, first, gypsum and then tartaric acid. To recover tartaric acid that still remained in solution after crystallization was complete, Wollaston added whiting to the solution to convert the organic acid once again to insoluble tartrate of lime (CaTar), which was filtered off and added to subsequent batches for re-acidification by vitriolic acid. Overall, in a sequence of steps that extended over six or seven days, Wollaston was able to isolate nearly fifteen lbs. of crystalline tartaric acid from each twenty-five lbs. of argol, a yield of product amounting to about 60 percent of the original weight of crude starting material. But, as previously noted, neither tartaric acid nor any of its salts were what Wollaston and Tennant had planned to sell, presumably because those compounds were commercially available at a price that would not give the partners sufficient profit. Their goal was to convert tartaric acid to the more valuable organic substances oxalic acid and salt of sorrel.

Wollaston began a set of experiments on April 30, 1802, to discover the best procedure for converting tartaric acid to oxalic acid (which he called "bricksalt").[8] The required reaction was well known at Wollaston's time, and involved the oxidation of tartaric acid by nitric acid to form oxalic acid, as shown in Equation 8.3. Because solid tartaric acid is not very soluble in nitric acid, the reagents must be heated to initiate the reaction. But, once the reaction begins, product formation releases a great deal of heat so that the rate of reaction increases with the rising solution temperature. Consequently, it is difficult to prevent the contents from overheating and boiling

$$\underset{\text{tartaric acid}}{H_2C_4H_4O_6} + \underset{\text{nitric acid}}{6\,HNO_3} \longrightarrow \underset{\text{oxalic acid}}{2\,H_2C_2O_4} + 6\,NO_2 + 4\,H_2O$$

Equation 8.3. Oxalic Acid Formation.

over and out of the reaction vessel. Wollaston was able to minimize (but not eliminate) this problem by reversing the order of addition of reagents and extending the reaction time, that is, by adding tartaric acid over a period of days to the nitric acid solution. By August 1802, he had settled on the following procedure. He began by adding two lbs. of tartaric acid to a large excess of nitric acid in a large carboy (a round glass vessel with a narrow neck), and left the solution to react and cool overnight. On the second day he added another three lbs. of acid and on the third day three lbs. more were added in the morning and two lbs. at night. On the fourth day, the solution had become too hot to add any more but on the fifth day the final two lbs. were added as the temperature remained steady at 92°F. Assuming that Wollaston allowed the solution to sit a day or two after addition of the last tartaric acid to allow the reaction to go to completion and cool to room temperature, the entire procedure would have taken six to seven days. Since the oxalic acid was the sole nonvolatile product of the reaction, he needed then only to boil off all the water in the reaction vessel to obtain the desired solid oxalic acid. Although the conversion of tartaric acid to oxalic acid was chemically simple, the reaction was very difficult to execute on a large scale because of the need to draw off the heat of the reaction. Wollaston could do little more than to allow atmospheric air to cool his reaction vessels, something that became especially challenging when multiple vessels, each releasing heat, sat side by side.

Not surprisingly, Wollaston attempted to establish a procedure for reacting two or three batches of tartaric acid concurrently, but without much early success. However, he was able to work around the problem in early 1803 by doing the reactions in longer-necked glass vessels known as "boltheads," and placing them in a specially-constructed "blowbox" in which multiple reaction vessels could be successfully worked in sequence. A small sketch of the blow box is shown at the end of a notebook entry for March 21, as reproduced in Figure 8.1.[9] As the small profile outline of the blowbox reveals, any material from the reaction solution that overheated and boiled out of the reaction vessel would strike the angled covering and settle into the bottom, from where it could later be recovered.

The last entry shown in Figure 8.1, dated April 2, indicates that Wollaston began to place two carboys and/or boltheads in the blowbox at one time,

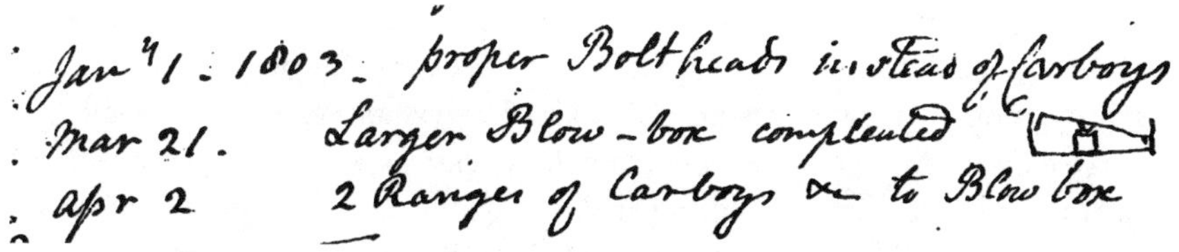

Jan 4 / 1803 – proper Bolt heads instead of Carboys
Mar 21. Larger Blow-box completed
Apr 2 2 Ranges of Carboys &c to Blow box

Fig. 8.1. Wollaston's Improvements to Oxalic Acid Production, 1803. Reproduced by kind permission of the Syndics of Cambridge University Library.

$$\underset{\text{oxalic acid}}{H_2C_2O_4} + \underset{\text{sal ammoniac}}{2NH_4Cl} \longrightarrow \underset{\text{oxalate of ammonia}}{(NH_4)_2C_2O_4} + 2HCl$$

Equation 8.4. Oxalate of Ammonia Formation.

$$\underset{\text{oxalic acid}}{2H_2C_2O_4} + \underset{\text{potash}}{K_2CO_3} \longrightarrow \underset{\text{salt of sorrel}}{2KHC_2O_4} + CO_2 + H_2O$$

Equation 8.5. Salt of Sorrel Formation.

and by 1804 he had adjusted the mixing of reagents and the placing of the vessels in the blowbox in such a way that three batches could be handled simultaneously. Each time a completed reaction vessel was removed, a new one was added so that a "range" of reaction vessels, each at a different stage of completion, was in the blowbox at the same time. By late 1808, Wollaston had slowed the rate of addition of tartaric acid to nitric acid from six days to twelve (to minimize overheating). This allowed him to double the number of reaction vessels in production at one time from three to six.[10] At that production rate, he calculated that he could produce about three lbs. of oxalic acid per day at a cost of approximately 3s. 8d. per pound. Small amounts of oxalic acid produced this way were sold to textile manufacturers but, as originally planned, most was treated with base to form salts.

A small amount of oxalic acid was converted to (the skin irritant and poisonous) oxalate of ammonia by simple treatment with sal ammoniac, as shown in Equation 8.4. Sales of oxalate of ammonia never exceeded several dozen pounds, which suggests that textile producers found little use for it. However, the potassium salt of oxalic acid did find a commercial use, and almost the entire quantity of processed argol was directed to its formation.

Wollaston converted oxalic acid to salt of sorrel by simple treatment with potash, as shown in Equation 8.5. Solid oxalic acid was added to a solution of potash in water until carbon dioxide ceased to bubble out of solution, indicating completion of the reaction. The product, salt of sorrel, precipitates out of solution and can be isolated by filtration. Curiously, there is no information in Wollaston's notebooks on how, when, or on what scale he carried out the conversions. Some notebook entries do show, however, that he ultimately converted about 8,400 lbs. of oxalic acid to 7,100 lbs. of salt of sorrel, all of which was sold to Manchester textile manufacturers.[11]

PRODUCTION AND SALE OF ORGANIC CHEMICALS

The production of organic chemicals required much more laborious and time-consuming processes than the purification and consolidation of platinum. Much of the work involved lifting and shifting heavy glass vessels, some holding 25 gallons of liquid and weighing over 200 lbs., handling multi-gallon volumes of aqueous and acidic solutions, boiling off water in large lead containers, and filtering precipitates from several gallons of liquid. Stoking the furnace to keep large vessels boiling in one area while trying to keep several carboys from overheating in another required constant vigilance and a tolerance for working in a hot environment with poor ventilation. Undoubtedly, John Dowse played a significant role in these laboratory procedures, although Wollaston himself must have been present most of the time to prepare the reactions, superintend the processes, and help with the most demanding tasks.

John's employment with Wollaston had begun in January 1800, when Wollaston still lived on Cecil Street and maintained a small medical practice. After the move in September 1801 to larger premises and a better equipped laboratory in the house on Buckingham Street, John's involvement in the laboratory work began to expand. By December 1801, when the first large-scale purification sequences of platina were underway, entries begin to appear regularly as debits in the financial account book for "John's book," itemizing reimbursement for laboratory supplies purchased on his own authority.[12] In addition, John's wages started appearing in Wollaston's financial records as a business cost, to be shared with Tennant. In 1803, when John began to play a larger role in the work required for the production of oxalic acid, entries appear in the account books for a supplement to his wages as a "bonus" and by the end of that year his wages had become a regular entry. By the end of 1805, his wages at £1.1.0 per week are recorded as a recurring business expense.[13] Interestingly, this is the same wage that Michael Faraday first received when he became laboratory assistant to Humphry Davy at the Royal Institution in 1813.[14] Wollaston obviously valued John's contributions highly for, after the chemical business began to show an overall profit in 1809, John's wages were increased to £1.4.0 per week and in 1820 his pay was increased once again to £1.6.0 per week.[15] As one of Wollaston's domestic employees, John also lived in the Buckingham St. house, which made it easier for him to tend to the laboratory furnaces around the clock.

Unlike the platinum work, which could have been conducted in a secure, dedicated room or two in the rear of Wollaston's house, the production of organic chemicals needed larger spaces with bigger furnaces and more exposure to the open air. Detailed London maps of the time show Wollaston's house as one of a triangle of residences enclosing a large open area containing some outbuildings, one of which appears to have been rented by Wollaston. Beginning in 1802, when the processing of argol began, Wollaston's account book has entries for "ground rent" or "yard" at £10 semi-annually. The entries continue until 1825, when he moved away from Buckingham St.[16] Coupled with these entries are additional ones for rent of a "laboratory" at £3 per year from 1802 until 1815, the years during which organic chemicals were produced. It is probable, therefore, that the rented building was one of those on the land behind Wollaston's residence. The rented building must have had significant floor space, because there is ample evidence in the account books of purchases necessary for the large scale of the organic chemicals operation. There are entries in 1802 and later for things like a shovel, a wheelbarrow, a pulley, jacks and winches, tubs, and more. The scale of the operations appears to be near the limit of what two men could do by themselves.

After Wollaston had established the basic procedures for the production of oxalic acid, the acquisition of argol began in 1802 with the purchase from three different vendors of 2,632 lbs. of raw material. The last purchase was made in 1811, by which time a total of 16,895 lbs., more than 8 tons, had been acquired. From this Wollaston was able to extract 9,878 lbs. of tartaric acid and convert it first to 8,441 lbs. of oxalic acid and then to 7,115 lbs. of salt of sorrel.[17] The labor involved is quite astonishing. At 25 lbs. per batch, initial digestion of the argol required more than 670 batch processes, each of which involved six washings of the tartrate cake with 25 gallons of water. Subsequent conversion of the tartaric acid to oxalic acid, which was carried out in 12 lb. amounts, would have required a total of about 820 batch processes, each of which involved the daily addition of tartaric acid to nitric acid over a period of several days, followed by the boiling down of the reaction solution to the concentrations needed for precipitation of the desired products, which in turn had to be collected by filtration. Finally the oxalic acid had to be converted to salt of sorrel. There are no notebook

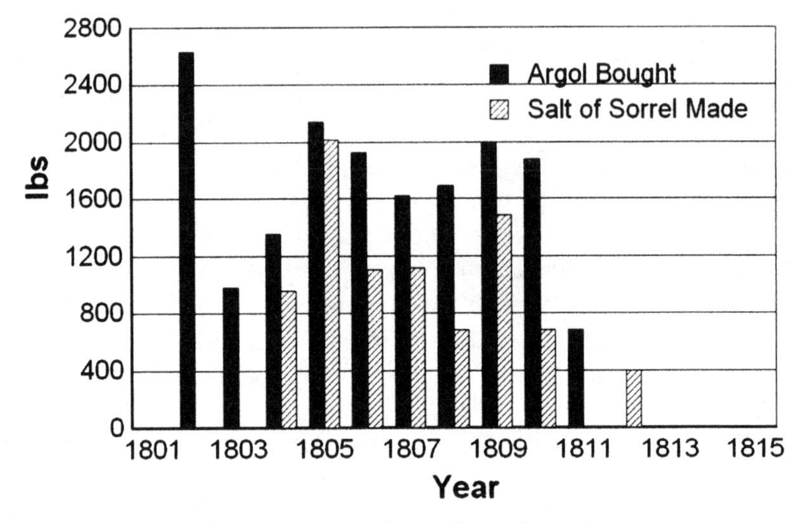

Chart 8.1. Organic Chemicals, Yearly Totals.

entries on the scale of that procedure, but if one makes a reasonable assumption that the conversion was done on a batch weight of 20 lbs. of oxalic acid, another 420 batch processes would have been required. All of this work was spread over a period of 10 years, with yearly totals illustrated in Chart 8.1.

The chart shows a pattern similar to that for the platinum business: an early large scale purchase of starting materials followed by a period of research on the chemical purification techniques and suitable production processes culminating in the production of marketable materials. But, in contrast to the success of the platinum business, the organic chemical business was doomed by shrinking profit margins. Argol was in plentiful supply when Wollaston and Tennant began to purchase their supplies. As markets for chemicals derived from argol grew, its price rose. In 1803, for example, 100 pounds of argol could be purchased for 67 shillings, but by 1810 the price had doubled to 130 shillings.[18] This increase did not fatally affect the costs of production because the price of argol still accounted for little more than one-quarter of the total expense of producing a pound of oxalic acid. But profits declined substantially when the market price of salt of sorrel began to fall. An entry in Tennant's account book reveals he purchased some salt of sorrel in 1802 for £1.16.0 per pound, the same year he and Wollaston began to acquire argol to make the salt.[19] Since the cost of producing one pound of the salt would not have been more than six shillings, the organic chemicals business initially promised a good financial

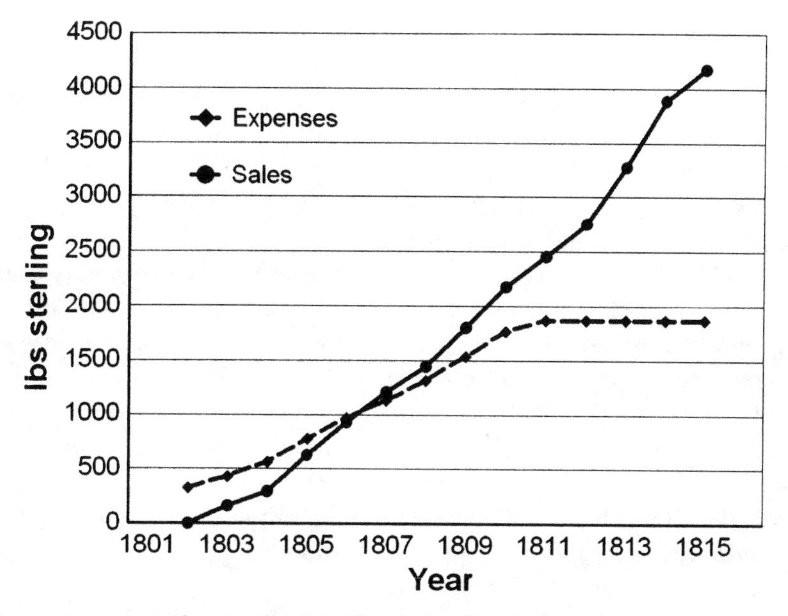

Chart 8.2. Organic Chemicals—Financial Totals.

return. But the price of salt of sorrel began to sink quickly. One entry in the notebooks from 1811 reveals that the partners were selling their product for only ten shillings a pound, which was a meager return for all the effort involved to make it.[20] A further entry for 1816, after peace in Europe, gives the market price for one pound as an even lower eight shillings.[21] Thus, even though Wollaston's process gave a good product, his two-man operation could not achieve the cost efficiencies of factory production, and, once such operations began to depress prices, Wollaston and Tennant got out of the business. Nonetheless, as Chart 8.1 shows, Wollaston continued to convert the last of the argol purchased in 1812 to oxalic acid, so that his labors in those processes continued until 1814.

Despite the ever-diminishing returns from the organic chemicals business, Wollaston and Tennant did manage to make an overall profit of about £2,000, as shown by Chart 8.2.[22] As shown, revenue from the sale of organic chemicals started in 1803, but accumulated revenue did not exceed expenditures until 1807. Then, as sales volume increased, profits began to grow, especially after purchases of argol ceased in 1811 and production and sale of products continued.

As mentioned in the last chapter, Wollaston drew up a new revenue-sharing agreement in February 1809, just at the time when the overall business partnership with Tennant began to move into the black. The first item mentioned in the agreement was that Wollaston would receive some unspecified share, before equal division with Tennant, for "every lb of Bricks [bricksalt, Wollaston's notebook term for tartaric acid] converted into SS [salt of sorrel] beyond what is sold & paid for." As shown in Chart 8.1, Wollaston did not produce any salt of sorrel in 1808, although he had prepared a large amount of the precursor substance, tartaric acid. One suspects that he believed he should receive more of the profits from the sale of organic chemicals because of his dominant role in their production, and therefore delayed any further synthesis of salt of sorrel until the terms of the partnership were renegotiated. As the evidence of workload in the notebooks suggests, such a proposal could not have been seen as unreasonable by any rational person. But the salt of sorrel component of the planned agreement was never implemented, and Tennant continued to receive a full half share of the profits from the sale of organic chemicals until his death in 1815. Why would Wollaston have agreed to such an inequitable situation?

THE CONTINUING PARTNERSHIP WITH SMITHSON TENNANT

There can be little doubt that Tennant's theoretical and practical chemical knowledge, enthusiasm for natural philosophy, engaging personality, long-term friendship and financial resources were key to the establishment in 1800 of the joint partnership with Wollaston. But Tennant's rooms at Garden Court, Temple, were unsuitable for large-scale chemical processes, so he could not execute anything other than benchtop experiments there. Although his contributions to the preparation of malleable platinum were negligible, there is some evidence that Tennant was a bit more involved in the production and sale of organic chemicals, at least in the early stages.

One undated entry in Wollaston's notebooks gives experimental results from Tennant (one of the very few times his name appears in the nonfinancial notebooks) for the conversion of cream of tartar to oxalic acid, but the process gave lower yields at a higher cost than Wollaston's method, and was not adopted.[23] In addition, Tennant's financial records reveal several purchases in April 1801 of materials that could have been used in preliminary experiments, although he played no role in the labor-intensive production of salt of sorrel.[24] Nonetheless, even the most generous allowance for the unknown cannot alter the fact that Wollaston did far more than 50 percent of the work. So the question remains: Why was the revised revenue-sharing agreement of 1809 not enacted for the organic chemicals part of the business? I believe the most plausible answer is that Tennant was dependent on revenues from the partnership for his continued financial and physical well-being and Wollaston could not bring himself to discontinue the 50:50 profit-sharing agreement.

Tennant, as previously noted, was an enigmatic person with a remarkable lack of resolve and intellectual focus. One of his many interests was rural economy, and in 1799 he purchased 500 acres of newly enclosed land near Shipham in Somerset, about 120 miles west of London.[25] He subsequently built a summer home on the property and spent the summer months there every year for the remainder of his life, far from the site of his partner's processing facilities. The contrast between Tennant and Wollaston during the years of their partnership is striking. Tennant was frequently unwell, spent several months each year out of the city, devoted much of his attention to agricultural practices and farm management, and fitted his daily activities around outings on horseback. Wollaston was physically active and generally in good health, applied himself diligently to a wide range of theoretical and applied scientific problems, and immersed himself in work except for the few weeks each year given to holidays. Although Wollaston was able to tolerate Tennant's eccentricities during the early years of their partnership, the draft version of the revised revenue-sharing agreement shows that he was not entirely happy with the original terms of the partnership. The platinum component was flourishing, and promised to provide significant profits well into the future, so the provision that Wollaston could keep 10 percent of that income before equal division was enacted in 1811. But, despite the preliminary statement of intention in the draft statement of agreement, no changes were made to the sharing of the argol profits.

Prior to August 1809, nearly all the payments for salt of sorrel (from a Manchester merchant named A. S. Burkett) were made to Wollaston. The primary reason for this, of course, was that Wollaston needed the money to offset his much greater production expenses. But after this date, nearly all of the payments were made to Tennant, and a few of them were from a new Manchester customer, Thomas Tatterthwaite. In addition, entries in Tennant's account book refer more frequently to things like packing cloth, pots, and barrels, indicating that he may have begun to package the organic chemicals for shipping. Such circumstantial evidence for Tennant's increased role in the organic chemicals business points to a more substantial contribution to that part of the business. And there is other evidence that he initiated a related business venture: the production of citric acid from lime juice.

In 1811, when it was becoming apparent that the argol business was no longer sufficiently profitable to justify the hours devoted to it, purchases of argol stopped. But in July of that year, Tennant began a new initiative by purchasing 805 gallons of lime juice at 3 shillings a gallon. The intent was to produce citric acid ($H_3C_6H_5O_7$, hereafter H_3Cit) from it, by a process similar to that employed to obtain tartaric acid from argol. Citric acid was employed in the textile industry to harmlessly remove unwanted coloration from the undyed portions of calico prints. But, because it could only be obtained in useful amounts from citrus fruits and was difficult to isolate in pure form, the acid sold at a higher price than oxalic acid. An entry in Tennant's account book gives the price of a pound of citric acid in 1813 as 17s. for the pure, colorless acid and 11s. 6d. for an impure, brownish-colored form.[26] Thus, if the cost of chemical extraction was not too high, a profit could be realized.

Immediately after the purchase, in a familiar routine, Wollaston set to work to determine the amounts of citric acid that could be extracted from the lime juice. He treated lime juice with whiting (calcium carbonate) to produce a precipitate of citrate of lime ($Ca_3(Cit)_2$), which he washed six times with water before adding a dilute solution of vitriolic acid to convert the solid citrate to an aqueous solution of citric acid. He then heated the solution to concentrate it to the point where citric acid precipitated out in crystalline form. By carrying out the process on a very small scale under ideal conditions, Wollaston found that it could be possible to obtain 10 oz. of citric acid from each gallon of juice.[27] If isolated and sold in an impure form, this would give a profit of about 3–4 s. per gallon, not a particularly rewarding return. Nonetheless, Wollaston scaled up the process to treat nineteen gallons of lime juice at a time, roughly similar to the volumes employed in the argol procedure. The process worked well until the final concentration step, when the very high solubility of citric acid in water made the evaporation needed to precipitate the solid acid especially difficult. The solution had to be heated for a very long time to drive off enough water for precipitation to begin in the very thick, syrupy solution. Wollaston found that use of a coal-fired furnace tended to overheat the solution in the final stages of concentration, so he had a special steam evaporator made that would hold the evaporating temperature at an appropriate level. The evaporator was finished on November 14, 1811, and the first citric acid for commercial sale was precipitated from it a few days later.[28] There are no production details in the notebooks on the overall production of citric acid, but all the lime juice must have been processed by early 1813, when Wollaston sold off all the empty casks.[29] All sales of citric acid were entered in Tennant's account book. Sales began to Manchester industrialists in April 1813 and continued until a few months after Tennant's death in

February 1815.[30] Since Tennant was abroad for much of 1814 and 1815, the sales must have been completed by Wollaston, although he entered them in the account books as credits to Tennant.

The organic chemicals portion of Wollaston's and Tennant's business was modestly profitable, bringing to each man a bit more than one hundred pounds sterling annually for a period of six or seven years, but the labor required to produce the commercial compounds was substantial. It is difficult to understand why the partners persisted when their financial prosperity had become assured as platinum sales surged. The most plausible explanation, I believe, is that the organic chemicals business had been conceived by Tennant, who viewed it as his principal and sustaining contribution to the partnership. If true, the premise would explain why Tennant's name is connected in the notebooks much more frequently with organic chemicals than with platinum, why he retained his equal share of profits from the sale of organic chemicals, and why the production of citric acid from lime juice was begun when there was no financial need for such an undertaking. But, as previously discussed, Tennant still did none of the work to produce the marketable goods and was not even in London for several months of the year. And so the question remains: Why did Wollaston continue with such an inequitable arrangement with such an unproductive partner? There must be several contributing factors: his loyalty to the man who helped him make the transition from miserable doctor to independent entrepreneur and experimental philosopher, his steadfastness to a mutually-plotted course, and his generosity to one who had no other source of income. Wollaston may have been reserved in most social environments and slow to form friendships but, once those friendships and relationships had been established, they survived all sorts of challenges, as his relationship with Tennant amply demonstrates.

MULTIPLE COMBINING PROPORTIONS IN THE SALTS OF ORGANIC ACIDS

The same eye for detail that served Wollaston so well in the platinum purification reactions enabled him to discover intriguing weight relationships among the salts of the organic acids he produced. Those novel observations drew him into the contemporary debates on the nascent atomic theory of John Dalton, and, for the first time in his career, Wollaston could not avoid the theoretical implications of his discoveries.

The investigation into the composition of salts began as soon as Wollaston began to study the cream of tartar (KHTar) present in argol. By October 1803,

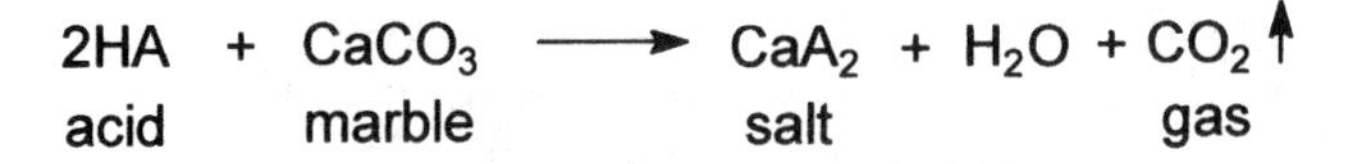

Equation 8.6. Acid Neutralization with Marble.

he had determined that the amount of potash present in 100 gn. of cream of tartar was equivalent to the neutralizing power of 27.4 gn. of marble ($CaCO_3$).[31] From the beginning of his chemical studies in 1800, Wollaston had always measured the strengths of his acids by the weight of marble they could neutralize. That reaction was a much-used method of determining the quantity of acid present in a solution because the neutralization reaction, shown in modern terms in Equation 8.6 for a generic monoprotic acid represented as HA, proceeds rapidly to completion (because of the loss of carbon dioxide) and the point of full neutralization is rendered visible by the cessation of effervescence. Moreover, powdered marble can be added in carefully weighed amounts so that the weight required to neutralize a given unit amount of acid can be accurately measured. As explained in Chapter 4, measuring the strengths of all of his purchased acids in terms of their marble equivalents allowed Wollaston to establish the most effective composition of aqua regia for the solution of platinum, and to maintain that exact composition from batch to batch when acids of varying strengths were used. It is not surprising, therefore, that he extended the concept to the salts of organic acids.

To quantify the amount of a base in a substance in terms of its marble equivalent, Wollaston chose to neutralize a fixed weight of acid first with marble and then with a different base, such as potash. The two measured base weights would then be deemed to be equivalent in their neutralizing power. Thereafter, any weight of muriatic acid could be expressed in terms of the equivalent weight of either marble or potash, and vice versa. This constancy of reaction equivalents had been recognized by many late in the eighteenth century and, by the opening years of the nineteenth century, tables (not all in perfect agreement) of equivalent combining weights of the common acids and bases were in general use. Wollaston was certainly aware of them, and his observation that 100 gn. of cream of tartar contained a weight of potash equal to 27.4 gn. of marble agreed well with similar analyses done by others.[32] But this was not the compositional aspect of cream of tartar that caught his attention.

To extract cream of tartar from argol in the process discussed earlier in this chapter, Wollaston converted it first to tartrate of lime by treatment

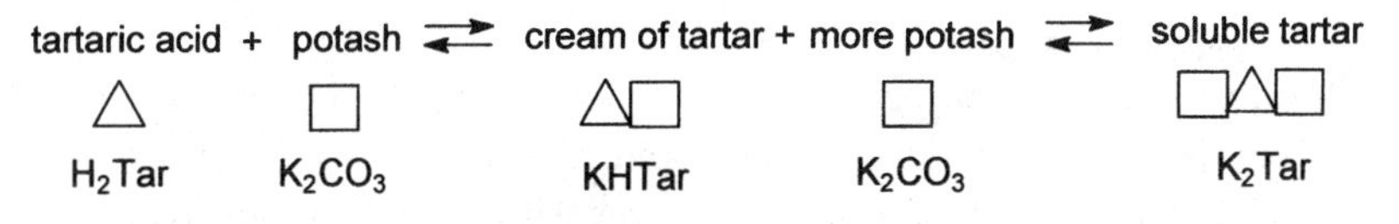

Fig. 8.2. The Salts of Tartaric Acid and Potash, as Understood in the Early Nineteenth Century.

with whiting. To determine the optimal amount of whiting necessary for complete reaction, Wollaston wanted to measure the amount of acid present in cream of tartar. He found that 100 gn. of cream of tartar could be completely neutralized by 26 gn. of marble.[33] This result intrigued him, for it meant that cream of tartar contained an amount of potash equivalent to 27.4 gn. of marble and an amount of acid equivalent to 26 gn. marble. Thus, it appeared that cream of tartar (KHTar) was a salt whose component tartaric acid was only half neutralized by potash. Further reaction of it with an added amount of potash equal to what was already in it would convert the acidic salt to the fully neutral one known as soluble tartar (K_2Tar).

The relationship among these three substances is shown schematically in Figure 8.2. To help understand the chemical conversions in Wollaston's terms, the group of elements that make up tartaric acid is represented simply as a triangle, and potash is represented by a square. The geometric symbols represent the package of elements intimately combined in the compound substance. Figure 8.2 helps us understand one of the key (but erroneous) beliefs of salt formation in Wollaston's time: the reaction of an acid with a base was understood to be a simple addition reaction, and the product salt contained all the elements (and weight) of the component acid and base. In theory, therefore, the full amount of acid and base could be recovered from the salt under appropriate decomposition conditions. We now know that salt formation is generally accompanied by the formation of coproducts, such as water, so that the weight of a salt is not normally a simple sum of the weights of acid and base used to make it. Thus the decomposition of a salt does not usually give the original weights of the acid and base used to make it. The water produced during the formation of most salts was unknown to early nineteenth century chemists because the neutralization reactions were nearly always done in water and the small weight of water formed as one of the products could not be perceived. Wollaston's discovery that cream of tartar contains amounts of acid and base nearly equal to the same equivalent weight of marble is not rendered invalid by what can be dubbed the "missing water" problem, because the equivalents measured are true combining proportions (the triangles and squares of Figure 8.2), not salt decomposition values. Wollaston did not have an understanding of salt formation that differed from his contemporaries. It was just that he had found that measurement of acid and base amounts in terms of their marble equivalents gave a more reliable measure of those quantities.

Below each symbol in Figure 8.2 is a simplified, modern formula of the compounds, although the interconversion of compounds is not presented in the form of a balanced equation. Chemists in the early nineteenth century often explained reactions of this type in terms of chemical affinity; in this case, tartaric acid united with potash because of an affinity of the acid for the base, and the combining weights necessary to form cream of tartar were those that allowed the base to mutually saturate the first affinity requirement of the acid. The partially neutralized cream of tartar was an acidic salt that could combine according to its own affinity requirements with an additional amount of potash to form the fully neutral salt known as soluble tartar. Although it was well known that the two salts of tartaric acid differed in the amounts of potash they contained, there was no expectation that the amount of potash required to saturate cream of tartar should bear any simple relationship to the amount required to partially neutralize tartaric acid. Thus it is not too surprising that Wollaston initially imparted no great significance to the near equality of the two proportions. Consequently he recorded the observation in 1803 as little more than a single curious observation nested within a network of several other salt analyses, which showed no similar combining ratios. Nonetheless, when he moved on to study oxalic acid and its salts, he was curious to see if a similar relationship existed there as well.

As soon as Wollaston began producing salt of sorrel from oxalic acid in 1804, he carried out a variety of analyses to determine the relative amounts of acid and base in both it and its neutral salt, known as oxalate of potash. While trying to obtain pure salt of sorrel by precipitating it from a solution containing an excess quantity of oxalic acid, he discovered sometime in 1804 a previously unknown salt of potash and oxalic acid. The new salt, which co-precipitated with salt of sorrel and was initially difficult to separate from it, contained even more acid combined with potash than salt of sorrel. This new salt, which later was found to be a common contaminant in commercial salt of sorrel, brought the total number of salts containing potash and oxalic acid to three, a phenomenon not previously observed in any other organic salts.

In his notebook entries from this time, Wollaston referred to the three salts as the neutral salt (K_2Ox), the salt of sorrel ($KHOx$), and the super salt of sorrel ($KHOx \bullet H_2Ox$). The relationship among oxalic acid and its three

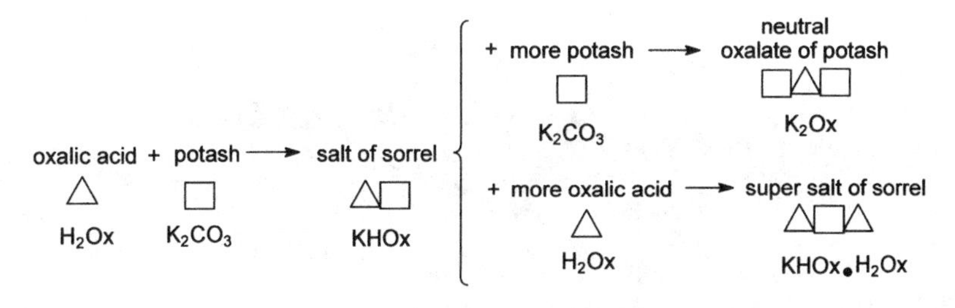

Fig. 8.3. The Salts of Oxalic Acid and Potash.

salts is shown schematically in Figure 8.3. He found that he could obtain each one of the three salts in an acceptably pure form by precipitating it from a solution to which appropriate amounts of acid and base had been added. After isolating small amounts of each salt, Wollaston heated 100 gn. of each to drive off all the contained acid, and weighed the amount of potash left behind. He determined that the neutral salt yielded 73.2 gn. of potash, salt of sorrel 54.5 gn. and super salt of sorrel only 25.3 gn.[34] These values are quite close to those predicted by modern stoichiometry (75.0, 53.9 and 27.2 gn. for the matching pure, hydrated salts) so there is no flaw in Wollaston's technique, nor much in the purity of the salts analyzed. He knew that the weights of base left behind by destructive heating of the salts were not the amounts that had been united to a fixed weight of acid, because each 100 gn. of salt analyzed contained a different weight of acid. Curiously, it does not appear that Wollaston at this time attempted to determine base equivalents using the same methods (conversion to marble equivalents) he had used earlier for the two salts of tartaric acid. But he soon changed tactics and proceeded to measure instead the number of acid equivalents in the three salts. That approach gave him a second set of integral combining proportions.

I mentioned earlier in this chapter that one of the products Wollaston sold in small amounts was the oxalate of ammonia, and he was investigating the composition of the ammonium salts of oxalic acid about the same time as the potash analyses described above.[35] He was able to discover by a very clever sequence of reactions and some good fortune that the potash and ammonium salts of oxalic acid contained weights of acid that stood in proportion to each other as the numbers 4:2:1, although the notebooks contain no details on how he did the experiments. The most definitive entry noting this result is shown in Figure 8.4.[36] Under the heading Oxalate of Ammonia, Wollaston listed the results of three experiments, each of which involved treating 100 gn. of starting compound with sal ammoniac

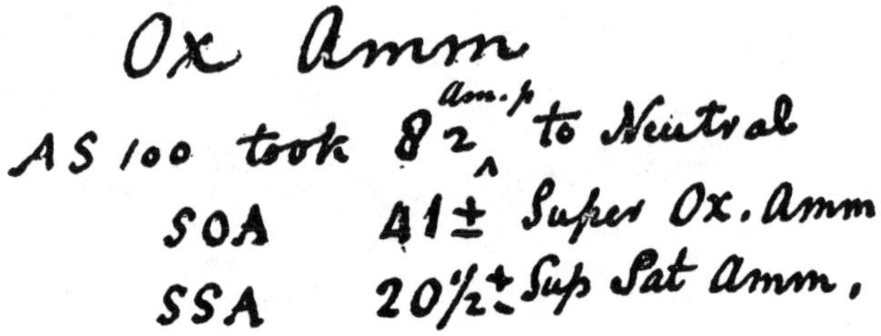

Ox Amm
AS 100 took 82 Am.p to Neutral
SOA 41± Super Ox. Amm
SSA 20½± Sup Sat Amm.

Fig. 8.4. Multiple Proportions in Oxalic Acid and Its Ammonium Salts. Reproduced by kind permission of the Syndics of Cambridge University Library.

(abbreviated Am.p., indicative of ammonia precipitate) to form different salts. This important notebook entry makes sense if interpreted in the following way:

EXPERIMENT 1:

100 gn. AS, acid of sorrel (H_2Ox), took 82 gn. of sal ammoniac to form the neutral salt ($(NH_4)_2Ox$).

EXPERIMENT 2:

100 gn. SOA, salt of oxalic acid ($KHOx$), took 41±gn. of sal ammoniac to form the super oxalate of ammonia ($(NH_4)HOx$).

EXPERIMENT 3:

100 gn. SSA, super salt of sorrel ($KHOx \bullet H_2Ox$), took 20.5±gn. of sal ammoniac to form super saturated ammonia (which Wollaston might have presumed to be $(NH_4)HOx \bullet H_2Ox$, although no such salt actually exists; the nature of the product formed is not important, as long as it is assumed that all the potash is replaced by sal ammoniac).

These experiments led Wollaston to conclude that the three salts of oxalic acid resulted from its combination with one, two, or four equivalents of base, that is, combination in exact whole-number ratios. It is clear from a number of related notebook entries that it took him several attempts to discover the reactions that best revealed the integral combining proportions, but he accepted the result as decisive, so much so that he transcribed the results onto the blotter facing the page on which he had earlier recorded the inconclusive thermal decomposition results for the potash salts.[37]

Once again, the combining weights measured by Wollaston for the amounts of sal ammoniac in the three salts of oxalic acid are in excellent agreement with those predicted by modern stoichiometry (which establishes the ideal combining weights for sal ammoniac in Wollaston's three experiments as 85:42:21), but those results are made possible by a most fortuitous coincidence. His values, we now understand, only represent true multiple proportions if the weights of base measured are all ones in combination with the same weight of acid. And this is not expected to be the case in those salts because of the missing water problem. However, as shown in Table 8.1, two of the three compounds used by Wollaston to react with sal ammoniac are isolated as dihydrates, and two contain a weight of potassium nearly equal to that of two water molecules. This means that all three reagents coincidentally contain very nearly the same weight of oxalic acid. As the table shows, 100 gn. of hydrated acid of sorrel contains only 71.4 gn. of actual acid, H_2Ox, after allowance is made for the water of hydration. Similarly, 100 gn. of salt of oxalic acid contain 69.5 gn. of oxalic acid after the weight of potassium is subtracted, and 100 gn. of super salt of sorrel contains 70.5 gn. of oxalic acid after the weights of both water of hydration and potassium are accounted for. The fact that the weight of actual oxalic acid contained in 100 gn. of each of the three salts is nearly the same depends on the fortuitous presence in them, and similarity in atomic/molecular weights, of potassium (39.1) and two water molecules (36.0). This means that Wollaston was, in fact, measuring the amount of sal ammoniac that reacted with nearly the same weight of oxalic acid, or potash combined with it, in each of the three salts and, consequently, the results he obtained manifested themselves as integral multiple proportions.

Wollaston recognized that his two good examples of multiple combining proportions, discovered in 1803 and 1804, were novel and unusual. But,

TABLE 8.1. Weights of Anhydrous Oxalic Acid in Analyzed Compounds

Compound	Weight Used (gn.)	Modern Formula	Weight of Contained Anhydrous Acid (gn.)
AS, acid of sorrel	100	$H_2Ox{\bullet}2H_2O$	71.4
SOA, salt of oxalic acid	100	KHOx	69.5
SSA, super salt of sorrel	100	$KHOx{\bullet}H_2Ox{\bullet}2H_2O$	70.5

as was common with him, he set the results aside until he could corroborate and make more sense of them. Three years later, in late 1807, he found that someone else was about to publish similar observations together with a supporting theoretical interpretation that involved the novel atomic ideas of John Dalton.

THOMAS THOMSON, DALTON, AND ATOMIC THEORY

Thomas Thomson was a Scottish chemist who, like Wollaston, had obtained a medical degree but became a devotee of chemistry. He made himself into a tolerable analytical chemist and the leading British textbook writer of his time. He was especially interested in chemical composition and familiarized himself with several examples of substances that combined with each other in more than one weight proportion to form compound substances, such as the elemental oxides and salts of acids. He had even published in 1804 some analytical results of his own on the oxides of lead, which showed that 100 parts of lead combined with 10.6 of oxygen in the yellow oxide, with 13.6 in the red deutoxide, and 25 in the brown peroxide.[38] But he saw the results as signifying nothing other than a progression of oxygen content in the three lead oxides, much like what Wollaston was doing about the same time for the potash content of oxalic acid salts. In August 1804, Thomson visited John Dalton in Manchester and learned for the first time of his host's developing ideas on the application of atomic theory to chemistry.

Dalton had just begun to publicize his hypothesis that the fundamental carrier of chemical properties was the differently-weighted atom of each chemical element. From this premise, he quickly foresaw how the fixed proportions by weight of the components of compound substances could be explained as a simple consequence of atom-to-atom combination. Thus, he envisioned the formation of binary compounds of units A and B to be the result of the combination of atoms of A with those of B, with a one-to-one combination being most stable. The stability claim was a consequence of Dalton's belief that like atoms repelled each other, a postulate required to explain the fact that gases of a single element (such as oxygen or nitrogen) expanded to fill the vessel that contained them. Further, Dalton believed that, if a second, or third, compound of A and B existed, it had to be a combination of two atoms of A or B with one of the other, as illustrated in Figure 8.5 for B combining with A in two different proportions. Figure 8.5 differs from Figure 8.2, which was drawn to illustrate the commonly accepted notions of combination according to degrees of affinity, in crucial ways. First the circles drawn around A and B represent the fundamental particles

Fig. 8.5. Dalton's View of Chemical Combination in Atomic Terms.

(either, according to Dalton, single atoms of elements or stable groupings of atoms called compound atoms) of chemical substances rather than the indefinite package of elements illustrated by triangles and squares in Figure 8.2. For the purposes of understanding Wollaston's contributions to the topic, it is helpful to consider the ramifications of such atomic thinking on instances of multiple proportions.

Although the same affinity forces act to bring A and B together in the concepts underlying Figures 8.2 and 8.5, Dalton's atomic hypothesis makes a specific prediction about the comparative amounts of B in compounds AB and AB_2. Affinity theory only recognized that AB_2 would contain more B than compound AB, since the affinity forces that regulated the formation of each compound were quantitatively unrelated to each other. Atom-to-atom combination, in contrast, requires that the weight of B in AB_2 be exactly twice the amount in AB, if the weight of A is the same in both. Dalton had himself realized the impact that instances of multiple proportions in fixed, integral ratios would have on support for his atomic hypothesis, and he believed the three oxides of nitrogen, the two hydrides of carbon, and the combination of nitrous air with two differing volumes of oxygen provided some fundamental confirmation of his views.[39] By 1807, he had begun to include the composition of salts among his instances of whole-number proportions. But, despite continuing efforts to build support for his ideas, Dalton's atomic hypothesis was not widely known outside of Manchester, and it was viewed with skepticism by many of those who were familiar with it, at least until Thomson became an adherent of the hypothesis.

In 1807, Thomson published the multi-volume third edition of *A System of Chemistry*, and included in volume 3 an exposition of Dalton's atomic hypothesis based on his notes from their meeting in 1804.[40] Thomson's work, as the first popular exposition of Dalton's ideas, brought the atomic hypothesis to a general audience and sparked debate on its merits and deficiencies. There was no doubt that Thomson had become an ardent supporter of the theory, and he used it to interpret the composition of salts, especially those in which an acid combined with a base in more than one proportion. He wrote:

> The simplest way of considering these bodies is, to conceive the supersalts to be compounds of two atoms of acid with one of the base, and the subsalts of two atoms of base with one of acid. Thus [as one example] the

> supersulphate of potash is composed of one atom of potash united to two of sulphuric acid.[41]

Then, shortly after bringing the new edition of his book to market, Thomson sent a paper to the Royal Society announcing his discovery of two new examples of integral combining proportions in the salts of oxalic acid. The paper was published in early 1808.[42] His study of oxalic acid, carried out in complete ignorance of Wollaston's investigation (and commercial preparation) of the same acid, began with an analysis of the several known salts of oxalic acid. For the potash salts, he reported: " This salt [neutral oxalate of potash, K_2Ox] combines with an excess of acid, and forms a super oxalate, long known by the name of *salt of sorrel* . . . . The acid contained in this salt is very nearly double of what is contained in oxalate of potash."[43] This observation is equivalent to Wollaston's measurement of the relative proportions of base in the same salts—the two men differed only in which component, acid or base, they took as the basis of comparison. Although Thomson gave no experimental details in support of his conclusion, his was the first published report of an exact 2:1 ratio in the components of a salt. However, his second example appeared to provide the requisite numerical precision.

Thomson's investigation into the combination of oxalic acid with the base strontia yielded an unusual result. Although others knew only of one salt, Thomson claimed that he had been successful in making two different salts, which had weights of base combined with 100 of acid in the proportion of 151.5: 75.7, exactly 2: 1.[44] Although the analysis of both salts was flawed (and found by a French chemist a few years later to be incorrect), Thomson believed his integral ratio to be a convincing example of atom-to-atom combination, and he proceeded to interpret all of his analyses of oxalic acid and its salts in atomic terms. Crediting Dalton for the insight that led him to his conclusions, Thomson brashly elevated combination in whole-number proportions to a law of nature, saying, "It follows equally from this law, that the acids and bases combine particle with particle, or a certain determinate number of particles of the one with a particle of the other."[45]

WOLLASTON'S INTEGRAL COMBINING PROPORTIONS

Although there is no evidence that Wollaston knew much of Dalton's hypothesis before reading of it in Thomson's 1807 book, he immediately recognized its relevance to his unpublished findings of multiple proportions in the salts of tartaric and oxalic acid. Despite his habitual caution at accepting theories that were not convincingly supported by experimental data, the atomic hypothesis appealed to him. In fact, Thomson tells us that Wollaston had embraced the hypothesis by the fall of 1807, when the two of them tried to convince a skeptical Humphry Davy of its merits after a dinner at the Royal Society Club.

> After dinner every member of the club left the tavern, except Dr. Wollaston, Mr. Davy, and myself, who staid behind and had tea. We sat about an hour and a half together, and our whole conversation was about the atomic theory. Dr. Wollaston was a convert as well as myself; and we tried to convince Davy of the inaccuracy of his opinions.[46]

Although Davy remained unconvinced and Thomson may have overstated the extent of Wollaston's conversion, there is no doubt that Wollaston recognized that multiple combining proportions were a logical consequence of atom-to-atom combination, and he began to reassess his own salt analyses in those terms. After learning of Thomson's analyses of oxalic acid and its salts, Wollaston moved quickly to expand and publish his own observations. The result was an enormously influential paper published in the *Philosophical Transactions* immediately after Thomson's.[47]

Wollaston's paper reported, not surprisingly, his own observation of whole-number combining proportions in the three oxalates of potash, together with similar results he had obtained for the carbonates of potash and soda, and the sulfates of potash. At the outset of the paper, after acknowledging Thomson's contribution, Wollaston put his own investigations into context.

> As I had observed the same law [integral multiple proportions] to prevail in various other instances of super-acid and sub-acid salts, I thought it not unlikely that this law might obtain generally in such compounds, and it was my design to have pursued the subject with the hope of discovering the cause to which so regular a relation might be ascribed.
>
> But since the publication of Mr. DALTON's theory of chemical combination, as explained and illustrated by Dr. THOMSON, the inquiry which I had designed appears to be superfluous, as all the facts that I had observed are but particular instances of the more general observation of Mr. DALTON, that in all cases the simple elements of bodies are disposed to unite atom to atom singly, or, if either is in excess, it exceeds by a ratio to be expressed by some simple multiple of the number of its atoms.[48]

TABLE 8.2. Wollaston's Multiple Combining Proportions, 1808

Compound	Base : Acid	Modern Formula (anhydrous)
oxalate of potash	1 : 1	K_2Ox
superoxalate of potash	1 : 2	$KHOx$
quadroxalate of potash	1 : 4	$KHOx \cdot H_2Ox$
sulfate of potash	1 : 1	K_2SO_4
supersulfate of potash	1 : 2	$KHSO_4$
subcarbonate of potash	1 : 1	K_2CO_3
carbonate of potash	1 : 2	$KHCO_3$
subcarbonate of soda	1 : 1	Na_2CO_3
carbonate of soda	1 : 2	$NaHCO_3$

Wollaston made no mention in his paper of the salts of tartaric acid that he had investigated a few years earlier, likely because he did not wish to let anyone know that he had commercial interests in those compounds. It is unfortunate that such obdurate commitment to secrecy cost him an even greater place of prominence in the discovery of multiple combining proportions. The several results he did publish are summarized in Table 8.2. The results for the three oxalic acid salts of potash are those he had obtained several years earlier, although the method of obtaining the combining proportions was different. Rather than measuring the amount of base in the salts by displacement with sal ammoniac (which gave integral results only because of fortuitous circumstances), Wollaston obtained his new whole-number proportions in a novel, simple, and compelling manner.

Building upon the insights gained from his first results for tartaric acid, that the amount of acid and base in a salt could be compared by expressing their amounts in terms of a common base equivalent, Wollaston devised the following general procedure. He took a fixed weight of an acidic salt and collected the base it contained by a decomposition reaction. He then measured how many multiples of that weight of base were needed to neutralize the acid that was present in the original fixed weight of the same salt. This procedure, likely inspired by thinking of salt formation in atomic terms, circumvents entirely the "missing water" problem, since the amount of base and its acid equivalent in a salt are measured relative to the same fixed weight of acid in the salt. We can understand his reasoning by reference to Figure 8.6.

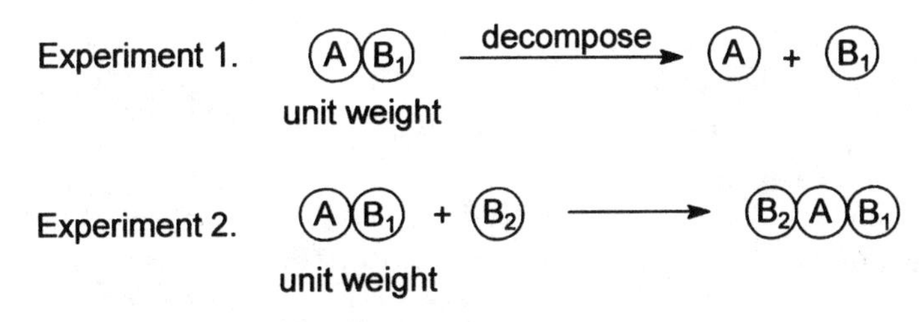

Fig. 8.6. Wollaston's Measurement of Multiple Proportions, 1808.

Decomposition of a fixed weight of an acidic salt AB, by heating, for example, leaves the critical part of component base, B_1, as residue. That same base B was then used in a second experiment to neutralize the acid in the fixed weight of salt AB, and the weight needed is represented as B_2. If B_2 turns out to be equal to B_1, then it follows that the acidic salt AB contains exactly half of the base necessary to fully neutralize the original acid A (to produce the salt AB_2).[49] For the previously discovered superacidic salt, here reported for the first time and unambiguously called the quadroxalate of potash ($KHOx{\bullet}H_2Ox$), Wollaston stated that three further weights of potash needed to be added to the one already contained in the salt to bring it to full neutrality.[50] In summary, Wollaston had shown, by simple and direct experimentation, that the two acidic salts of potash and oxalic acid contained potash equal to one-quarter and one-half of the weight needed to completely neutralize the acid. Put another way, as he elected to do to correlate with the names of the salts, the proportions of acid in the three salts united to a fixed weight of base were in a weight ratio of 1: 2: 4.

Wollaston tried to make a salt with three proportions of acid united to one of base by mixing the components in a 3:1 ratio of acid to base, and evaporating the solution to force precipitation. But he found that the precipitate contained two distinct crystalline substances, one being salt of sorrel and the other being the quadroxalate of potash. He concluded that, even under conditions favorable for the formation of a salt containing a 3:1 ratio of acid to base, such a compound did not form; instead the reagents partitioned themselves equally into salts with a 4: 1 and a 2: 1 ratio.[51] He wrote, in words carefully crafted to avoid a first-person declaration of his own acceptance of the hypothesis,

> To account for this want of disposition to unite in the proportion of three to one by Mr. DALTON's theory, I apprehend, he might consider the neutral salt as consisting of

- 2 particles potash with 1 acid, $[AB_2]$
- The binoxalate as 1 and 1, or 2 with 2, $[AB]$
- The quadroxalate as 1 and 2, or 2 with 4, $[A_2B]$

in which cases the ratios which I have observed of the acids to each other in these salts would respectively obtain.[52]

The three salts of oxalic acid were clearly the centerpiece of Wollaston's paper, but the evidence for integral combining proportions was strengthened by his description of simple procedures for measuring those proportions in the well-known salts of carbonic and sulfuric acids. The examples given by Wollaston, three of which were salts of well-known acids and the fourth exhibiting a double instance of multiple proportions in an organic acid, elevated the concept of integral multiple proportions from a curious, perhaps atypical, observation to an inductive generalization, perhaps even to a law of nature.

Even though Wollaston accepted Dalton's atomic hypothesis as the best available explanation for the integral proportions he had so convincingly established, he recognized that atomistic explanations were still in their infancy. Nonetheless, in a passage near the end of the paper that explored the consequences of an architecture of substances composed of atomic units, Wollaston brilliantly foresaw developments that were decades in the future.

> I am further inclined to think, that when our views are sufficiently extended, to enable us to reason with precision concerning the proportions of elementary atoms, we shall find the arithmetical relation alone will not be sufficient to explain their mutual action, and that we shall be obliged to acquire a geometrical conception of their relative arrangement in all the three dimensions of solid extension.
>
> For instance, if we suppose the limit to the approach of particles to be the same in all directions, and hence their virtual extent to be spherical (which is the most simple hypothesis); in this case, when different sorts combine singly $[AB]$ there is but one mode of union. If they unite in the proportion of two to one $[AB_2]$, the two particles will naturally arrange themselves at opposite poles of that to which they unite. If there be three $[AB_3]$, they might be arranged with regularity, at the angles of an equilateral triangle in a great circle surrounding the single spherule; . . . but when the number of one set of particles exceeds in the proportion of four to one $[AB_4]$, then, on the contrary, a stable equilibrium may

> again take place, if the four particles are situated at the angles of the four equilateral triangles composing a regular tetrahedron.[53]

The foregoing passage, which would not look out of place in a modern introductory chemistry textbook, is a good example of the power of Wollaston's deductive abilities. But, as we have seen with so much of his scientific thinking, deduction from hypothesis was consistently subjugated to experiment and observation. So, after tantalizing his readers with a vision of polyatomic structures in three-dimensional space, Wollaston adds a cautious, less visionary, statement.

> But as this geometrical arrangement of the primary elements of matter is altogether conjectural, and must rely for its confirmation or rejection upon future inquiry, I am desirous that it should not be confounded with the results of the facts and observations related above, which are sufficiently distinct and satisfactory with respect to the existence of the law of simple multiples.[54]

This declaration reiterates Wollaston's strong and unwavering belief that experimental and observational "facts" formed the foundation of science, and hypotheses and theories derived from them were not to be used to shape or "confound" those facts. One might argue that Wollaston had himself exploited the deductive consequences of atomic theory to devise the conditions under which he was able to obtain multiple combining proportions. But, without specific dates in his notebooks accompanying the analyses reported in his paper, it is impossible to know for sure the full intellectual context of his discoveries. Although he may not have conformed steadfastly to his stated beliefs, there is no doubt about the nature of those beliefs—experiment and observation were the bedrock of his science. I have no doubt that Wollaston thoroughly understood the nuances of atomic theory, and if he was not at this time ready to admit the uncontested existence of atoms, he was certainly able to foresee as well as anyone the potential an atom-based metaphysics would have for a fundamental comprehension of chemical phenomena. His intellectual reach, coupled with the train of circumstances that led him to the observation of multiple proportions, render invalid the retrospective, superficial judgment of William Whewell that "the scrupulous timidity of Wollaston was probably the only impediment to his anticipating Dalton in the publication of the rule of multiple proportions."[55] Timidity had nothing to do with Wollaston's intellectual stance.

THE IMPACT OF WOLLASTON'S PAPER ON MULTIPLE PROPORTIONS

Wollaston's results were quickly disseminated throughout the chemical community and acted in concert with Thomson's advocacy to bring atomic theory to the attention of continental chemists. One of the first to take note was the Swedish analytical chemist Jöns Jacob Berzelius, who first learned of Dalton's atomic theory by its mention in Wollaston's paper and was later to become one of the most influential proponents of atomic theory. He, like nearly all of his contemporaries, usually expressed the composition of analyzed substances in percentages, quantitative values that did not reveal multiple proportions. In his autobiography, he recalled the impact of Wollaston's paper:

> Among other things, I had analyzed the three oxides of lead, but had arranged the results in percentages. Now, after the direction afforded by Wollaston's paper, I compared the quantity of acid for a similar weight of the metal and found to my glad surprise that the acid quantities were in the proportion of 1,1 ½ and 2. . . . From all this resulted my first paper on definite proportions. . . .[56]

Another renowned chemist who was impressed with Wollaston's results was Claude Louis Berthollet. In 1808 he was composing a lengthy introduction to the French edition of Thomson's 1807 edition of *A System of Chemistry*. In the Introduction, Berthollet praised the ingenuity of Dalton's hypothesis and acknowledged the rationale it provided for combining proportions, but tempered his assessment with a call for more compelling experimental support.[57] He published some analyses of salts himself in 1809, but assigned a wider investigation to a younger associate at Arcueil named Jacques Étienne Bérard. In a well-designed series of experiments carried out with great care and accuracy, Bérard analyzed the carbonates and sulfates of potash and soda, and all the known salts of oxalic acid. In the second of two papers on the topic, Bérard confirmed all of Wollaston's results and added three new instances of integral combining proportions. He was, however, unable to replicate Thomson's work on the oxalates of strontia and concluded that the Scottish chemist's analyses were incorrect and that only one oxalate of strontia existed.[58]

Thus, although historians of chemistry have consistently cited Thomson's and Wollaston's publications on multiple combining proportions as jointly providing crucial experimental support for the atomic hypothesis,

Wollaston's results actually had a much greater impact since they were more extensive, more carefully reported, and could be replicated. Bérard's confirmatory results, of course, added to Wollaston's reputation as a reliable and imaginative analytical chemist. Consequently, while Thomson continued to be an active champion of atomic theory, Wollaston's views on the topic were seen by most of his contemporaries to be of more significance. Unintentionally, and certainly reluctantly, Wollaston was thereafter drawn into the debate on the physical reality of atoms and, as I will relate in Chapter 9, he reacted in two seemingly incompatible ways. On the one hand, he tried to obtain compelling evidence for the physical existence of atoms while, on the other, he worked to insulate his (and others') analytical results from any prejudicial dependence on a hypothetical, invisible world of atoms.

CHAPTER 9

Crystals and Atoms 1803–1818

> This simple, cheap, and portable little instrument [the reflective goniometer], has changed the face of mineralogy, and given it all the characters of one of the exact sciences.[1]

Much of the work Wollaston carried out to develop chemical products for commercial applications was generally unknown to his contemporaries, so it appeared to them that his published discoveries emanated instead from short-term bursts of inspiration and experimentation. Wollaston saw no reason to contradict the prevailing view of his experimental talents, for he was content to have others believe great discoveries could be made by small means while maintaining his competitive advantage through secrecy. Consequently most nineteenth- and early-twentieth-century biographical essays present Wollaston as a solitary, quite private character who jumped from one research area to another with little intellectual continuity. The reality, of course, is quite different. The intensive private research and process development needed to produce commercial chemical goods, coupled with extensive interaction with craftsmen and manufacturers, provided the foundation for many of the publications that appeared to have come from nowhere. Moreover, Wollaston never did insulate himself from society. As the chemical processes evolved and John Dowse became more capable of handling the routine aspects, Wollaston began to renew the active social life he had enjoyed as a provincial physician in Bury St. Edmunds.

YEARLY EXCURSIONS

After receiving a portion of the Hyde inheritance in 1797, Wollaston started to take yearly summer vacations, such as those to the Isle of Wight (1799) and the Lake District (1800), as described in Chapter 2. In 1801 and 1802, however, he remained in London while he concentrated on developing the platinum and organic acids processes. Those were the only years from 1797 until his death when he did not get out of the city for at least a few weeks in the summer. In 1803, while the controversy over the elemental nature of palladium engaged the chemical community, he resumed his annual summer tours, traveling with younger sisters Anna and Louisa for ten days to the Dover area of southeast England.[2] In 1804 he embarked in mid August on a twenty-eight-day tour of Scotland,[3] and in 1805, he spent about six weeks in Ireland. There, by chance or design, he met up with Smithson Tennant in Belfast and the two visited the basaltic columns of the Giant's Causeway on the northern coast not far from Coleraine.[4]

In 1806, he traveled through the Midlands with Anna. It was on this trip that he first began to make sketches with a prototype of the camera lucida, which he was to improve and patent a year later. During each of the vacations, Wollaston would visit friends and acquaintances, cathedrals and castles, areas of historic and geographical interest, factories and mines, while taking copious notes of all that he visited. Although Warburton's notes indicate that Wollaston kept detailed summaries of each excursion in individual notebooks, none of them are present in the Cambridge archives, and it does not appear that any still exist. However, succinct itineraries of a few of the trips were transcribed by Warburton, and all show the same pattern of meandering tours through hamlets, towns, and cities, by carriage or boat. In 1807, Wollaston spent all of August on a second tour of Ireland, largely the southern regions that he had not covered in 1805.[5] This trip came at the insistence of Dublin resident John Brinkley, his old Caius College friend who had become the Royal Astronomer of Ireland. One year later he was accompanied by his younger brother Charles on a three-week exploration of southeast England, including Hastings and Brighton.[6] In August 1809, Wollaston traveled north to Hull, and thence to Manchester and Liverpool via York, Harrowgate, and Leeds.[7] In Liverpool, he dropped in on John Bostock, a physician to the General Dispensary with whom he had exchanged letters on ways to detect poisons in suspected murder victims. Wollaston's reputation had spread widely by this time, and Bostock was thrilled to find him at his door. Soon after the visit, Bostock wrote to their mutual friend Alexander Marcet about the encounter:

> I was sitting some time ago in my study, when a gentleman walked in, whom to my surprise, I recognized to be Dr. Wollaston . . . I think that I should set him down as the most acute man of the age on all

> philosophical subjects. His conversation is rich in instruction, but his great accuracy keeps you in a kind of aweful agitation, for if you let drop a single syllable that is not to the point, he sees it instantly, & nails you down to your argument.[8]

The Wollaston that Bostock describes is one whose character is becoming familiar to us by now. He was one who observed with minute care and thought deeply about the very many things that interested him. He interpreted natural phenomena with a rigorous logic and questioned every instance of what he perceived to be sloppy thinking. To those who had enough honest curiosity, or strength of character, to engage him in intellectual discussion, Wollaston could become an engaging teacher and, sometimes, a close personal friend. But to those who took umbrage at having their beliefs examined and found wanting or, even worse, who feigned knowledge on topics they knew little about, Wollaston was a feared conversationalist. His personality was not one that appealed to all.

This brief synopsis of his summer travels to the year 1809 has been introduced here to emphasize that, even in the most labor-intensive years of the chemical business operations, Wollaston was no recluse. In addition to the annual excursions, he was very active in the meetings and governance of the Royal Society and was frequently invited to serve on government committees, as I will discuss in Chapter 12. He continued to maintain close connections with his father and siblings, as his trip summaries attest, and he frequently hosted dinner parties at his home. The convivial atmosphere of Wollaston's dinner parties was later captured by the young geologist Roderick Murchison, who was a guest at several of them.

> Wollaston's little dinners of four or five persons were most agreeable, and you were sure to come away with much fresh knowledge. A good dish of fish, a capital joint and some game, followed by his invariable large pudding, filled in with apples, apricots, or green-gages, all served on plain white porcelain by two tidy, handsome women, was the bill of fare.[9]

The frequent attendance at such dinners of geologists, such as Murchison, reflects the growing interest of Wollaston in that new science, and especially in one of its subdisciplines, mineralogy. True to form, Wollaston was quick to make himself into a mineralogical expert.

About 1808, entries in Wollaston's daybook begin to reveal purchases of minerals from various London dealers. These purchases correlate with his growing interest in the crystal structure and chemical analysis of common minerals. Identification of very small crystalline substances and determination of their chemical composition had been, as has been frequently noted, a key objective of Wollaston's from his earliest studies in Huntingdon. His connection of crystalline regularity with chemical identity had served him well in the isolation of palladium and rhodium as well as in the discovery of multiple combining proportions in salts. So it was a natural progression for him to seek the same geometrical regularities in large crystals that he had been so careful to identify in small ones. Consequently, his wish to measure accurately the angles at which the planar surfaces of crystals intersected soon led him to design an instrument that was to revolutionize the study of crystals.

THE CONTACT GONIOMETER

The scientific study of crystals began to be regularized in the late eighteenth century, and the angles between crystal faces started to become an important physical characteristic of a mineral species. The pioneering French crystallographer René Just Haüy developed an extensive mathematical interpretation of crystal structure, based largely on the angles between the crystal's axes and each of its faces. For Haüy, every crystalline form of a mineral could be reduced by mechanical cleavage to a simple geometric form he called its "primitive form."[10] The primitive forms of all minerals were a small number of geometrically regular solids such as the tetrahedron, the triangular prism, or the parallelepipedon.[11] Consequently, once the primitive form of a mineral was determined and the angle of inclination of a crystal axis to its cleavage faces was measured (defined as the angle of intersection between the crystal axis and a line drawn perpendicular to a crystal face), all values of its surface angles could be calculated by plane trigonometry. Haüy's method can be illustrated by reference to Figure 9.1, in which an axis for a representative crystal is shown by the line xy. Once the angles between the axis xy and crystal faces such as ABCD and ABEF are measured then, for example, the angle of intersection of face ABCD and face ABEF could be calculated. Measurements on a number of crystals seemed to give facial angles in agreement with such calculations, and the paradigmatic example was Iceland spar, the well known rhombohedral form of calcium carbonate. Haüy (and others) had observed that the faces of Iceland spar were inclined at an angle of 45° to the crystal axis. Calculations then gave a value of 104° 28′ 40″ for the angle at which adjacent faces met.

Haüy and his contemporaries measured crystal angles with a simple device invented by Carangeot in 1780, known as a contact goniometer.[12]

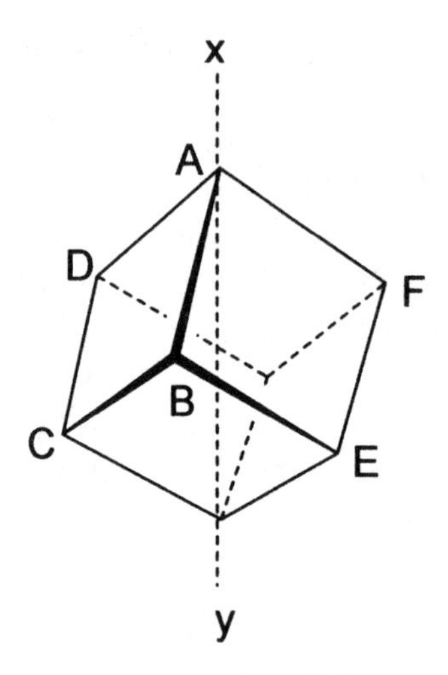

Fig. 9.1. Representative Crystal Cleavage Faces.

This simple measuring device, shown in Figure 9.2, consisted of a hinged pair of arms attached to a protractor that could be placed on the two faces of a crystal to give the angle of their juncture. Under ideal conditions, the contact goniometer could give readings accurate to about half a degree. Using it, Haüy measured the external angle of two faces of Iceland spar and found the angle to be close to 104°, near enough to the calculated value to convince himself that his mathematical construction of the crystal was legitimate. He found similar correspondences for a number of other crystalline minerals, and gradually became convinced that all crystals formed according to the geometrical dictates of a few simple primitive forms. Haüy's ideas were widely disseminated, well received, and enormously influential.

Such a fundamental idealization of crystal architecture would have been, of course, anathema to Wollaston, who was always wary of theoretically-biased observations. And in this case, he had good reason to doubt the accuracy of Haüy's values for Iceland spar. Wollaston had, as noted in Chapter 3, used his newly invented refractometer in 1802 to investigate double refraction in that mineral.[13] As part of that study, he had "measured with care, an angle at which two surfaces of the spar are inclined to each other, and found it to be 105° 5′."[14] No mention was made at that time about the difference between his and Haüy's calculated value of 104° 28′ 40″. It is possible that Wollaston was unaware of Haüy's crystallographic ideas, for in the paper he compared his own results only with the similar, but much earlier ones of Huygens. It is certain, however, that Wollaston came to know of Haüy's speculations a few years later, for an entry in his daybook mentions

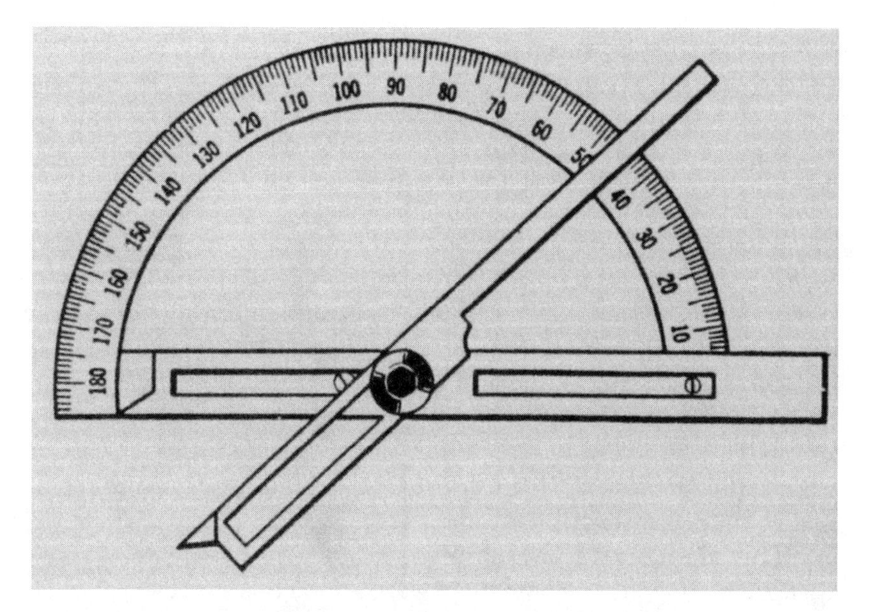

Fig. 9.2. A Contact Goniometer.

the purchase of Haüy's 5 volume *Traité de minéralogie* in August 1805, just prior to his first trip to Ireland. That purchase likely coincided with his growing fascination with crystallography, which was supported by subsequent purchases of mineral samples, including a large batch bought from Jacob Forster's shop in 1808 (the same shop he had selected in 1803 to sell samples of palladium).[15] About this same time Wollaston also began to interact extensively with the Professor of Mineralogy at Cambridge, Edward Daniel Clarke.

Clarke had graduated from Cambridge with a BA in 1790 and had amassed an extensive mineralogical collection during his travels around the United Kingdom, continental Europe, and Siberia. In March 1807 he began a very popular series of lectures on mineralogy, which led the university in late 1808 to create a professorship for him in the subject. Although Clarke possessed no great intellectual depth, he was a natural history enthusiast who acquired some expertise in mineral identification and classification. From 1808 on, to gather content for his mineralogical lectures, he began to exchange minerals and letters with Wollaston, and the letters from Wollaston that still exist reveal much about the characters of both men.[16]

The letters reveal that Clarke in 1808 was a novice at the chemical analysis of minerals and regularly sought confirmation or correction from his good friend in London. Clarke's initial identification of minerals was frequently erroneous, and his unskilled chemical analyses were generally unreliable, but Wollaston remained a patient and valuable mentor. Clarke was not the only one to benefit from his friend's growing mineralogical expertise, for Wollaston provided similar services to many others, and gradually became a well-known and trusted analyst of minerals. There still exists in the Science Museum of London a small chestful of minerals that Wollaston received from correspondents throughout Europe, over a period of many years. It is likely that the frequently-witnessed micro methods he developed for the qualitative analysis of such minerals contributed to their recognition by his contemporaries as the epitome of his research style.

The very first letter to Clarke in 1808 confirms that Wollaston was then using a contact goniometer to measure crystal angles, and that he introduced Clarke to its use. But he clearly was not satisfied with the accuracy attainable with the device—it required clearly delineated crystal planes in relatively large crystals, a steady hand to orientate it at right angles to the crystal surfaces and, even then, yielded values accurate only to half of a degree. His solution to its limitations was the invention of a new type of goniometer, one that was to change the course of nineteenth-century crystal measurement.

THE REFLECTIVE GONIOMETER

In June 1809, Wollaston read a paper to the Royal Society describing an instrument he had invented that allowed the angles of crystal faces to be measured to a precision of 5 minutes, or about 1/12 of a degree.[17] Wollaston's innovation, the basic idea of which had been contemplated (but not implemented) by others, was to measure the angles between crystal faces by means of a light ray reflected from their surfaces. The reflective goniometer, as he called the instrument, is shown in Figure 9.3. The crystal to be studied was attached with wax to the small mounting plate (shown on the left of the diagram, labeled "*n*"), such that the reflecting faces of the crystal could be brought horizontal and parallel to the axis of rotation of the large measuring wheel by appropriate adjustment of the mounting plate. A distant image, of a building's horizontal edge for example, was then brought by reflection from a crystal face to the observer's eye. Then, with the eye held in the same place, the crystal was rotated on the axle of the measuring wheel until the same reflection from a second crystal face was brought to the eye. The angle through which the wheel had rotated was measured in degrees marked on the circumference of the large wheel, and to increments

Fig. 9.3. Wollaston's Reflective Goniometer.

of 5 minutes by the attached vernier. The angle of rotation of the measuring wheel actually gave the internal angle between the two faces, but the wheel was inscribed to give the supplement of this angle, the desired external facial angle. In the hands of a careful and experienced user, the reflective goniometer could give, with unprecedented accuracy and precision, the facial angles of crystal planes that were no wider than 1/50 inch, a measurement impossible with a contact goniometer.[18]

Wollaston predicted (because he had already found such to be the case) that the new goniometer would yield results different from many of the commonly accepted ones. His illustrative example, not surprisingly, was carbonate of lime, Iceland spar. Wollaston made it clear that measurements with the reflective goniometer spelled trouble for Haüy's mathematical idealization of crystal structure. He wrote

> The inclination of the surfaces of a primitive crystal of carbonate of lime is stated [by Haüy], with great appearance of precision, to be 104° 28′ 40″: a result deduced from the supposed position of its axis at an angle of 45° with each of the surfaces, and from other seducing circumstances of apparent harmony by simple ratios . But however . . . I find the inclination of the surfaces to each other is very nearly, if not accurately 105°.[19]

We know that Wollaston had for some time been skeptical of Haüy's values for the crystal's facial angles. In his personal copy of the 1802 paper on the refraction of calcium carbonate, beside the paragraph where he reported his own measurement of the crystal's definitive angle, Wollaston inserted a comment that predated his goniometer paper: "May 5 1809. I think Haüy is wrong & that the angle of inclination of the planes is nearer to 105° than any other assignable."[20]

Despite his certainty on the matter, Wollaston was judicious in his criticism of Haüy's result and remained reluctant to have his name attached to any concept that he had not fully studied and adopted. In fact he could be quite irascible to those who too carelessly cited his opinions as authoritative. For example, after Clarke had named him as an authority on a certain mineralogical specimen, Wollaston quickly rebuked him.

> I find that you . . . have quoted authorities which do not exist—for instance what has Dr. Wollaston written about pyroxene? & if he has written nothing you certainly make a most unwarrantable use of his name—I have it moreover on very good authority that Dr. W knows little or nothing of the matter & does not recollect the specimen alluded to, & threatens, if any misrepresentation of a mere conjecture has been made, to roast you alive the first time he catches you in London—however I will do my best to pacify him.[21]

Such a stern admonition to a good friend over a relatively minor indiscretion, even if deflected by a humorous aside, seems exaggerated. Another tamer, but more illustrative, example appears in a letter written in 1812 to the Edinburgh mineralogist Thomas Allan.

> I certainly do object to being quoted as a geological authority till I think myself qualified to write upon such subjects. And this objection applies

> generally to all other subjects—When I publish let any one who pleases quote & let any one who can detect errors.—My first experiments may differ from subsequent results. I may be quoted as having freely communicated the former, & not quoted as having freely acknowledged the correction of my own error as well as that of my neighbours.[22]

This comment tells us much about Wollaston's stance on intellectual property. Only after he had published a well-considered opinion and/or experimental observation did he become comfortable at having his ideas acknowledged. Any preliminary conjectures or opinions given to others in conversation were to be recognized as mere working hypotheses, subject to confirmation, modification, or refutation as time and experiment played out. If the comments to Allan are taken at full value, we can better understand why Wollaston was sometimes slow to publish and occasionally quick to anger. One cannot help but suspect that he could take offense at the slightest misuse of his name or authority, even when done by a close associate. It is easy to see why many found intellectual exchanges with him to be unsettling.

Despite such intellectual caution, Wollaston's stature had, in fact, grown year by year as the quality and breadth of his papers became appreciated throughout the scientific world. Unbeknownst to him when he composed his goniometer paper, Étienne Malus in France had just completed his own thorough study of the double refraction of Iceland spar, a study prompted by Wollaston's 1802 paper on the topic.[23] To investigate more precisely the phenomenon of double refraction, Malus invented an instrument that employed an optical sighting tube attached to a graduated circle marked in tenths of a degree.[24] The instrument allowed him to measure accurately Iceland spar's refractive index at different angles of incident light rays. Consequently, he was able by 1809 to confirm Wollaston's (and Huygens's) findings, as mentioned in Chapter 3. In the course of his experiments, Malus also found the external facial angle of the crystal to be 105° 5′, exactly the same value that Wollaston had published in 1802.[25] When Malus's observations on double refraction and his related discoveries (such as the polarization of light by reflection) came to Wollaston's attention, he checked and verified the results for himself. Impressed by the Frenchman's acuity, he praised Malus's talents in a letter to Hasted, announcing that he could "not recollect any one discovery in optics which evinces more accuracy of observation & acuteness of discrimination."[26] Clearly, although Wollaston was reluctant to have his own work praised excessively, he had no qualms about giving plaudits to others for work well done, especially when that work was marked by fastidious observation.

Malus's measurement of the external facial angle of Iceland spar came to the attention of Haüy in time for him to include it in a publication dated 1809.[27] Haüy there acknowledged that Malus's value was the "same as that obtained by M. Wollaston, the famous English physician,"[28] but he did not believe the half degree discrepancy in observed angles of one particular crystal was large enough to warrant a revision of his crystallographic ideas. However, the reflective goniometer was soon to throw up more serious challenges to Haüy's interpretations.

Although Wollaston's goniometer paper contained no new data on crystal measurement, the instrument (which he did not patent) quickly became the working tool of mineralogists. In a letter to Hasted describing it, Wollaston referred to the goniometer as "my present hobby horse, & I flatter myself that crystallographers in general will find it a most excellent hackney, travelling well upon the road without ever stumbling or tripping."[29] The reflective goniometer was adopted quickly in Britain and in time supplanted the use of other measuring devices on the continent, reaching France in 1813 and Germany in 1815.[30] By the early 1830s it had advanced the fields of mineralogy and crystallography to such an extent that John Herschel cited it as a prime example of the effect that an instrument can have on the advance of science. He wrote

> What an important influence may be exercised over the progress of a single branch of science by the invention of a ready and convenient mode of executing a definite measurement, and the construction and common introduction of an instrument adapted for it cannot be better exemplified than by the instance of the reflecting goniometer. This simple, cheap, and portable little instrument, has changed the face of mineralogy, and given it all the characters of one of the exact sciences.[31]

Goniometry of crystals by reflective methods continues to modern times, and Wollaston's instrument was the first to do so with accuracy and convenience.

As soon as he had perfected the design of the reflective goniometer, Wollaston began to use it to measure the facial angles of all the minerals he could get his hands on, including multiple measurements of angles of the same mineral obtained from different sources. Most of his measurements agreed with those published by Haüy, except for a few anomalies, such as the unusual differences in the supposedly identical rhombohedral angles of carbonate of lime ($CaCO_3$), bitter spar ($Ca(Mg)(CO_3)_2$) and iron spar ($FeCO_3$).

Wollaston's measurements gave different results for each: a facial angle of 105° 5′ for carbonate of lime, 106° 15′ for bitter spar, and 107° for iron spar.[32] These three minerals all shared the same primitive form, according to Haüy, and were therefore expected to have the same facial angles, which measurement with a contact goniometer had seemed to confirm. Wollaston used the more discriminating reflective goniometer to show conclusively that such was not the case. The three crystalline forms were slightly, but unmistakably, different.

Wollaston shared Haüy's fundamental belief that identity of crystalline form was evidence for identity of composition (a theoretical commitment that had helped him recognize and separate different salts in many of his small-scale chemical analyses), and he therefore expected that the three chemically different carbonates must differ structurally from each other in some way. This expectation was happily confirmed when he discovered the differences in their facial angles. This meant that the three minerals no longer appeared to be exceptions to the prevailing rule that each chemically different crystal exhibited its own unique crystal structure.

After announcing the revised rhombohedral angles for two of the carbonates, Wollaston speculated on the structures that might result from an intermixture of two or more of the carbonates, as was possible under differing conditions of geological formation. He proposed that if a mineral could form from intermixture of two closely similar crystalline substances, such as the carbonates, the resultant mineral would have a form closely similar to, but not identical with, its constituent minerals. And the final form was fixed, however indeterminately, by the organizing "power" of the primitive forms of the constituent minerals. Wollaston was unprepared to venture any guesses as to the origins or nature of such crystal "powers," but he did predict that minerals formed by intermixture of different crystals should occur frequently in nature. This suggestion prompted others to search for more examples of crystals of mixed composition and, once found, to determine their distinctive external facial angles and, finally, to speculate further on the relationship between external form and chemical composition. A comprehensive explanation was offered a few years later by Eilhard Mitscherlich, who introduced the concept of isomorphism to explain the relationship among substances of different composition that adopted closely similar crystal forms.[33] Mitscherlich's explanation of the phenomenon depended on an atomic interpretation of crystal form, something that Wollaston could well have anticipated if only he had allowed himself to commit wholeheartedly to atomic theory.

CRYSTALS AND ELEMENTARY PARTICLES

As noted in Chapter 8, Wollaston believed in early 1808 that the integral combining proportions he had discovered in salt formation were best explained by atom-to-atom combination, and he had even speculated at the time on the three-dimensional shapes that would result from stable arrangements of neighboring atoms. As others confirmed and extended his chemical discoveries, they too began to view Dalton's atomic hypothesis more favorably, and Wollaston began to be seen as an adherent of the atomic hypothesis. In 1812, he appeared to take atomistic explanation in a new direction by proposing a role for atoms, now more ambiguously referred to as "elementary particles," in crystallography. When selected to give the Royal Society's Bakerian Lecture for that year, he chose to discuss some ideas he had been ruminating over for "more than three years," which places their origin at the time he began studying minerals with his goniometer. The lecture was read to the Society on November 26 and was published early in 1813.[34] In it, Wollaston discussed how the concept of primitive forms fundamental to Haüy's crystallographic theory could be simplified by considering the primitive structures themselves to be assemblies of spheres and spheroids.

Wollaston began by illustrating, as shown in Figure 9.4, how one sphere placed upon a triangle of three others gave a structure whose faces constituted a tetrahedron (sketch 5) and how one sphere above and one below a square of four spheres gave a structure with octahedral faces (sketch 6). Thus, two of Haüy's primitive forms could be reduced to stable arrangements of simple spherical particles. Going further, by adding two more spheres into the pockets of opposing faces of the octahedron in sketch 6, Wollaston was able to make an acute rhomboid, thus demonstrating that "the simplest arrangement of the most simple solid that can be imagined, affords so complete a solution of one of the most difficult questions in crystallography."[35]

When he first thought about explaining crystal structure by the close-packing of spheres Wollaston was unaware that Robert Hooke had much

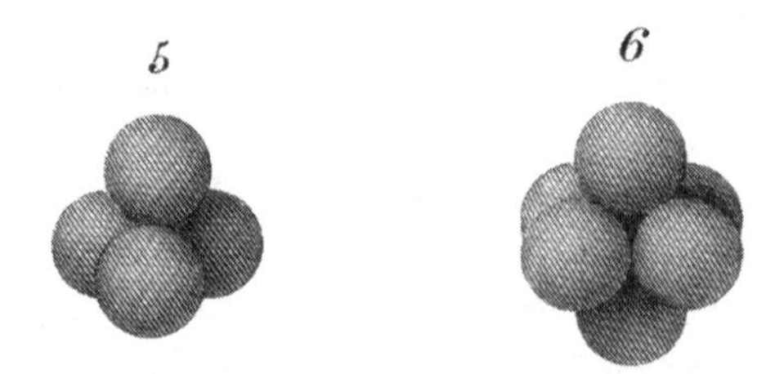

Fig. 9.4. Construction of a Tetrahedron and an Octohedron from Spheres.

earlier advanced a similar explanation for the structure of quartz in his famous *Micrographia* of 1665. Hooke's hypothesis was mentioned to him by a friend prior to the Bakerian Lecture, and Wollaston consequently inserted Hooke's relevant statements into his paper, adding that his predecessor's idea had "been totally overlooked, from having been thrown out at a time when crystallography, as a branch of science, was wholly unknown."[36] But Wollaston did not mention another person who had entertained similar thoughts, John Dalton. In Part 1 of his *New System of Chemical Philosophy*, published in 1808, Dalton had written

> Crystallization exhibits to us the effects of the natural arrangement of the ultimate particles of various compound bodies. . . . The rhomboidal form may arise from the proper position of 4, 6, 8 or 9 globular particles, the cubic form from 8 particles . . . but it seems premature to form any theory on this subject.[37]

The speculation was placed, without emphasis, in a section of the book on the constitution of solids, but there can be no doubt that Wollaston was aware of it, for Dalton had pointed it out to him. While in London giving a course of lectures at the Royal Institution in December 1809, Dalton had several conversations with Wollaston, whom he considered to be "one of the cleverest men I have yet seen here."[38] In January, Dalton visited Wollaston at his home, where the host mentioned his emerging ideas on the particulate structure of crystals. Dalton later mentioned what then transpired

> When he mentioned those ideas to me in conversation, . . . He could scarcely credit that I had entertained and published the same, until he brought my book out of his library, and I shewed him the page. He had probably seen it before, but had forgotten it.[39]

Dalton was probably correct that his words had been forgotten for, until the reflective goniometer began to shape Wollaston's thinking about the primitive forms of crystals, a construction of crystal forms from close-packed spheres would have appeared to be little more than a fanciful hypothesis, unsupported by any essential mineralogical observations. So, even though Wollaston's ideas on the architecture of packed spheres were not completely original, he can be credited with developing them to a greater extent than any of his contemporaries by broadening the concept to include both non-spherical, and non-identical, particles. That portion of the paper marked a new departure for the hypothesis.

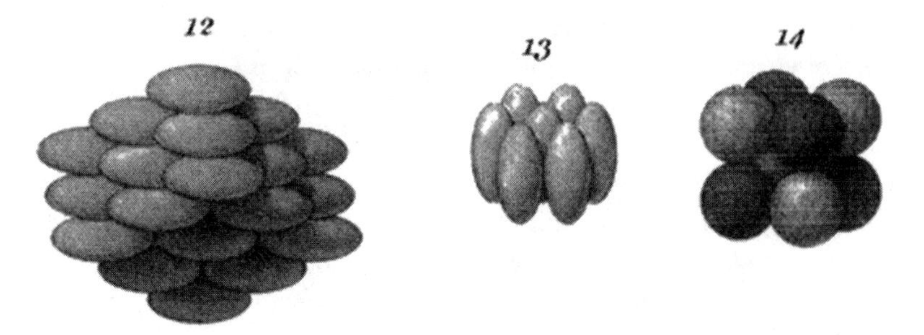

Fig. 9.5. Wollaston's Construction of Crystals from Differently Shaped Elementary Particles.

His 1802 paper on double refraction of light rays in Iceland spar had led Wollaston to agree with Huygens that the path of the differently-refracted "extraordinary ray" could be understood as the transmission of an elliptical wave front through the crystal, and Huygens had postulated that spheroidal particles in the mineral might be responsible for the phenomenon. In his Bakerian Lecture, Wollaston considered in greater detail the crystalline forms that could be constructed from what he called "oblate spheroids." He recognized that oblate spheroids could give rhombohedral crystals of differing facial angles, depending on "the degree of oblateness of the primitive spheroid." Wollaston believed this might be the case in doubly-refracting Iceland spar.[40] Another possibility would be the mutual attraction of so-called oblong spheroids, which would align themselves with their longest axis parallel, as shown in sketch 13 of Figure 9.5. A crystal made this way "would be liable to split into plates at right angles to the axes, and the plates would divide into prisms of three or six sides with all their angles equal, as occurs in phosphate of lime, beryl, etc."[41]

Wollaston's last hypothesis in the Bakerian Lecture dealt with the construction of a cube from two different types of spherical particles, both of the same size, as shown by the differently shaded spheres of sketch 14 in Figure 9.5. A cubic crystal having cleavage planes at right angles to each other requires all constituent spheres to be placed one on top of another, a less stable form of packing than that shown in sketches 5 and 6 of Figure 9.4. Wollaston's premise, which presages the modern sodium chloride crystal structure, was stated in the following way,

> Let a mass of matter be supposed to consist of spherical particles all of the same size, but of two different kinds in equal numbers, represented [in sketch 14] by black and white balls; and let it be required that

> in their perfect intermixture every black ball shall be equally distant from all surrounding white balls, and that all adjacent balls of the same denomination shall also be equidistant from each other. . . . The four black balls are all in view. The distances of their centres being every way a superficial diagonal of the cube, they are equidistant, and their configuration represents a regular tetrahedron; and the same is the relative situation of the four white balls.[42]

AN OPPORTUNITY MISSED

Wollaston's speculations on the construction of Haüy's primitive forms from spheres and spheroids opened up new territory for atomistic thinking, and some of his contemporaries moved tentatively in that direction. John F. Daniell, for example, published several experimental studies, which he believed provided support for Wollaston's particulate interpretation of crystal structure.[43] But Wollaston himself did not pursue further the ramifications of his own imaginative thinking. Yet again, his wariness about theory-driven observations continued to limit his intellectual reach. As he told the readers of his paper,

> though the existence of ultimate physical atoms absolutely indivisible may require demonstration, their existence is by no means necessary to any hypothesis here advanced, which requires merely mathematical points endued with powers of attraction and repulsion equally on all sides, so that their extent is *virtually* spherical, for from the union of such particles the same solids will result as from the combination of spheres impenetrably hard.[44]

These words reveal the broad outlines of Wollaston's evolving attitudes to atomic theory in 1812, much influenced by his crystallographic thinking. He continued to believe that particle-to-particle combination (a retreat from the atom-to-atom terminology, used four years previously in the multiple proportions paper) continued to be the best explanation for the combining proportions observed in chemical compounds. But the logical connection between the particles of chemical combination and the ultimate physical atoms presumed to be the last point of division of matter had not been satisfactorily established. So Wollaston held fast in the safe harbor of what he believed to be neutral descriptive language, and the word "atom" is used in the Bakerian Lecture only for the ultimate physical units that, for example, provided mass to objects and were acted on by gravity.

The relationship between those physical atoms, and the character-bearing units of different chemical elements, and the presumed primitive forms of minerals, was unknown and (in Wollaston's opinion) unknowable with the experimental techniques available at the time. He therefore reverted to usage of the generic term "particle," which, as his wording suggests, could be identified as a chemical atom, but need not necessarily be that very thing. In addition, he recognized that it was not even essential that the particles from which he assembled the primitive forms of minerals had to be made of solid matter. Point sources of forces that manifested themselves with spherical or elliptical field boundaries could give the same results.

Clearly, Wollaston had taken heed of some of the objections to a comprehensive atomic theory voiced by many of his contemporaries, chief among which were the lack of compelling evidence for indestructible chemical atoms, and complete ignorance of how characterless, presumably fundamental, physical atoms could acquire the chemical properties that defined the chemical elements. Nonetheless it is evident that, however carefully Wollaston expressed himself, he believed that the primitive forms of crystallography so central to Haüy's mineralogy could be nicely explained by the ordered assembly of differently-shaped particles. The particles could be single or compound atoms, or just vacuous force fields with appropriate geometric boundaries. Of course, solid bodies of the postulated shapes provided the simplest explanation and one that "had appeared satisfactory to various mathematical and philosophical friends to whom [Wollaston] had proposed it."[45] Unfortunately (we can say in hindsight), Wollaston did not seize the opportunities his hypothesis presented for a comprehensive interpretation of form based on chemical composition, as becomes apparent in another paper on crystalline form, published six years later.

In 1818 Thomson, editor of the journal *Annals of Philosophy*, asked Wollaston for his opinion on the contents of a paper he intended to summarize in a forthcoming issue. The original paper, from the French mineralogist François Beudant, had been published in France the previous year. It reported that sulfate of iron crystallized in the same form even when it was heavily contaminated with sulfates of different crystal structure.[46] In the publication, Beudant first confirmed an earlier observation by Berthollet that a crystal containing equal amounts of the sulfates of iron and copper gave the same rhomboidal crystals as pure sulfate of iron. The same rhomboidal form was maintained until the sulfate of copper exceeded 91 percent of the crystal content; only then did the crystalline form change to that characteristic of pure sulfate of copper.[47] Similarly, in crystals containing the sulfates of iron and zinc, the sulfate of iron form was maintained until the amount of the zinc salt exceeded 85 percent of the total. Most surprisingly of all, a crystalline mixture of the sulfates of copper and zinc containing only 3 percent of the sulfate of iron gave the same crystal structure. From these results Beudant concluded that sulfate of iron was able to impose its crystalline form on mixed crystals even when present in very small amounts. Curiously, Beudant gave no facial angles to confirm his claim of identicalness, but if his observations were correct, they provided compelling examples of crystals of differing chemical composition with the same geometric shape, something that could not easily be accommodated in Haüy's theory.

Wollaston complied with Thomson's request, and his commentary on Beudant's findings was published in the *Annals of Philosophy* a few pages after the Frenchman's results.[48] He began, as we might expect, by repeating Beudant's synthesis of the relevant sulfate crystals and comparing their crystal structures. Wollaston confirmed Beudant's more important observations and stated,

> But though I thus differ from M. Beudant with regard to the primitive form of the common sulphate of iron, I must admit the justness of his remark, that the forms assumed by mixed sulphates of copper and iron, of zinc and iron, or of copper, zinc, and iron, appear the same as that of simple sulphate of iron alone.[49]

Intrigued by Beudant's results, Wollaston went one step further by preparing a crystal containing a 4:1 ratio of sulfate of zinc to sulfate of copper that was a rhombic prism indistinguishable from sulfate of iron, even though it contained none of the iron salt itself. In addition, Wollaston stated that all transparent crystals that analysis showed to be a mixture of sulfates had to be homogeneous solids. This conclusion stemmed from his conviction that any inhomogeneous mixture of sulfates, each of which had a different refractive index (such as the sulfates of copper and zinc) could not form a transparent solid. So Wollaston concluded that, for transparent crystals at least, the chemical components identified by analysis were united in an intimate way, likely in the form of a chemically distinct triple salt with a characteristic refractive index. And, as Beudant had discovered, some of these distinct chemical compounds did crystallize in forms that appeared to be identical. Wollaston could not, however, abandon the principle that equated compositional differences with crystallographic ones. He was therefore forced to conclude that the observed identity of crystal forms in the sulfates could not be sustained. At some time in the future, he speculated, differences would be found. But he did not himself wish to invest the time that resolution of the relationship between crystal structure and chemical composition would require. Consequently, in the last paragraph of his commentary he conceded:

> It must be owned that the foregoing remarks leave the subject involved in difficulty; but it is to be hoped that they may at least serve to excite the industry of others, and answer the purpose for which they are designed, by suggesting to chemical and crystallographical inquirers a train of curious and useful investigation.[50]

And that is where Wollaston left the debate. Other than a few other minor reports in Thomson's journal on the primitive forms of some other crystals,[51] Wollaston published nothing further on the subject, even though he continued to carry out private researches on the chemical and crystallographic characteristics of minerals.[52] His interest in the theoretical underpinnings of the subject had run their course, and he contented himself after 1818 with collecting accurate observational data only for minerals that caught his interest. He had the luxury of being a fully independent philosopher (and by that time a fairly wealthy one) with an ever-broadening range of interests and commitments; he could do what he wanted. But his hope that others would exploit his and Beudant's observations was quickly realized, most strikingly by the German mineralogist Eilhard Mitscherlich.

In brief, Mitscherlich accepted as a fact that several chemically-distinct and well understood compounds did crystallize in nearly identical crystalline forms, as Beudant had discovered.[53] He also accepted that chemical compounds were made up of atoms, or groups of atoms, in strict stoichiometric ratios. By integrating these two concepts, Mitscherlich concluded that atoms could substitute for one another in a chemical compound to produce a set of compositionally different crystals all of the same form, a phenomenon he named "isomorphism" in 1822. Mitscherlich's interpretation made clear how the crystalline form of sulfate of iron, for example, would be relatively unchanged by the replacement of iron atoms by copper or zinc atoms if one accepted that the substitution of one atom for another did not alter the position of those atoms in the crystal structure.

Many historians have retrospectively given Wollaston credit for many of the antecedent principles of isomorphism,[54] but I think it is clear that he had such a commitment to the fixed elemental composition of each chemical substance and the (often subtle) distinctiveness of the crystalline form of each pure substance that he could not have articulated a theory of isomorphism equivalent to Mitscherlich's. Certainly, Wollaston had all the key ingredients in hand: he had designed the instrument that enabled the accurate measurement of crystal angles, he had provided new and compelling examples of fixed and multiple proportions in salts, he had published a paper on a particulate interpretation of crystal structure, and he had even made crystals that were, as Mitscherlich was to call them, isomorphous. He was even capable of bold and imaginative thinking on the structural consequences of a particulate chemistry. But all of these talents were steadfastly subjugated to an overarching commitment to meticulous observation and incontrovertible (in his opinion) experimental facts. Consequently, it was to be Mitscherlich, directing a "train of curious and useful investigation" that Wollaston encouraged, who laid claim to the discovery. Although Wollaston failed to conceive of isomorphism, his contributions to crystallography were significant enough that a French mineralogist, identified only by the initials LN, in 1818 named a mineral (calcium silicate, $CaSiO_3$) associated with meionite as Wollastonite "in honor of one of the most respected chemists of this century."[55]

Wollaston's reluctance to place full confidence in atomic theory manifested itself in another influential paper published about this time. It presented chemists with a labor-saving device that was to become an essential tool of practicing chemists, while simultaneously provoking widespread reevaluation of atomistic explanations. These developments will be discussed in Chapter 10.

CHAPTER 10

More Practical and Conceptual Innovation 1809–1822

> I have not been desirous of warping my numbers according to an atomic theory, but have endeavoured to make practical convenience my sole guide.[1]

In 1809, the same year Wollaston introduced the reflective goniometer, the Council of the Royal Society selected him to give the Croonian Lecture, an annual award lecture that had begun in 1738 with an endowment from the anatomist William Croone. The lecture was meant to focus on the nature and laws of muscular motion, and Wollaston stayed true to the benefactor's wishes by combining three ideas of his relevant to the topic: the frequency of muscular contractions, an antidote for motion sickness, and a reason for the purported beneficial effects of motion on blood flow.[2]

THE SOUNDS OF MUSCULAR CONTRACTION

Part I of the paper, titled "On the Duration of Muscular Action," is an instructive example of Wollaston's ability to discover something new in common occurrences, and to quantify it in simple ways. He reported observations first made when he was completing his medical studies in London prior to beginning his medical career in Huntingdon. He observed then that when muscles contracted they did so in a series of very short sequential contractions instead of a long, continual one. He described how he made this discovery in the following words.

> I have been led to infer the existence of these alternate motions from a sensation perceptible upon inserting the extremity of the finger into

> the ear. A sound is then perceived which resembles most nearly that of carriages at a great distance passing rapidly over a pavement.[3]

Combining his own observations with those of a few friends trained to make similar ones, Wollaston found the frequency of muscular contractions ranged from a low of 14 to a high of 36, with a mean of 20–30 beats per second.[4] How did he measure the frequency of such muscle contractions? By the following clever method.

> While my ear rested on the ball of my thumb, my elbow was supported by a board lying horizontally, in which were cut a number of notches of equal size, and about 1/8 of an inch asunder. Then, by rubbing a pencil or other round piece of wood with a regular motion along the notches, I could imitate pretty correctly the tremor produced by the pressure of my thumb against my head, and by marks to indicate the number of notches passed over in 5 or 10 seconds, observed by my watch, I found repeated observations agree with each other as nearly as could be expected. . . .[5]

Wollaston was able to get comparable results for foot muscles by leaning his ear on a small cushion placed on a notched stick resting on one foot. When he contracted the muscles in his foot, the vibrations conveyed to his ear by the crude stethoscope-like device were matched by rubbing a piece of wood against the notched stick. Since the frequencies compared favorably with those of finger and thumb contractions he was able to conclude that they were common to all muscle groups.

A frequency of twenty-four cycles per second is at the very lowest range of human hearing, so many would have found Wollaston's observations difficult to replicate, but there were a few nineteenth-century investigations that yielded similar results. The most convincing was carried out many years later by Helmholtz, who used different techniques to obtain a value of 18–20 beats per second. He surmised that the audible frequency was the first overtone of that value, that is, about 36–40 vibrations per second.[6] Modern measurements using an electronic stethoscope have confirmed Wollaston's value of 25 hertz (cycles per second) as the main frequency of muscle sounds, originating in the action of single muscle fibers.[7] While not of great medical value, Wollaston's detection and measurement of the frequency of muscle contraction is a good illustration of his observational and inventive talents, and of the close monitoring of his own sensory information.

The other two parts of the Croonian Lecture were undistinguished by comparison. Part II, "On Sea-Sickness," contained Wollaston's hypothesis about the causes of that affliction, which he had endured on one long sea voyage. He suggested that the nausea was caused by the fluctuating pressure of blood on the brain occasioned by each sudden dip of the boat. His proposed remedy was to inhale each time the ship's deck descended, on the belief that increased blood flow to the chest on inhalation would counteract its movement to the brain. Part III of the lecture, "On the Salutary Effects of Riding, and other Modes of Gestation," sought to give a rational explanation for the widely-held belief that rocking motions, such as those experienced during horseback riding, carriage rides, or even boat trips, provided beneficial health effects. Without questioning the beliefs themselves, Wollaston suggested that such external motions accelerated the passage of blood through the one-way valves in veins, thereby easing the load on one's heart. Such consequences, he suggested, might be especially valuable for those who were suffering from illness or cardiac weakness, and who could not derive similar benefits from active exercise. Unremarkable as Wollaston's animadversions on motion sickness and the beneficial effects of rocking motions were, they were consistent with contemporary thought. They certainly help us to understand the widespread contemporary medical practice of prescribing travel as a restorative and also why Tennant believed it essential to his health to go for long horseback rides each day, whatever the other demands on his time.

MICROANALYSIS

Having made himself into an expert in the microanalysis of chemical substances and minerals, Wollaston was frequently asked to determine the composition of substances of interest to others. He agreed to most such requests, and if he believed the analysis to be of broad interest, he published the result. For example, about 1809 the Portugese ambassador resident in London gave him a few grains of a platina ore that had been discovered in the gold mines of Brazil. Wollaston's analysis led him to conclude that the Brazilian ore was very different from that of New Granada—it was nearly pure platinum with intermixed grains of quite uncontaminated palladium.[8] Unfortunately, the ore was not in sufficient abundance to be commercially useful, so neither Portugal nor Wollaston was able to benefit from its discovery.

Later the same year, Wollaston sought to determine whether or not the recently-discovered elements columbium (later to be renamed niobium) and tantalum were what their discoverers had claimed them to be. In 1801, the

London chemist Charles Hatchett had identified the oxide of a new metal, which he named "columbium," in a rare mineral (later called columbite) of American origin present in the British Museum.[9] One year later, the Swedish mineralogist Anders Ekeberg discovered another new metal, which he named "tantalum," in a Finnish mineral he named tantalite.[10] The properties of the two newly discovered elements were so similar that many believed them to be one and the same thing. Because both source minerals were quite rare, chemists could not easily resolve the issue, and belief in each discoverer's claims fell mostly along national lines.[11] The problem intrigued Wollaston, who, shortly after he had begun to take a serious interest in mineralogy and collect his own samples, was able to get a piece of the mineral tantalite from a Swedish source and a few grains of columbite from the same British Museum sample Hatchett had analyzed. He found the two minerals to be physically indistinguishable, apart from their very different specific gravities, and after submitting a few grains of each to a battery of qualitative tests, he observed that they appeared to be chemically identical as well.[12] On the dubious assumption that their puzzling density difference could be attributed to distinctive modes of aggregation, he therefore concluded that both minerals contained the same metal.

Wollaston's conclusion was generally accepted until the 1840s when the German chemist Heinrich Rose subjected greater quantities of the two minerals to analysis and concluded that they contained tantalum and two new metals, one of which he named niobium. Several years after that, niobium was conclusively shown to be identical to columbium.[13] The metals discovered by Hatchett and Ekeberg were in fact different. Wollaston's error, as later events determined it to be, cannot be linked to faulty analysis. The minerals columbite and tantalite have the same crystalline form, and both contain oxides of tantalum and niobium/columbium that could not be distinguished by the chemical tests he used. But the density differences between the two minerals which he attributed to aggregation peculiarities were a clue to the eventual solution, as tantalum is a much heavier metal than niobium/columbium and is more prevalent in the denser tantalite.

THE CRYOPHORUS AND FINE PLATINUM WIRES

Wollaston's talent at illustrating scientific principles with small, simple devices was a characteristic much admired by his contemporaries. I have already described his miniature galvanic cell of 1801 and the thimble battery of 1815. In 1811, he was developing two other initiatives that further demonstrated his characteristic ingenuity. The first was a heat transfer device

Fig. 10.1. Wollaston's Cryophorus.

he named a "cryophorus," and the second was a novel method of producing platinum wires with diameters far smaller than had ever been made previously, which have since come to be known as "Wollaston wires."

The glass cryophorus, a name meaning "frost-bearer," was described in a *Phil. Trans.* paper of 1812, just a few pages after his paper on the elementary particles of crystals.[14] The simple glass apparatus, shown in Figure 10.1, consisted of two bulbs, each with a diameter of about one inch, connected by a 2–3 foot length of thick glass tubing with an internal diameter of 1/8 inch. One of the bulbs was partially filled with water, the other was left empty. The whole apparatus was freed of air by sealing off a capillary tube on the bottom of the empty sphere after the water had been boiled for "a considerable time." When the empty bulb was then immersed in a freezing mixture of salt and snow, the water in the half-filled bulb would freeze solid in a few minutes. Wollaston did not use the operation of the cryophorus to promulgate any theory of heat—he only noted that the freezing power of the salt/snow mixture was transferred through the glass tube to the water in the distant bulb. Thus the apparatus served to transfer the "frost" from one location to another or, phrased differently, the heat extracted from the empty bulb by the external freezing mixture was withdrawn in equal amount from the liquid water in the other bulb, which froze as a result. Wollaston conceded that his invention functioned as little more than an impressive demonstration device and that he could foresee no immediate practical applications. However, a century later, analogous heat transfer devices did become important commercially. The cryophorus is a heat engine in its simplest form, and a generally unrecognized precursor to modern heat pipes.[15]

Another investigation begun in 1811, aimed at producing extremely thin wires to be used as a reference grid in the eyepiece of telescopes, had a more lasting impact. Wollaston's sequence of "trial-and-error" experiments are fully recorded in his research notebooks. He initially sought to obtain gold wires with diameters approaching 1/5,000 inch by drawing a silver wire with a gold core through a series of dies with progressively narrower openings. The silver coating on the composite wire was then dissolved off by immersion in nitric acid, leaving the central gold wire behind.[16] Even when successfully done, the best he could do with that method was to obtain a gold wire with a diameter of about 1/1,000 inch. However, the wire so made was found to be "broken and irregular," and quite unsuitable for its intended use.[17] He then modified the process in two ways. First, he poured molten silver into a cylindrical mold around a central steel wire and then made a hollow core by pulling the steel wire out of the silver rod. Then, he inserted a right-sized platinum wire instead of a gold one into the silver cavity and drew out the composite wire to the desired diameter. The silver coating was then removed from its platinum core by solution in nitric acid, leaving the platinum wire intact. Unfortunately, the wire so obtained was not much better than the earlier one made of gold, for it too was found to be "interrupted and irregular."[18] But the second process led Wollaston to realize that there was no need to begin with a cored-out silver wire at all. The process could be both simplified and improved by beginning with a relatively thick platinum wire and surrounding it with molten silver to make the composite wire in one step. Subsequent drawing out and dissolving off of the silver exterior left behind a thin platinum wire. In the first trial of this novel process, carried out in late 1811 or early 1812, Wollaston was finally able to prepare a few short lengths of platinum wire as narrow as 1/5,000 inch.[19]

Not surprisingly, Wollaston's process required platinum of quite high purity for success, since impurities would lead to breaks in the continuity of the very fine wires he wished to make. The first platinum wire he made with a diameter of 1/1,000 inch came from a sample of his purest consolidated platinum. Wires of even smaller diameter required more ductile platinum, which could only be obtained by melting metal made by his powder-compaction process. Because of its very high melting point (1,768°C), bulk platinum could not easily be melted, but Alexander Marcet discovered that he could melt a short length of normal platinum wire in the flame of a spirit lamp fed by a stream of oxygen.[20] After using Marcet's method to form a tiny fused globule of platinum, Wollaston was able by December 1812 to draw wire to a diameter of 1/4,000 inch and, in one instance of a particularly ductile globule, to 1/30,000 inch.[21]

By 1820, Wollaston had accumulated a selection of fourteen composite wires that could be made into thousands of inches of ultra-thin platinum wire, a supply that far exceeded any contemporary need.[22] By that time he had added a new and unpublished wrinkle to the process, which enabled him to make usable wires of diameters ranging from 1/30,000 to 1/50,000 inch. He did this by employing a variation of one of the production steps he had originally discarded. He folded a layer of silver around a steel wire and drew out the composite wire to a diameter of 1/10 inch. Then he pulled out the steel wire and replaced it with one of his previously made silver-coated platinum wires. The resulting wire with a double thickness of silver surrounding the narrow platinum core could then be drawn through a series of ever narrower dies to give the smallest diameter wires he ever succeeded in making.[23] Wollaston's technique for making fine wires continues to be used in modern times and, in one instance, a group of researchers has extended his methods to make ultra-pure platinum wires three-millionths of an inch thick.[24]

Wollaston generally sent free samples of the fine wires to all who requested some, but even the giveaways did not much diminish his supply. After his death in 1828, the collection of his prepared wires passed on to Henry Kater, and from him they found their way successively to Michael Faraday, Thomas Graham, and finally to the latter's assistant James Young, whose descendants ultimately bequeathed them to the Science Museum, where they now reside.[25] My colleagues and I have analyzed a selection of wires in this collection to determine the purity of the platinum in them. We found that he was able to produce some platinum of an average purity of 99.3 percent, a remarkable achievement for an early nineteenth-century purification and powder-compaction process.[26] Wollaston's best platinum is comparable to that marketed today as Grade 3 metal, suitable for use in crucibles and other laboratory ware.[27]

As he made ever thinner wires, Wollaston became interested in the relationship between tensile strength of the wires, "tenacity" as he termed it, and their diameters. It was well known to those involved in wire drawing that wires became stronger in proportion to their diameter as they became thinner, a phenomenon resulting from what is now known as "coldworking." He found by experiment that fine platinum wires exhibited the expected increase in tenacity, even in wires drawn down to diameters of 1/18,000 inch.[28] He compiled a table of measured tenacities for a series of wire diameters from 1/500 through to 1/30,000 inch. The table was included in the draft version of the 1813 paper, but Wollaston decided to exclude it from the published version.[29] What reason might there be for this decision? We know that Wollaston guarded the crucial details of his platinum process and never published any physical parameters of his purified platinum that would allow others to compare its purity with platinum produced elsewhere. The two parameters he held to be most indicative of the metal's purity were its density and its tensile strength, and so he decided to keep those numbers to himself.

Only in his posthumous paper did Wollaston reveal the key chemical and technical details of his platinum process. In it, for the first time, he reported that his best, fused platinum had a specific gravity of 21.5 (the modern value is 21.45), and its tenacity was 590 (pounds supported by a wire of 1/10 inch diameter).[30] The surprising aspect of these numbers is that they were not exceptional. Lavoisier had, for example, in 1789 reported platinum's specific gravity to be 22.1,[31] and Wollaston reported that the accepted tenacity value for gold was 500 and iron 600. He must have feared that, if the unexceptional values for his platinum's physical properties were to be published, others would realize there was no arcane secret in his manufacturing process that would prevent them from producing platinum of similar quality. His contemporaries believed that he had discovered some completely novel processing techniques that could not easily be accessed, and Wollaston was not inclined to correct that (mis)perception. Full disclosure of scientific details was normal procedure for a chemist, but not for a chemical entrepreneur. Wollaston was both, and he behaved accordingly as circumstances dictated. The withholding of the tenacity data for his platinum wires illustrates how deftly he needed to function to be successful in both roles.

THE LOGARITHMIC SCALE OF CHEMICAL EQUIVALENTS

Both in the chemical purification of platina and the preparation of organic chemicals, Wollaston paid close attention to the weight relationships that established the combining proportions of reagents. Once those proportions became known, the calculations required to determine the optimum weights of reagents to be used for each reaction involved frequent multiplication and division of large numbers. Circumstantial evidence suggests that Wollaston used a calculating device known as a slide rule for most of his calculations. We know, for instance, that he designed and printed novel logarithmic scales for a variety of slide rules intended for mercantile applications before producing his most famous one in 1813, the scale of chemical equivalents.

The slide rule was a simple hand-held device in general use in the early nineteenth century that allowed the multiplication and division of numbers to be done rapidly and accurately. It was based on the fact that the operations of multiplication and division are reduced to the simpler ones of addition and subtraction if the numbers in question are first converted to their logarithmic equivalents. If one set of logarithms are aligned on a scale, then multiplication (or division) can be carried out by the movement of a secondary logarithmic scale, which was designed to move along the side of the first. Calculations made with a slide rule of normal dimensions were limited to operations involving numbers of three significant figures or less, a limitation that was of little consequence in chemical applications, where three-figure accuracy was sufficient for most purposes.

Although the slide rule was first designed for basic arithmetical calculations, it could be adapted for a wide variety of practical applications that required conversion of one value to another by multiplication or division, and this is what Wollaston did for several different purposes. There is evidence in his notebooks and elsewhere that, as early as 1802, he had designed a specialized slide rule for the use of his younger brother Henry, who had hoped to establish himself as a merchant banker in Amsterdam.[32] That slide rule was inscribed with the exchange rates for a number of European currencies, and one could use it to quickly convert the currency amount of one country to that of a number of others. One notebook entry mentions delivery to Wollaston in December 1805 of 100 paper scales (for gluing onto a wooden slide rule) suitable for use in Hamburg, and another dated March 1806 mentions a new copper printing plate and 50 printed scales for use in Amsterdam.[33] The currency exchange sliders are another of Wollaston's little known inventions, and they demonstrate more of his practical inventiveness and broad range of interests. There are several engraved copper plates and many of the printed scales themselves in the collections of the Science Museum of London, having been placed there by some of Wollaston's descendants, but they have been little studied and largely forgotten.

Even though his currency exchange sliders were highly valued by their users, Wollaston did not wish to be associated with them by name until he judged the time to be right. Apparently, he did not even want the brother for whom they were first designed to acknowledge his inventiveness. In a remembrance written many years later for the benefit of his children, Henry recalled that "My brother William is so very reserved in his communications that to this day [sometime after 1813] I scarcely know whether or not I am at full liberty to shew the ruler lest he should take me to task for developing his ingenuity."[34] This baffling aspect of Wollaston's character, so "reserved in his communications" as Henry vaguely described it, was an innate trait that confounded many of Wollaston's contemporaries. Why, one wonders, would he react so negatively to being associated with ideas and inventions that, however incompletely developed, would do little other than earn him the admiration of his peers? Even after studying Wollaston and his works for many years, I have no compelling answer to this question, although I believe one of his biographers (oddly enough one who never met him) provided an explanation that sounds right.

> It was untruth that Wollaston so greatly dreaded; and the fear of it made him prone to under-estimate the positive worth of any fact. An inquiry thus became for him a very tedious and protracted affair. It was not sufficient that a fact, perhaps quite incidental to the main object, and what other men would have called trivial, was true enough for the use he had to make of it. It must be true enough for every purpose it could be applied to: in a word, positively and absolutely true.[35]

Although Wilson overstated the issue, for Wollaston certainly knew that absolute truth in science was an elusive goal, he was certainly right to conclude that Wollaston feared being wrong. Consequently, he was more cautious in making inductive generalizations than almost every one of his contemporaries. And nowhere does Wollaston's theoretical circumspection appear more strikingly than in the opinions he expressed in the chemical equivalents paper I will now discuss.

In early 1814, Wollaston published a paper containing a comprehensive list of the combining weights of common chemical elements and compounds, together with an arrangement of them on a logarithmic scale suitable for gluing on a standard slide rule such that the combining weights requisite for their interconversion could be easily calculated.[36] He opened the paper by listing the challenges faced by an analyst who wished to determine the constituents, by weight, of a compound such as crystalline sulfate of copper ($CuSO_4 \bullet 5H_2O$). The first point of analysis was to determine how much sulfuric acid, copper oxide, and water are in the salt; the second might be to ascertain the amounts of the constituent elements copper, sulfur, oxygen, and hydrogen. To carry out such an analysis, a chemist would need to employ a number of quantitative tests for each component under investigation, each of which required specific amounts of other reagents. Wollaston estimated a total of twenty individual chemical reagents might be involved in the complete analysis of sulfate of copper.[37] To provide a comprehensive, or "synoptic," overview of the weight relationship of all these reagents, and a means to facilitate calculation of ideal combining weights, Wollaston designed his scale of chemical equivalents. He described it in these words:

> The scale . . . is designed to answer at one view all these questions, with reference to most of the salts contained in the table, not merely

> expressing numerically the proportions by which the desired answers may be calculated, but directly indicating the actual weights of the several ingredients, contained in any assumed weight of the salt under consideration, and also the actual quantities of several reagents that may be used, and of the precipitates that would be obtained by each.[38]

Wollaston, we know, was familiar with the traditional use of the term "equivalent weight" to denote the weights of two bases that would react completely with a specific weight of acid. Long before 1814, he had understood Richter's law of neutrality, which stated that neutral salts always contained the same equivalent proportions of acid and base (later to become known as the law of reciprocal proportions), and he credited Richter for "having first observed that law of permanent proportions on which the possibility of this numerical representation is founded."[39] But Wollaston broadened the concept of chemical equivalents to include combining proportions, the relative amounts in which all components of a compound substance, not just acids and bases, united with each other. After scouring the literature for the most reliable and mutually consistent analyses of a large number of chemical elements and compounds, and adding a few observations of his own, Wollaston compiled a list of about 100 "equivalent weights" with specific literature references for each value.[40] Instead of setting the lightest element hydrogen = 1 as the basis of his scale (as most of his contemporaries did for their lists of atomic weights), Wollaston selected oxygen = 10 as the reference weight, because that element was present in more of the compounds under study than any other.

The decision to make oxygen the reference point for tables of combining weights had also been made independently by Berzelius, who had himself discovered many examples of multiple proportions in the oxides of metals. The two chemists spent much time together in the summer of 1812 when the Swede visited England. Berzelius had first learned of Dalton's atomic theory by reading Wollaston's 1808 paper on multiple proportions and had become a staunch proponent of the theory by the time he traveled to England. After meeting with Berzelius on the morning of August 12, Wollaston (who was set to depart the city a few days later on a holiday to the south of England) mailed Berzelius an advance copy of his list of equivalent weights, with one column based on H = 0.5 and another based on O = 10.[41] Sometime after receiving Wollaston's list, Berzelius wrote to Davy,

> I have spoken much with Dr. Wollaston on definite proportions and on the atomic way of considering the composition of bodies. He agrees

with me that oxygen must be taken as unity, as much for its chemical properties as for the convenience of calculation.[42]

It is interesting that Berzelius in this letter refers to composition in "atomic" terms, a designation Wollaston took care to avoid in his paper. Also, Berzelius implies that he was the first to consider setting oxygen as the reference value (he made it "unity") and he shortly thereafter published a footnote in a paper co-authored with Marcet stating that "the best way to form a system of definite proportions, and to make it harmonize with the general views of chemistry, would be to take oxygen as the base of the scale."[43] On reviewing the paper before publication, Wollaston informed Marcet that he regretted that Berzelius had published the idea before him, since he had decided on the same role for oxygen long before he had spoken of it with Berzelius.[44] Fortunately, through the intermediacy of Marcet, the two chemists recognized that each had come to the same decision independently and did not let the priority issue affect their publication plans or their long-term friendship.

Having settled on oxygen = 10 as the reference value for his scale, Wollaston then placed the equivalent weight of every other substance on one side or another of a central, logarithmic sliding scale. The full sliding scale included equivalent weights ranging from 10 to 300; the top portion of the scale is reproduced in Figure 10.2. That figure shows the central slider (the only moving part of the device) placed at its "home" position, with the numerical values on the central slider aligned with the corresponding equivalents engraved on each side. Thus we can see at a glance, for example, that oxygen = 10.0, water = 11.3 (since hydrogen = 1.3, not shown on the scale) and sulfur = 20.0. Several of Wollaston's equivalent weights are only half of the modern values for the elements because he, in harmony with most of the contemporaries whose analyses he cited, assumed that water was composed of one equivalent each of hydrogen and oxygen.[45] With the central slider in its home position, Wollaston's scale was a very handy, concise table of the equivalent weights of the best characterized elements and compounds. But the practical value of the scale came from readjustment of the numerical slider to reveal experimental combining weights.

In a second diagram in the paper, a portion of which is reproduced as Figure 10.3, Wollaston demonstrated how a user could at a glance determine the constituent weights of a substance such as muriate of soda (sodium chloride, NaCl). By moving the slider so that the number 100 aligns with muriate of soda on the right-hand listing, one can read from the scale that 100 parts of the salt contain "46.6 dry muriatic acid, and 53.4 of soda, or 39.8 sodium, and 13.6 oxygen; or if viewed as chloride of sodium, that

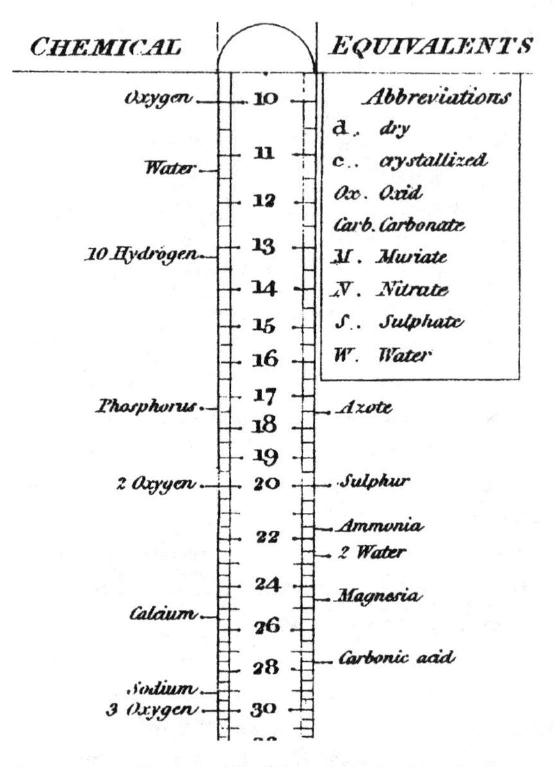

Fig. 10.2. Top Portion of Wollaston's Scale of Equivalents.

it contains 60.2 chlorine, and 39.8 sodium."[46] Furthermore, if the goal was to convert muriate of soda to sulfate of soda, one could read from the scale that 84.8 of oil of vitriol (s.g. 1.85) would be required for each 100 of salt, since that amount of vitriolic acid is equivalent to the muriatic acid in the original salt. This is the same result one could find by consulting any one of a number of published or personally-compiled lists of equivalent weights calculated on a "parts per hundred" basis. But by moving the central slider so that any real, experimental weight was aligned with muriate of soda, a chemist could just as easily read off the constituent weights for that specific amount of the salt. It was for such experimental applications that the chemical slide rule demonstrated its real worth—it was an inexpensive, portable table of the most common chemical equivalents, mated with a calculational function that relieved analysts from the repetitive (and error-prone) tasks of multiplication and division. Not surprisingly, in a few short years the scale of chemical equivalents became a commonplace accessory to chemical experimentation everywhere.

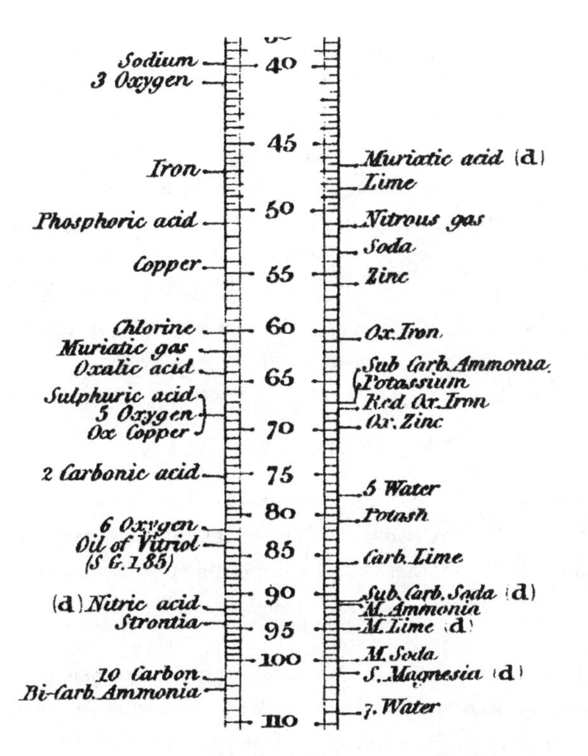

Fig. 10.3. Equivalent Scale for Muriate of Soda.

Many chemical textbooks and manuals began to include descriptions of, and instructions for the use of, either Wollaston's original sliding scale or improved versions of it. A typical comment is that of the Scottish chemist, D. B. Reid, who wrote in 1825

> I have been thus particular in describing Dr. Woolaston's [*sic*] scale, because a student who knows how to use it, will derive more information from it, and with much more facility, than he can acquire in the same time from any other source. Moreover, another great beauty of the scale is, that the student not only gets acquainted with individual facts, but likewise becomes familiar with the most difficult and important laws of chemical action; for the scale is founded on them, and every example on it affords an illustration of one or other of these.[47]

Wollaston published the scale as a practical aid to chemical investigation and encouraged its development by others. Consequently more complex versions soon appeared. For example, a chemist based in Leeds, William

West, designed a scale containing equivalents for more than 200 substances, and others appeared that contained as many as 500. In addition, larger scales of a size that permitted reading some values to four figures, or that used two sliders, or opened on hinges, were made in different places at different times.[48] It was not until the latter half of the nineteenth century, as the number of well-characterized substances grew too large for convenient inclusion on a scale of compact size, and the need for quantitative precision beyond three figures increased, that use of the chemical scale gradually declined to the point that it became primarily of historical interest.

ATOMS OR EQUIVALENTS

To make the sliding rule of greatest practical use to the working chemist Wollaston extended the concept of chemical equivalents to the combining weights of chemical elements and compound substances. Of this decision, Wollaston simply said

> In the formation of this scale, it is requisite in the first place to determine the proportions in which the different known chemical bodies unite with each other, and to express these proportions in such terms that the same substance shall always be represented by the same number.[49]

This is a curious statement for a person who had a few years earlier discovered multiple proportions in a number of salts and consequently knew full well that many substances combined in more than one proportion. In such cases it was not possible to assign a single equivalent weight to the substances involved. Wollaston circumvented the issue by interpreting multiple proportions as successive instances of fixed proportions, stating that

> when a base unites with a larger portion of acid than is sufficient to saturate it, the quantity combined is then an exact multiple of the former, thus exhibiting a new modification of the law of definite proportions, rather than any exception to it.[50]

To illustrate the assumptions Wollaston had to make to construct his list of chemical equivalents, I will summarize his treatment of the two known oxides of carbon, remembering that he had set oxygen = 10 as the numerical reference point. He explained

> The first question . . . to be resolved is, by what number are we to express the relative weight of carbonic acid [carbon dioxide, CO_2], if oxygen be fixed at 10. It seems to be very well ascertained, that a given quantity of oxygen yields exactly an equal measure [volume] of carbonic acid by union with carbon; and since the specific gravities of these gases are as 10 to 13.77, or as 20 to 27.54, the weight of carbon may be justly represented by 7.54, which, in this instance, is combined with 2 of oxygen forming the deutoxide, and carbonic oxide [carbon monoxide, CO] being the protoxide will be duly represented by 17.54.[51]

Wollaston was doing nothing more here than repeating the consensus view that the lowest oxide of carbon (dubbed the protoxide in Thomson's nomenclature) consisted of one equivalent each of carbon and oxygen, and the next-higher oxide (the deutoxide) had one equivalent of carbon combined with two of oxygen. Wollaston made no mention of other possibilities, such as carbonic oxide being 2 carbon + 1 oxygen and carbonic acid being 1 carbon + 1 oxygen, an alternative that contained multiples of carbon united with one of oxygen. The alternative would give different equivalent weights for one element or the other, but would fit less well with measured gas densities. John Dalton, who had confronted a similar dilemma when using combining weights to compile his first table of relative atomic weights in 1803, resolved the issue by assuming the simplest, and most stable, atomic configuration for the lowest oxide to be CO, relegating the higher oxide to the next-most-stable grouping, CO_2. Although Wollaston was thoroughly conversant with Dalton's reasoning, and had accepted it as the best explanation for his previously published instances of multiple proportions, he was at this later date no longer willing to identify his list of combining weights as atomic weights. In a paragraph that struck a chord with many of his contemporaries, but misrepresented the ontological status of his own equivalents, Wollaston wrote:

> According to this [atomic] view, when we estimate the relative weights of equivalents, Mr. DALTON conceives that we are estimating the aggregate weights of a given number of atoms, and consequently the proportion which the ultimate single atoms bear to each other. But since it is impossible in several instances, where only two combinations of the same ingredients are known, to discover which of the compounds is to be regarded as consisting of a pair of single atoms, and since the decision of these questions is purely theoretical, and by no means necessary to

> the formation of a table adapted to most practical purposes, I have not been desirous of warping my numbers according to an atomic theory, but have endeavoured to make practical convenience my sole guide, and have considered the doctrine of simple multiples, on which that of atoms is founded, merely as a valuable assistant in determining, by simple division, the amount of those quantities that are liable to such definite deviations from the original law of RICHTER.[52]

It is certainly true that Wollaston constructed his scale of equivalents for practical use, but "practical convenience" was not, nor could it have been, his sole guide. He had to make the very same assumptions as Dalton about the constitution of binary compounds.[53] One constituent had to be arbitrarily chosen (in a way consistent with a whole network of analyses) as the unit component of a compound, and there was no way to do so in a manner that could be fully validated by experiment. So Wollaston's equivalent weights had no more fundamental connection to analytical results than did Dalton's atomic weights: the two terms were essentially synonymous in all ways but one. Combining proportions interpreted as atomic weights invited a belief in fundamental chemical atoms, whereas their interpretation as equivalent weights allowed one to be agnostic about the nature of the weight-bearing matter. As we have seen in Wollaston's earlier musings about a particulate explanation of crystal structure, he preferred to be seen publically as an atomic agnostic. He could well have said as much in his paper, but instead he stated, in a way that invited scrutiny, that he was not desirous of "warping" his numbers to fit atomic theory. What Wollaston meant by such choice of words is difficult to discern with confidence: there is nothing in his calculation of equivalent weights that would have required change if they were to be presented instead as atomic weights. I think we must assume he simply did not want his numbers to be discredited by those who opposed, for one reason or another, the interpretation of chemical phenomena in atomic terms.

Wollaston's reinterpretation of combining weights in the seemingly neutral language of equivalents was seen as a significant advance by those who had reservations about the reality of chemical atoms and/or the utility of atomistic explanations of chemical phenomena. Consequently, his work became the starting point for a flow of studies on "equivalent weights" that became such a resilient, and anti-atomistic, theme of much nineteenth century chemistry. In contrast, several others ignored or misinterpreted Wollaston's cautionary words and took his equivalent weights to be synonymous with atomic weights, and therefore good evidence for the fecundity of

Dalton's atomic theory. Whatever their views on the philosophical ramifications of the scale of chemical equivalents, however, chemists in general gave unanimous praise to the practical benefits of the sliding scale.

Some chemists even used the sliding scale to generate molecular formulas for compounds based on the weights of its constituent elements listed on the scale. Thomas Thomson frequently did so, as did William Prout, who exploited Wollaston's values to derive constitutional formulas for several organic compounds. His debt to Wollaston and the scale of equivalents was made clear in the following passage:

> On the supposition that this instrument be correct, or nearly so, which no one can doubt, and that organic substances be really formed on the principles of definite proportions, we are enabled by its means to approximate in most instances, with almost absolute certainty, to the number of atoms of each element entering into the composition of a ternary or quaternary compound.[54]

Both Thomson and Prout interacted frequently with Wollaston, both at the Royal Society and elsewhere, and their facile use of his equivalent weights for atomic purposes suggests that they, like Berzelius, did not judge their interpretations to be contrary to his opinion. We know that Wollaston, on several occasions, spoke favorably of atomic theory in conversation and may well have expressed himself in words with less caution than he was prepared to do in print. It certainly appears to be the case that several of his contemporaries did not believe Wollaston's numbers would have had to be "warped" to make them conform to atomic theory, and there is no evidence that Wollaston objected to such atomic applications of his equivalent weights.

Wollaston's scale also found favor with many skeptics of atomic theory, such as Humphry Davy, who believed that the majority of the known chemical elements would one day be reduced by novel methods to a much smaller number of fundamental substances. Davy had himself found evidence of integral multiple proportions in his analysis of the oxides of nitrogen and had published what he called the 'proportional weights' of thirty-seven elements in his *Elements of Chemical Philosophy* of 1812. Even as late as 1826 when, as president of the Royal Society, he presented the first Royal Medal of the Society to John Dalton for "the Development of the Chemical Theory of Definite Proportions, usually called the Atomic Theory," Davy could not bring himself to accept without qualification the central claim of Dalton's atomic theory. In the award address, he made sure to emphasize the contributions of the author of the scale of equivalents.

> [Dalton] first laid down clearly and numerically, the doctrine of multiples, and endeavoured to express by simple numbers, the weights of the bodies believed to be elementary. His first views, from their boldness and peculiarity, met with but little attention; but they were discussed and supported by Drs. Thomson and Wollaston; and the table of chemical equivalents of this last gentleman separates the practical part of the doctrine from the atomic or hypothetical part, and is worthy of the profound views and philosophical acumen and accuracy of the celebrated author.[55]

Davy's comments must have lit a fire under one of the Fellows, for a short time later the following bombast appeared in the journal of the Royal Institution.

> I next ask, whether Mr. Dalton has materially contributed to its [atomic theory's] development and extended application? Whether we owe to his suggestions those "Tables of Equivalents," which are so useful in the laboratory, and so important to the manufacturer of chemical products?—No: all this is due to Dr. Wollaston, whose *logometric scale of chemical equivalents* brought the theory into practice, and rendered that, which was a mere abstract subject of chemical inquiry, little understood and less investigated—an instrument of the utmost usefulness and value. He did that for the theory of definite proportions which Mr. Watt effected for the steam-engine.[56]

Even after the hyperbole is muted, the core message remains: several of his contemporaries believed Wollaston's scale made the premises of atomic theory useful to chemical practitioners of all kinds, and his argument for fixed combining weights as the practical foundation of chemical reactivity had a profound influence.

Despite the caution expressed in his published papers, there is good reason to believe that Wollaston privately accepted atomic theory as the best available explanation for the fixed, multiple, and reciprocal proportions revealed by chemical analysis. It also appears he believed atoms (possibly of different shapes) could account for some of the properties exhibited by crystalline minerals. Nonetheless, he found it difficult to accept unequivocally the reality of chemical atoms. This was a metaphysical commitment that confronted all chemists in the early nineteenth century. Daltonian atomism had the potential to explain many (but not all) chemical observations, and it had no serious explanatory rival, but there appeared to be no way to prove that atoms of the elements were the ultimate units of chemical phenomena. In the absence of such proof, it was possible for many to remain opposed, or at least agnostic, to the theory. However, a few years later Wollaston proposed a solution to the problem by reporting a study that he believed yielded evidence for the reality of indivisible atoms.

THE UPPER LIMIT OF THE ATMOSPHERE

Building upon the observation of a friend that the moon had no measurable atmosphere, Wollaston reasoned that a gaseous atmosphere consisting of mutually repulsive and indivisible particles would have an upper limit where the inter-particle repulsive forces in a rarified atmosphere would be exactly balanced by the attractive force of gravity. Alternatively, if matter was infinitely divisible, the attractive force exerted on it by gravity would become infinitely small at some great height in the atmosphere, and the matter would diffuse throughout space and collect itself in atmospheres around other large heavenly bodies. Building on this premise Wollaston concluded that, if massive bodies of the solar system had no atmosphere, gaseous particles in the earth's atmosphere could not be infinitely divisible. Astronomy, Wollaston hoped, could provide a solution to a metaphysical problem that was thought to be beyond experimental reach. His resultant search for atmospheres surrounding the sun and Jupiter was described in a bold and highly speculative paper on the subject, published in 1822.[57]

The opportunity to test the premise presented itself in May 1821, when Venus passed behind the sun when viewed from earth. Wollaston calculated that, if the sun had an atmosphere, it would refract light from Venus during the periods just before and after the planet disappeared behind it, and the apparent irregularities in planetary motion could be observed telescopically. The required observations made on his behalf by Captain Henry Kater, together with those he collected himself, showed no evidence of a solar atmosphere. Thus encouraged, Wollaston sought further evidence for his hypothesis from the orbits of Jupiter's moons. He calculated that, if Jupiter had an atmosphere appropriate to its mass, light from its fourth moon would remain visible to a viewer on earth even when the moon was directly behind the planet. Observations showed that such did not occur, so Jupiter did not appear to have an atmosphere either. Based on these results Wollaston bravely concluded that "all the phenomena accord entirely with the supposition that the earth's atmosphere is of finite extent, limited by the weight of ultimate atoms of definite magnitude no longer divisible by repulsion of their parts."[58] He thus unambiguously declares his belief in the existence of indestructible chemical atoms. The atoms (as they were then assumed to be) of the atmosphere, principally oxygen and nitrogen, were not divisible into infinitely smaller portions. And what was so for those two elements must be so for all the others. He had come to believe in 1822 that Dalton's hypothesis had been proven correct: each chemical element was made up of its own unique and indestructible atom, characterized by a distinctive atomic, and combining, weight.

Wollaston's paper was widely reprinted in foreign journals, was much cited, and was generally praised for several years after its appearance, especially by those who already had an affection for atomic theory. Four years later, John Dalton used an argument similar to Wollaston's to estimate the extent of the earth's atmosphere according to the law of partial pressures.[59] Michael Faraday, despite his skepticism of corporeal atoms, observed that low temperatures set a limit to the evaporation of solid bodies in a vacuum, and drew analogies to Wollaston's "proof" of a similar limit to the earth's atmosphere by "a beautiful train of argument and observation."[60] However, objections to Wollaston's argument also began to surface. Thomas Graham, then a young aspiring chemist in Edinburgh, questioned whether gravity was the only factor to be considered. He calculated, in a line of reasoning consistent with Faraday's observation, that the decreasing temperature of the atmosphere would also limit its extent to a height of about twenty-seven miles.[61] The great polymath William Whewell viewed Wollaston's argument as untenable because it was not known how gas density varied with the compressing force of gravity at the far reaches of the atmosphere.[62] Later in the century, as it gradually became recognized that the known gases of the atmosphere were actually polyatomic molecules, the chemist George Wilson was able to claim that the best Wollaston's argument, which he categorized as "one of the most interesting physical essays on record," could do was to demonstrate that an atmosphere of molecules had an upper limit.[63] And that conclusion said nothing about the divisibility of those molecules, which were known to be composed of smaller units.

The impact of Wollaston's argument for the indivisibility of chemical atoms waned with the passage of time and the increasing sophistication of chemical knowledge. By the middle of the nineteenth century, the professor of chemistry at Oxford, Charles Daubeny, initially an adherent of Wollaston's atmospheric argument for atomism, was forced to reconsider the hypothesis. In the second edition of his treatise on atomic theory, Daubeny still summarized the argument, but for a different purpose.

> as the idea appears to have emanated from no less a man than Dr. Wollaston, it would be fitting here to introduce some mention of it, were it only as an illustration, of that marvellous acuteness of mental vision, . . . which peculiarly characterized that eminent philosopher in all departments of physical research.[64]

Wollaston's argument for the indivisibility of atoms had proven to be untenable, and evolved instead, somewhat ironically, to become an illustration of his "marvellous acuteness of mental vision."

CHAPTER 11

Changing Priorities
1809–1815

> In his manner he is simple and unaffected, in conversation both modest and instructive, but so sure of what he says positively that they have a proverb here among chemists that whoever argues with Wollaston is wrong.[1]

To his contemporaries who had little knowledge of the extent of his secret chemical business, the years from 1807 to 1814 marked the most scientifically productive period of Wollaston's career. Over that brief time frame, he published a total of eighteen papers, covering topics such as fairy rings, super- and sub-acid salts, and a particulate view of crystal structure. In addition, he introduced devices such as the camera lucida, the reflective goniometer, the cryophorus, and the scale of chemical equivalents, plus the technique of making fine platinum wires. There is a reason those years were so rich in discovery. Before 1807 Wollaston had little free time to do much other than bring his commercial enterprises to profitability and it was only after 1809 that he had the time to give freer rein to his scientific curiosity and expand his social life. Surprisingly, remnants of his medical interests resurfaced, such as the sounds emitted by contracting muscles. But that was not Wollaston's only contribution to scientific medicine at this time.

ELECTROCHEMICAL SECRETIONS AND BLOOD SUGAR

A few months prior to the Croonian lecture, he had published the results of his observation that electricity had the power to move chemicals through biological membranes.[2] His premise was that weak electrical forces might be responsible for the movement of substances through the cell membranes of animals, a process he called secretion. Unwilling to test his hypothesis on living creatures, he did his experiments on a simple model system. He placed a dilute solution of salt into a glass tube that had one end covered with a membrane of cleaned animal bladder. The tube was then placed upright with the moistened bladder end on a silver coin. A zinc wire was immersed into the solution to a depth of about one inch and bent around outside the tube to come into contact with the silver coin. The zinc-silver interface acted as a small galvanic battery similar to the small cell Wollaston had used to decompose water in 1801 and, after a few minutes of action, he was able to detect the deposition of soda on the external surface of the bladder. To use modern terminology, Wollaston had shown in a very simple way that an electric potential could act to move ions through a membrane that was impermeable to those ions in the absence of an applied voltage. Of course, we now know that a host of physiological processes, including nerve signal transmission, involves similar voltage-regulated ion transport through cell walls.

Wollaston's experiment attracted much attention on the continent, where physiological chemistry (like most other chemical disciplines) was under more intensive investigation. In France, for example, de Morveau repeated the experiment with cork in place of the bladder to show that animal matter itself was not the source of the effect.[3] In time, however, Wollaston's demonstration of chemical transport across biological membranes faded into obscurity, and it is little recognized today as a precursor of a fundamental biochemical process.

The animal secretion paper was followed a year later by Wollaston's description of cystic oxide, a new type of urinary calculus (discussed in Chapter 2), and shortly after, in 1811, by a fourth paper on a physiological topic—the sugar content of the blood of diabetics. This cluster of medically-oriented papers in the years 1809–1811 marks his last publications on animal chemistry. Although he was never to be reckless in his published hypotheses, Wollaston was certainly less cautious in these physiological papers than he had been in earlier years. He was a different man in 1809 than in 1805, the first year of platinum sales: he was more intellectually confident, on sounder financial footing, and better connected to the London social and scientific scene. It was an appropriate time to become bolder and to publish less compelling results, and he certainly did so with his paper on blood sugar levels.[4]

It was known by the late eighteenth century that the urine of diabetics contained sugar, but how it got there was a puzzle. Some believed it was absorbed from the stomach into the blood stream and from there passed through the kidneys into urine. Others suggested that diseased kidneys themselves produced sugar, which they subsequently discharged into urine, a hypothesis that did not require transport of sugar through the blood. A few even believed that it was transported from stomach to bladder by some unknown physiological pathway. In 1797, John Rollo, a Scottish military surgeon, published a treatise on diabetes in which he suggested that blood did contain sugar, even though it could not be detected by taste.[5] In the second edition of the work, published in 1798, he reported William Cruickshank's method of detecting sugar in diabetic blood. Cruickshank treated dried blood with nitric acid to convert the sugar to oxalic acid, which he was able to detect at abnormally high levels in samples from persons with diabetes. Upon learning of this test, the London anatomist and physician Matthew Baillie, knowing of Wollaston's chemical skills, asked him to confirm the test results. In fact, Baillie and Wollaston did some of the early testing together in late 1798 or early 1799, as a letter from the latter to the former attests.[6] But Wollaston abandoned that research when he left medicine in 1800 and did not publish the results until asked to by Marcet, who wished to couple them with some of his own on a similar topic.

Wollaston did not believe Cruickshank's method could give reliable results because he knew that some components of blood other than sugar could give oxalic acid by treatment with nitric acid.[7] Consequently he developed what he believed to be a more reliable test. To samples of healthy blood serum provided by Baillie, Wollaston added some diluted muriatic acid to coagulate any albumin that remained in it, a process that was brought to completion by heating to the boiling point of water for four minutes. After removal of the coagulated material the remaining aqueous solution was slowly evaporated until salts began to crystallize from it. Wollaston easily identified the principal crystalline product as common salt. But he observed that the final crystallization step did not occur as cleanly in samples to which either common sugar or urinary sugar had previously been added.[8] Although an imperfect crystallization of salts did occur when very small amounts of sugar had been added to the blood serum, the co-precipitating sugar could be detected by reaction with added nitric acid, which resulted in a foaming and blackening of the precipitate. So Wollaston concluded that a blood or serum sample that contained diabetic sugar would not, by his process, give a clean precipitate of salt. Those that contained very low levels would give a precipitate containing sugar that could be detected by reaction with nitric acid. He applied his test to dried diabetic blood and diabetic blood serum, both supplied by Baillie. Wollaston was unable to detect sugar in any of the blood samples, although he did obtain a positive test if he added some urinary sugar to the samples before the analysis. He was therefore led to conclude that sugar in the urine of diabetics did not get there by transport through the blood.

Wollaston's test for blood sugar, we now know, did not give correct results and his conclusion that ingested substances could pass to the bladder by a route other than through the bloodstream was erroneous. But he was not the only one misled by the challenges of blood chemistry. Others in the early nineteenth century failed to find sugar in diabetic blood, and it was not until mid century that the French physiologist Claude Bernard discovered that sugar could only be detected in fresh blood serum, or serum stored at cold temperatures, since the small amounts present in blood are quickly degraded by red blood cells.[9] Animal chemistry, its practitioners would discover in later decades, presented much greater experimental difficulties than the inorganic, aqueous chemistry that Wollaston understood so well.

THE ATTRACTIONS OF GRAVITY, FRANCE, AND ENGLISH INSTITUTIONS

Not all of Wollaston's experimental investigations worked out well, and his lengthy study of the effects of gravity on different substances, carried out in 1809 and recorded in four notebooks, was perhaps the biggest disappointment of all.[10] His interest in the topic was probably aroused by a letter from the astronomer John Pond, who informed him in early 1808 that the observed position of the moon was very slightly different from what would be expected from rigorous application of the law of gravitation.[11] Wollaston must have placed great faith in the accuracy of Pond's observations because he soon began a series of experiments with pendulums to seek experimental support for an unconventional hypothesis.

In the early years of the nineteenth century, it was generally accepted, in accord with Newton's gravitational law, that equal-length pendulums with bobs made from different materials of the same weight had the same periodicity of swing. Thus, to discover if Pond's observations about the moon's position in the sky could be explained by varying gravitational effects upon it by different substances in the earth, Wollaston had to show that it was possible to detect very small differences in the period of a pendulum caused by bobs made from different materials. So, in May 1809, he embarked on an extended series of experiments to see if such was the case.

Although the notebook records of his pendulum experiments do not provide enough information to decipher exactly how he constructed his pendulums or conducted the experiments, there is enough data to understand roughly what he did. He obtained cylindrical bobs of equal diameter and weight, varying only in their thicknesses, from fine craftsmen such as Cary and Troughton, from a variety of materials, such as nitre, sulfur, alabaster, zinc, lead, and tin.[12] He then set pendulums with bobs of two different materials into simultaneous motion, and repeatedly measured the period of their oscillations after several time intervals of a few hours each. His first sequence of experiments led him to suspect that "there does exist matter which does not follow the common law of gravitation. The difference between zinc and nitre seems about 1/8,700 part, [between] zinc and alabaster about 1/21,000."[13] Such very small differences could be attributed to experimental error, so Wollaston pressed on with further measurements. After hundreds of experiments that lasted until December 1809, Wollaston found that he could not discern a consistent variation in the periods of pendulums that could be attributed to the substance of the bobs. He dejectedly entered the marginal comment in the third notebook of results "Fallacy of N [the hypothesis] fully proved & the subject abandoned."[14] Newton had been right after all—the force of gravity depends only on the mass of a material, not its chemical type. The fruitless pendulum experiments do show, once again, Wollaston's aptitude for exhaustive experimentation and careful observation. The failure to discover something new is not always a failure of the experimenter or his methods; sometimes there is just nothing new to discover.

Many of Wollaston's publications were reprinted in the *Annales de chimie*, the premier chemical journal in France, and the French science he encountered as a secretary of the Royal Society had exposed him to the works of the several of the country's leading practitioners. He was greatly impressed by their scientific acumen and, probably encouraged by some of those who had been there, hoped to have the opportunity of visiting Paris himself one day. The Napoleonic wars prevented most travel between France and Britain, but passports could be obtained by a select few when the timing was right. So, in 1810 the three mutual friends Wollaston, Tennant, and Edward Howard sought to obtain passports, the first two for scientific, the third for health, reasons. In February of that year, the itinerant Geneva-born geologist and physician Jean-François Berger offered to carry their passport requests with him on his upcoming trip from London to Geneva via Paris.[15] Once in Paris, Berger wrote several letters to Marcet informing him of the promising progress of the negotiations with French authorities. In one letter, he stated that

> Messrs Laplace, Berthollet, Malus, etc., appear especially desirous to make the acquaintance of Dr. Wollaston: they consider him to be the

> first natural philosopher [*physicien*] of England, and have just made him Corresponding Member of the Institute.[16]

Marcet quickly conveyed the good news and complimentary remarks to Wollaston, who was initially taken aback by the information, but shortly thereafter replied by letter.

> I wished to have seen you on the subject of the very agreeable intelligence conveyed by Berger previous to your writing to him in order that my silence on Tuesday night not be represented as utter insensibility to the greatest honour that ever was conferred upon me. —Indeed the expressions of wishes to cultivate one's acquaintance are also highly flattering; but our good friend Berger would admit that a man under these circumstances may feel a certain apprehension sometimes called mauvaise honte [bashfulness] at the thought of encountering the civilities of such men in whose estimation one could not but fall 50 per cent at first interview.[17]

The reply shows both the tremendous admiration Wollaston had for his French counterparts and the excessively modest opinion he had of his own eminence. Unfortunately, however, Berger's report was not fully accurate. The information on Wollaston's reputation in Paris was correct, but that on his election to the Institut de France was not—he was not to receive that "greatest honour" until a few years later, in 1816. But the political situation in 1810 ultimately turned out to be unfavorable for passport requests by British nationals and, despite support from Berthollet and Laplace, the applications were not approved.[18] The next opportunity for travel to France would only arise after the first Peace of Paris in 1814, and Wollaston was then to take advantage of it.

Although Wollaston did not get the chance to travel to France in 1810, he was courted by London's Royal Institution in the same year. After the reorganization of the Institution to make it more financially viable and publically accountable for its scientific initiatives, Wollaston was proposed as a member on August 6 and elected on November 5.[19] In December he paid a one-time membership fee in lieu of annual payments,[20] but he never became actively involved in the activities of the Institution. He declined an invitation to lecture there in 1812,[21] and at some time turned down an opportunity to serve as one of the Institution's managers, saying he would "certainly not attend any of the Meetings of the Managers, & be wholly inefficient in that capacity."[22] He had obviously become sufficiently independent that he was unwilling to sacrifice time and effort in support of the Institution's objectives.

As mentioned in Chapter 6, Wollaston had explored the possibility of succeeding his brother Francis as the Jacksonian professor of natural philosophy at Cambridge in 1807, but held back when he learned that he had little chance of winning the position in competition with William Farish, who was professor of chemistry at the time. However, a second opportunity arose in the spring of 1813, when Francis decided to vacate the natural philosophy chair to become rector of Cold Norton in Essex. Once again, the leading local candidate for the Jacksonian vacancy was the well-qualified Farish. By this time in his life, however, Wollaston had relinquished any interest he once had in gaining a Cambridge professorship. Even though he was encouraged to enter the contest by his mineralogical friend, E. D. Clarke, Wollaston quickly poured cold water on the idea in reply, saying that "I have long since decided not to think of it & I will not."[23] As Wollaston anticipated, Farish did become the new Jacksonian professor, opening up a spot for someone to succeed him as the university's eighth professor of chemistry. Interestingly, there was significant support, from both his Cambridge and London friends, for Smithson Tennant to gain the position.[24] The members of the Chemical Club, including Wollaston, banded together to urge Clarke to campaign on Tennant's behalf.[25] Wollaston's approval of Tennant's candidacy confirms that he had no objection to his partner becoming even more distant, intellectually and geographically, from their chemical business. It seemed to Tennant and his friends that the demands of mounting an annual series of lectures with experimental demonstrations might serve to augment his diligence.[26] Tennant won election in May 1813, and thereafter turned his full attention to preparing the course of lectures he was to deliver for the first, and only, time in 1814.

Surprisingly, Tennant's changed circumstances were not accompanied by any alterations to the 1809 financial agreement with Wollaston. Ironically, income from the joint venture probably gave Tennant the resources needed to procure the instruments, chemicals, and lab ware for his Cambridge lectures. The appointment to the chemistry professorship made little difference to the London partnership with Wollaston. Tennant's contributions were just reduced from negligible to nil.

THE VISIT OF BERZELIUS

Jöns Jacob Berzelius arrived in England in June 1812 as an ambitious man, thirty-one years old, with a growing reputation as an experimental and theoretical chemist. He was a prolific letter writer, and accounts of his travels and those he met are a source of much useful information. He met Wollaston at several events in and around London, including meetings of the Chemical Club, and spent a few days in mid July with him in Buckingham Street, where he was undoubtedly shielded from the platinum works in the rear of the house and the organic chemicals production in the building out back. He was, however, invited to observe, and be impressed by, Wollaston's microscale techniques for the detection of nickel in a speck of meteoric iron. He described those analytical methods in a letter to Johan Gahn, himself an accomplished analyst.

> The whole of his apparatus for these experiments is some bottles with stoppers, drawn out to a point so that they reach down into the liquid and collect one drop, which is the quantity he needs to extract. In these he keeps the commonest acids, alkalis and a few reagents. The solutions are made on a narrow glass strip, and he uses a small lamp for his blow-pipe and evaporation experiments on the piece of glass. Everything stands on a small wooden board with a handle, and is taken out or put away all together.[27]

Wollaston never published any comprehensive description of his micro methods of qualitative analysis, but the reliable description by Berzelius illustrates why Wollaston was held in such high esteem as an analyst, and why his laboratory "on a small wooden board" became iconic of his style, even though he was equally, but secretly, competent at the bucket-sized scale of his chemical business.

Berzelius was as much impressed with the man as he was with his experimental skills. In his letter to Gahn, he continued,

> Wollaston is about 45 years old and in face and build so like the statue of Newton at Cambridge that one can almost take that to be him. In his manner he is simple and unaffected, in conversation both modest and instructive, but so sure of what he says positively that they have a proverb here among chemists that whoever argues with Wollaston is wrong. But even when he is talking positively everything comes out so gently that Wollaston has no jealous rival and is still looked upon without exception by London chemists as their chief. . . . In his way of handling scientific matters and his zeal to investigate everything, which moreover he does with the greatest skill, he resembles you in a very striking way.[28]

Additional assessments were expressed in a letter to Berthollet:

> My stay here [in London] has been most interesting and instructive. . . . But what I value most of all is the personal acquaintance of the admirable Wollaston and the brilliant Davy. I am sure that among the chemists who are at present in the prime of life there is none that can be compared with Wollaston in mental depth and accuracy as well as in resourcefulness, and all this is combined in him with gentle manners and true modesty. I have profited more by an hour's conversation with him than frequently by the reading of large printed volumes. . . . Simplicity, clarity, and the greatest appearance of truth are always the accompaniments of his reasoning.[29]

Berzelius's comments are valuable as independent and objective assessments of Wollaston's science, character, and standing among his peers after he had established himself in London. But they also show that Wollaston was not perceived as an imposing, intellectually-intimidating person by those who approached him with an open mind and without pretension. Clearly, Berzelius and Wollaston established a good relationship in the summer of 1812, and they would continue to remain in contact for many years after.

A MAN AT THE PEAK OF HIS POWERS

In the paper describing his method of producing very fine platinum wires, Wollaston could do no more than estimate the diameter of the wires by assuming that the diameter of the wires was reduced proportionally as the length of the drawn wire increased. He then set his mind to inventing a device that could measure the diameters of the wires directly. Thus, he designed and constructed a small microscope-like instrument that brought a magnified image of a very narrow wire to a focus that could be viewed by eye simultaneously with a direct image of a calibrated scale that he had made from a parallel array of several very short wires of stepped lengths.[30] The scale itself was attached to the end of a sliding telescopic tube so that it could be brought closer to, or further from the eye, through a distance which allowed dimensions as small as 1/10,000 inch to be measured. Wollaston claimed his instrument could measure such narrow diameters accurate to within 2 percent of their true value, an accuracy far better than that obtainable with other micrometers of the time.

Although Wollaston's micrometer had limited utility, it was another of his inventions that added to his reputation as one who was consistently able to apply the fruits of experimental philosophy to practical applications. If we focus only on the seven publications that appeared in the two-year period around 1813, we can gain an appreciation of Wollaston's intellectual and inventive breadth as he approached 50 years of age: practical and speculative crystallography, atomic theory, discovery of combining proportions and design of a sliding scale of chemical equivalents, invention of three novel optical devices and one for the transference of heat, and a completely new method of fabricating fine platinum wires. The Wollaston of 1813 was a man with an enviable scientific reputation, in spite of his independent bent and idiosyncratic discomfort with public acclaim. He was seen by many, including Berzelius, as England's leading chemist. Not surprisingly, Berzelius proclaimed his English friend's merits to his Swedish colleagues, which led to Wollaston being elected as a foreign member to both Sweden's College of Medicine (May 1813) and its Royal Academy of Sciences (January 1814).[31] These were the first of many foreign societies to which Wollaston was to be elected in the following years.

Another example of Wollaston's standing in England's scientific community at this time is the leading role he played in the experiments done with the enormous voltaic battery constructed at John Children's estate at Tonbridge, Kent. There Children constructed a battery composed of 21 pairs of zinc and copper plates, each of which measured 6 feet long and 2.7 feet wide. All of the pairs of plates were connected in sequence and could be lowered by pulleys into a vast sulfuric acid bath of 945 gallons.[32] Children invited a large group of philosophical friends to the first trials of the battery in early July 1813. At those trials Wollaston was able to pass a large electrical current through some of the platinum wires he had recently fabricated, heating some to brightness and melting others.[33] After finding that the battery was not as powerful as he had hoped, Children acted on Wollaston's advice to redesign the plates such that each zinc plate was located between a pair of copper plates, effectively doubling the current emanating from each voltaic cell.[34] After alterations to the great battery had been made, a group of philosophers met once again at Tonbridge in March 1815, at which time Wollaston acted as the chief operator of the battery.[35] Several experiments were conducted to compare the effects of passing a large current through metallic wires of different diameters and, in the paper summarizing the results, Children described one of the experiments recommended by Wollaston.

> In an experiment in which equal lengths of two platina wires, of unequal diameter, (the larger being 1/30, the smaller 1/50 of an inch,) were placed together in the circuit *parallel* to each other, the thicker wire

> was ignited, because it conveyed more electricity without proportional increase of cooling surface. When connected continuously, the order of ignition was reversed. These two results were foreseen by Dr. WOLLASTON, who suggested the experiments.[36]

It is amusing to speculate that the same person who conducted the experiments with the largest battery ever made to that time in Britain probably had in his pocket the miniature thimble battery he had first designed in 1812, which even more dramatically demonstrated the ignition of a platinum wire. One can only imagine the delight with which the Tonbridge assembly of philosophers would have viewed the comparative effects of a thimble battery and one thousands of times larger, and the inner satisfaction the inventor of the smaller device must have felt. In fact, Children was prompted to compare the two batteries in his paper.

> It is known, I believe, to almost every member of this society [the Royal Society] that Dr. WOLLASTON has shown, with the delicate apparatus invented by him, that a platina wire, of the same dimensions as that just mentioned [1/5000 inch in diameter], is instantly ignited by a single pair of plates one inch square, on being immersed in a diluted acid. The ratio of the areas of the plates of the respective batteries [Wollaston's and Children's] is as 1 to 48384.[37]

Perhaps it was the impact Wollaston's thimble battery made at Tonbridge that finally prompted him to publish details of its design and function in 1815, three years after its invention.

THE RESURRECTION OF EUROPE

Wollaston lived during a period of great political and economic turmoil, and the funding of Britain's war effort against Napoleon had become an enormous drain on the country's finances. Wollaston himself did not suffer greatly from the consequences of war, and it is even possible that the conflict with France was beneficial for his platinum and organic chemicals business, mainly by isolating the markets for his products from continental competition. On the other hand, Wollaston's small chemical ventures, well protected by a wall of secrecy, could well have thrived even in a Britain at peace, then growing rapidly in both population and manufacturing innovation. But Wollaston, like everyone else, followed the progress of war carefully.

In 1808, after invading Portugal, Napoleon installed his brother Joseph as king of Spain, but the French armies sent there to expand the Continental blockade against Britain began to encounter fierce resistance throughout the country, which continued throughout the six years of the conflict known as the Peninsular War. In July of that year, Spanish action forced the surrender of an entire French army corps at Bailén, marking the first significant defeat of one of Napoleon's many armies. After learning of the outcome, Wollaston wrote to Hasted:

> Let us rather congratulate each other upon the present appearance of Spanish determination to resist Bony—the Great rascal has never before been engaged in a contest with a whole people & one cannot help hoping they may be successful although neither their government nor their religion for which they are so zealous are really worth contending for, & in fact the very first act of King Joseph in abolishing the tax upon produce is perhaps the very best thing that could be done for them.[38]

In this candid letter, Wollaston reveals that he was no revolutionary in political, religious, or mercantile matters. He was as opposed to traditional monarchies as to unfettered republicanism, and to Catholicism (but not Catholics), and to taxation upon the products of manufacture: in short, to many of the things that ran counter to British establishment thinking in the early nineteenth century. But the war with France did limit the possibility of travel to the continent, and prevented face-to-face meetings with the Frenchmen whose scientific work Wollaston so admired. So, when the opportunity to travel to Paris arose, Wollaston was quick to seize it.

After a string of decisive battles in early 1814, the combined forces of Russia, Prussia, Austria, and Great Britain moved triumphantly into Paris on March 31. Eleven days later Napoleon was forced to abdicate and remove himself to the Mediterranean island of Elba. Word of the promising events quickly reached England, and as early as March 3, Wollaston had entered the optimistic words "Resurrection of Europe" into his daybook.[39] Acting promptly to the changed circumstances, Wollaston, William Blake, and Alexander Marcet decided to make a trip to Paris within a few days of Napoleon's abdication, arriving there even before the return of King Louis XVIII from his exile in England.

They sailed from Dover to Calais on April 15, arrived in Paris three days later and remained there for fifteen days, returning to Dover on May 6.[40] Much information about this remarkable trip has been preserved in letters by the travelers, and one from Wollaston to Hasted shortly after his return to London is especially informative about the surprisingly pleasant journey to Paris through a scarred landscape populated by war-weary inhabitants.

> But now that I have taken up my pen . . . Paris you will say should be the order of the day, . . . [after obtaining passports] from Louis XVIII by his Sec'y La Chartre . . . I must say our confidence increased as we proceeded, for we soon found that being English was itself almost a passport. —Think what a change in a short 3 weeks! . . . We jogged merrily, very merrily on, I infinitely amused with the everything of novelty in the towns; all infinitely gratified with any discussion we could get at the posthouses, all perhaps over vain of the apparent effect of an English carriage full of Englishmen, when in fact the real cause of the commotion we witnessed was probably their expectation of the King, & their imagination that we might be precursors. The only doubt that beset us was respecting accommodations in a part of the country that would be occupied by a hundred thousand others.
>
> But it is really marvellous how little show such engines make. We met here 100, & there 100, here 50, 40 or 30, mostly Prussians & some few Cossacks & thus perhaps, I have named all we saw. We passed thru a part of the last scene of action before Paris, & saw lying perhaps 6 dead horses, saw wounded perhaps 15 or 20 trees, and one cut in two by a cannon shot, & no other vestige of the havock on that very ground not 3 weeks before when the fate of Europe was decided . . .[41]

Wollaston's letter ends without describing the group's time in Paris, but the daily activities of the three Englishmen while in the city were summarized by Marcet in a letter to Berzelius.

> We arose at 7 o'clock in the morning, hastily gulped down our breakfast, and left to examine without respite the scientific, political and dramatic sites, the artistic masterpieces, the scenes of the last battles, the Emperors, the Kings, and all the frivolities of this astonishing city. We almost never returned before midnight, and as we fell asleep we always asked ourselves if everything was only a dream.[42]

The three visitors met frequently with Berthollet, who impressed them with his candor, modesty, and paternalistic demeanor, as well as Vauquelin,

Thénard, Clément, Desormes, Collet-Descostils, Ampère, and Gay-Lussac, the last of whom Wollaston found especially interesting.[43] But Wollaston was disappointed that his 1812 work on the crystal angles of the mineral carbonates, which he had hoped would be highly appreciated by French crystallographers, was instead almost unknown to them.[44]

Although the trip to Paris was hastily-conceived, it turned out to be a rewarding one, and Wollaston appears to have thrived in the company of those he had previously admired from across the Channel. No longer did he feel that his scientific merits would be diminished upon conversation with the best French natural philosophers.

DEATHS OF A BUSINESS PARTNER AND A FATHER

In early August of 1814, just a few months after Wollaston's return from France, his eighty-three-year-old father suffered a stroke which left him temporarily paralyzed on one side. The son hurried to provide assistance at the family home in Chislehurst and quickly recognized the symptoms, which did not appear to be life-threatening. By September he was able to report to Hasted that his father was "really wonderfully well, recovering fast both hand & foot, clearing his faculties from the first, & tho' not yet clear in articulation, certainly mending considerably."[45] This was a promising turn of events but, about six months after his father's stroke, Wollaston had to deal with the loss of his long-time friend and business partner.

After becoming professor of chemistry at Cambridge in 1813, Tennant presented his first lecture course there in April and May of 1814, at the same time Wollaston and friends were exploring Paris. After the course was completed, Tennant set out himself in September for a lengthy tour of France, intending to acquaint himself with the latest advances in continental science in preparation for his 1815 series of lectures.[46] On February 22, 1815, he boarded a vessel at Boulogne for his return to England, but the lack of a favorable wind delayed his departure. Seizing the opportunity to invigorate himself with another of his customary horseback rides, he engaged a fellow traveler to ride with him along the seashore, where they attempted to explore an old fort by crossing the drawbridge to its entrance. Tragically, the unfastened bridge gave way and Tennant fell into the gulley below where he was crushed when his horse landed on top of him. He was taken to a nearby hospital but died there soon after admittance and was later buried in the Boulogne public cemetery.[47]

Wollaston first learned of the tragedy from Tennant's close friend, the lawyer John Whishaw, and shortly thereafter received confirmation from Marcet. To the latter, Wollaston replied on February 25

> I cannot but thank you for your intelligence the most unwelcome - Were it not for the decided tone of Whishaw's note, I should still retain a hope that report had made the event more serious than was really fact. It is too evident however that he has further particulars, & it would be absurd to entertain a doubt.[48]

And that was all. Other than a promise to relay the news to a mutual friend, Wollaston expresses no sense of great personal sorrow or loss. If this short note can be taken as a reliable measure of Wollaston's grief, it appears that the long-term friendship between the two men had ceased to be an emotionally strong one. Tennant's death had little impact on the chemical business other than to make Wollaston the sole recipient of profits just at the time when platinum sales were surging. But before he could consolidate the business in his own hands, Wollaston had to settle financial details with Tennant's estate, and he prepared a summary statement for George Pryme, the husband of one of Tennant's cousins, who superintended the dispersal of Tennant's assets.[49]

The summary confirms in monetary terms the great disparity in the contributions of the two partners. Total revenues from the beginning of the joint endeavor in December 1800 to its conclusion on the day of Tennant's death amounted to £14,351, against expenses of £8,334, for a total profit of about £6,000. Tennant's investment in the partnership amounted to little more than the purchase of raw starting materials, and had been mostly reimbursed before his death. Consequently, Wollaston calculated that a payment of £304.16.6 would bring the partners' shares into balance. Pryme must have been well informed about the workings of the business, for he countered that the estate should also receive payment for half of the accumulated platinum scraps that had been bought back from Cary, and the impure organic salts that had not been fully processed. Wollaston believed that such unprocessed materials should be his alone as the partnership survivor, but he struck a deal with Pryme to reimburse the estate for one-quarter of the value of the scraps, an amount calculated as £58.14.0.[50] Consequently, the settlement to Tennant's estate was completed on June 27, 1815, by payment to Pryme of £363.10.6.[51] Nonetheless, something about the settlement rankled Wollaston, for beside the notice of payment, he added the remark, "I paid too much forgetting the property tax which I should have to pay on joint account," a somewhat petulant remark given the small amount involved.

Becoming sole proprietor of the chemical business had many benefits for Wollaston. First, it allowed him to shut down for good the labor-intensive organic chemicals venture. This freed him and John from hours and hours of heavy labor, a welcome outcome for Wollaston as he neared his fiftieth birthday. The lifting and shifting of heavy carboys, even if shared with his assistant, must have demanded considerable physical exertion. Although the research notebooks make no mention of serious accidents or injuries, the work may well have contributed to the attacks of lumbago that began to flare up in late 1814. On December 24, 1814, for example, Wollaston described the debilitating effects of his lower back pain to Hasted.

> It would be a sin to sit still a whole day & write nothing . . . but if you ever felt the twinges of a thorough lumbago, the pangs that almost force a scream, with the almost ludicrous intervals of perfect ease, you may imagine the state I have been in for some days past, & conceive the effort sometimes necessary to fetch a dip of ink.[52]

While Wollaston convalesced, he remained at home, and the pain limited his attendance at meetings of the Royal Society. It was not until February 1815, only a week or two before he learned of Tennant's death, that he was able to return to his normal routine.

The second benefit arising from the end of the business partnership was a financial one. The greatest profits from platinum sales accrued after 1815 and, being his alone, made Wollaston into a quite wealthy man. His decision to abandon medicine for a career in the "business of chemistry" had ultimately turned out to be an overwhelming success. A third benefit was related to the other two. Without worrying about the need to enhance his scientific standing, his financial situation, or commitment to a business partner, Wollaston was able to return to many of the interests he had set aside to concentrate on business. He began to partake once again in the outdoor excursions he had enjoyed years earlier in Bury, and added partridge shooting and fishing to his pursuits. He also began to travel more extensively, taking advantage of enduring peace on the continent after the final defeat of Napoleon at the battle of Waterloo in 1815. But before settling into a more relaxed lifestyle, Wollaston had to fend off one last attempt to lure him to Cambridge.

Not surprisingly, Wollaston's friends in Cambridge encouraged him to seek election as Tennant's successor as professor of chemistry. After learning of the campaign on his behalf, Wollaston quickly wrote to his supporters to withdraw from the contest, claiming that he did not need the money, did not wish to lecture to students, and did not want to diminish the chances of any other candidate.[53] He never again sought out, or was proposed for, other positions at Cambridge, even though he continued to visit there frequently and retained his Caius College fellowship until his death.

The first several months of 1815 were tumultuous for Wollaston, but worse was to come in the closing months of that year. On October 31, seventeen years after the death of his wife, Francis Wollaston died, perhaps from a second stroke, although there is no confirmation of that in any of William's letters. The death of this good patriarch, Anglican priest and astronomer, who had provided moral and financial support to all of his and Althea's many children, left a huge void in the close-knit Wollaston family. After the customary period of mourning for himself and his household staff, Wollaston attempted to maintain his normal routine, including the near daily processing of his twelfth solution series of platinum purification,[54] together with regular attendance at meetings of the Royal Society.

Francis's will, finalized in July 1815, provided for all but one of his remaining family: sons, daughters and granddaughters alike.[55] The youngest son, Henry Septimus, who had struggled to establish a career in the financial world, received nothing other than unspecified amounts "already given" him in previous years. William was given a 430-acre farm near Wingfield in Suffolk. But it was not unnatural for many outsiders, overlooking the family's size and unaware of the fairness of the dispersal, to assume that William benefitted even more substantially from his father's estate. Even a close friend like Marcet could claim that "The excellent Wollaston has just lost his father who left a large fortune which, I venture to say, will not spoil our friend."[56] Of course, Wollaston no longer had any need of a generous inheritance from a father who had done so much to shepherd his development from a Cambridge medical student to a prosperous chemical entrepreneur.

A MORE RELAXED LIFE

After losing the two men to whom he owed the greatest personal and professional debts, Wollaston began to rekindle his love of the outdoors, in part by angling and hunting, but also by the study of geology. The first indication of his sporting interests had appeared in his daybook as an entry in June 1813 recording the purchase of fishing tackle, just two days after a visit to the Beechwood estate of Sir John Sebright, located about 20 miles northwest of London near Hemel Hempstead.[57] Visits to Beechwood became regular occurrences over the next few years and continued until a few months before his death, as Wollaston became very fond of the shooting excursions and the convivial environment.

Sebright, nearly the same age as Wollaston, had left the army in 1794 when he succeeded his father as baronet and inherited his estates. He was elected Member of Parliament for Hertfordshire in 1807 and continued to represent that county until 1834. He was notorious in parliament for the oddness and bluntness of his speeches and at home for dominating his large family (one son and eight daughters, some of whom died young) and for his love of animals.[58] It is not surprising that Wollaston would be drawn to the vibrancy of a large family, especially one located in a verdant country setting not far from London, or to a man with a passion for science and the outdoors, but how he was able to establish a strong friendship with a man of such a differing personality is something of a mystery. The association of the two men may have begun at the weekly meetings of the Chemical Club, which Sebright regularly attended. Certainly, Sebright would have been keen to advance his scientific interests through discussion with such a well-informed philosopher, and it was he who dubbed Wollaston "the pope" for the certainty with which he expressed his scientific opinions.[59] It is also possible that Sebright wanted to expose his children to the rational instruction that Wollaston delighted in delivering to enquiring minds. And there is good evidence that the eldest Sebright daughter, at least, became an apt student. In early 1816 Wollaston informed Berzelius, who had himself visited Beechwood in 1812, "You will be gratified to hear that the eldest Miss SEBRIGHT is really becoming an expert analytic chemist with considerable skill and ingenuity in operating upon microscopic quantities."[60] It is clear that Wollaston was a much-welcomed guest at Beechwood for his conversation and scientific expertise, perhaps even for his matrimonial eligibility, although there can be no doubt that he went there primarily for the sport.

In September 1813, Wollaston returned to Beechwood with Marcet for partridge shooting and successfully shot a bird on his second attempt.[61] At the hunt, Wollaston soon became a good marksman and often returned to London with much game for the dinner table. After one outing at Beechwood in the fall of 1815, for example, he wrote to Hasted in characteristically cheerful and self-deprecating prose

> Would you believe that I have turned sportsman in my old age and in one of three days shooting have just killed a leash [three] of pheasants

> and a leash of hares, & all by mere chance, for on the following 2 days I missed almost everything, mortified certainly to find that skill had no concern in the business in proportion to the claims that vanity had set up on the score of the 1st day's success. But when a man returns to sports that had been relinquished for a score of years & finds himself still equal to 3 days' good work with temperate weather & cheerful company, provided he sees sport, he may be very well satisfied without dealing death & destruction at every blow.[62]

By 1816, Marcet was able to inform Berzelius that "The dear Doctor, Pope that he is, has begun to hunt in earnest, and has already acquitted himself with great success. The fact is that he does not know how to do anything poorly."[63] We find in the Beechwood visits the same sort of scientific enthusiasm that emanated from the young Wollaston during his time in Bury: his wonder at the natural world and curiosity about its workings had not diminished over time.

The ever more frequent trips out of London that Wollaston began to make after 1815 allowed him to gain better firsthand knowledge of the geology of the British Isles. He had become a member of the Geological Society in 1813, and thereafter cultivated friendships with several of the members who were geological enthusiasts. For example, in August 1815, he set out on a nineteen-day excursion to North Devon and southern Wales with the express purpose of studying geological formations and exposed strata. His notes on the strata suggest that he was well informed on the nature and extent of significant outcrops of geological formations in the region.[64] Fortunately, at Swansea, he encountered an old friend who had years earlier ordered some of the first platinum crucibles made by Cary, the Scottish chemist and geologist James Hall. Hall escorted Wollaston around southern Wales for a few days, pointed out a few of the more interesting features and answered many of his questions.[65] From trips such as this one, together with the purchase and perusal of geological treatises and meetings with prominent geologists, Wollaston quickly became competent in the science, although not to the extent that he was ever to publish anything beyond the mineralogical realm.

CHAPTER 12

Service to Government and the Royal Society 1803–1820

> You figure to yourself a solemn PRS in great cocked hat with mace before him in his chair of state, surrounded by a set of grave Philosophers; I think of bobtailed sportsmen each with his gun, each with his dog.[1]

As someone with recognized expertise in both science and commerce, Wollaston was frequently selected to serve on government committees, often through the agency of the Royal Society. Three that we know of were short-term projects that did not require much in the way of experimental investigation, but two others occupied Wollaston for many years. The first long-term project, the details of which are recorded in government records in the National Archives of the UK, began in 1803 and ended fifteen years later. That project, which began as a relatively straightforward comparison of hydrometers for the measurement of alcohol in spirits, evolved into a technically- and politically-charged introduction of a national standard for alcohol measurement and taxation.

EXCISE TAXES AND SIKES'S HYDROMETER

Excise taxes, originally intended as taxes on domestic goods, were first introduced by England's parliament in 1643 to finance its military campaign against the king, Charles I, during the English Civil War. The taxes were collected from the primary producers of a wide range of high-volume commodities including beer, cider, wine, and distilled alcoholic spirits.[2] Despite widespread opposition to the tax and its officers by both producers and consumers, its success at generating revenue for a government that was involved in a succession of wars until the early nineteenth century ensured its continuation and expansion. By the early eighteenth century, for example, excise taxes had surpassed traditional land taxes as a proportion of total government revenue, and, as that century advanced and alcohol consumption increased, the tax on alcoholic spirits became an ever more lucrative component of the excise.[3]

The excise tax on spirits was based on the amount of alcohol contained in each type of product, although the methods of gauging alcohol content were varied and contentiously qualitative. One test that came to be of importance in the eighteenth century in "proving" the relative strength of spirits was based on the flammability of alcohol dissolved in water. If a small amount of gunpowder soaked with the spirit burned steadily, the alcohol/water solution was said to be of a strength known as "proof." If the soaked gunpowder was difficult to ignite or burned poorly, the spirit was classified as "underproof"; alternatively, if it burned with what was judged to be excessive violence, the spirit was "overproof."[4] Because the tax on overproof spirits was quite a bit higher than that on proof spirits, producers often challenged the judgment of the excise officers when a product thought to be of proof strength was measured, and consequently taxed, as overproof.

A partial solution to the vexing problem of measuring the alcohol content of spirits arose with the introduction of an instrument known as a hydrometer, which was a deceptively simple device consisting of a relatively buoyant bulb attached to a graduated upper stem, which sank to a certain depth in spirits in a way dependent on the liquid's specific gravity. The point at which the liquid surface intersected the graduated stem then revealed the specific gravity, and therefore the alcohol content, of the liquid in which the hydrometer floated. Although the principle of such a specific gravity–measuring device had been established by others, a London engine maker named John Clarke introduced the first useful hydrometer for measuring spirit strength in 1725.[5] The earliest version of "Clarke's hydrometer," as it became known, simply had a mark on the stem that indicated whether the spirit being measured was below, above, or at proof strength. Nonetheless, as its ease of use and apparently objective measurement became widely known and valued, improved versions followed. By the end of the eighteenth century, the hydrometer's usefulness had been extended by a design that allowed weights to be added to the submerged end. For illustration, a diagram of a hydrometer made by a competitor named Bartholomew Sikes, but similar in construction to Clarke's, is shown in Figure 12.1.[6] The most durable hydrometers were made from brass and consisted of three parts: the central hollow brass sphere imparted buoyancy, the lower fixed brass weight was attached in such a way that disks of different sizes and weights, such as the four shown, could be placed on top of it, and

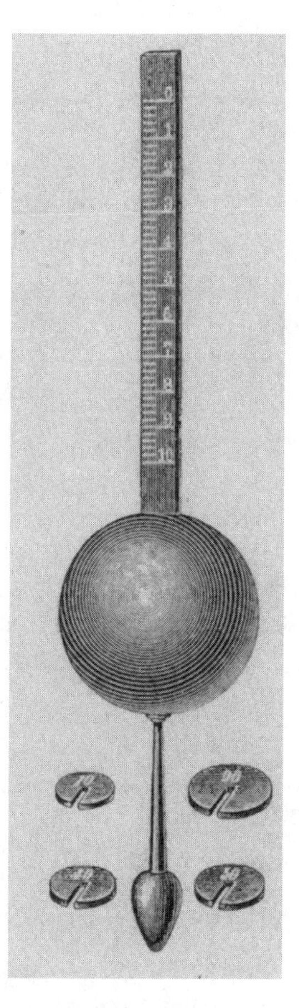

Fig. 12.1. A Sikes's Hydrometer.

an uppermost stem that intersected the liquid layer somewhere along the calibrated range. Addition of appropriate weights to the submerged stem made the hydrometer capable of measuring the specific gravity of spirits from the least to the most alcoholic, over a range of temperatures. As both the producer and the government were keen to have the strength of spirits measured more accurately (the former to avoid excessive tax and the latter to maximize it), the number of weights continued to increase throughout the eighteenth century. It was not until 1788, however, that Clarke's hydrometer received parliamentary approval as the official instrument for gauging alcoholic spirits for the purpose of the Excise.[7]

In the same year that Clarke's hydrometer was given official status, parliament asked the president of the Royal Society, Joseph Banks, to seek a scientifically sound method of measuring the alcoholic content of spirits. The task was assigned to Charles Blagden, who concluded in 1790 that the specific gravity of an alcohol/water mixture was the physical property that could be measured with greatest accuracy.[8] A few years later, his assistant, George Gilpin, published a comprehensive table of specific gravities of such mixtures varying by parts per thousand from pure water (s.g. 1.000) to the strongest spirit commercially available (s.g. 0.825), at every degree of temperature from 30 to 80°F.[9] Gilpin's tables provided the data for converting measured specific gravities of spirits to weight or volume percentages of alcohol content. Bartholomew Sikes, an experienced Excise officer, constructed hydrometers on this principle, such as the one illustrated in Figure 12.1.

The Act of Union between Ireland and England in 1800 made life difficult for administration of the excise tax. Ireland taxed spirits at a graduated rate based upon the percent content of alcohol with a Sikes-type instrument, while Britain continued using the proof system and Clarke's hydrometer.[10] After being informed of the inconsistencies in, and the possible challenges to the collection of, the tax on spirits, the joint secretary of the Treasury (later to become Chancellor of the Exchequer), Nicholas Vansittart, received parliamentary approval in 1802 to assemble a committee of experts to examine the relative merits of the various hydrometers available for the measurement of alcohol strength. Vansittart began by asking the Commissioners of Excise to seek input from excise officers in England, Scotland, and Ireland, and by placing advertisements in the public press for all makers of hydrometers to submit instruments for trial. Probably through the intermediacy of the Royal Society, Wollaston and the Cambridge professor of chemistry William Farish were chosen to test the submitted hydrometers and submit a report on their findings.[11] The whole investigation was coordinated by Vansittart, who relied on Wollaston for the scientific details of specific gravity measurements.[12]

In trials carried out at the Excise Offices from January until the end of April 1803, Wollaston and Farish assessed the construction and mode of use of nineteen submitted hydrometers, and selected the best seven for extensive testing.[13] After due consideration for the accuracy of the measurements, the speed with which they could be made, the sturdiness of construction, and the quality of the conversion tables that came with each instrument, Wollaston and Farish concluded that "they are of the opinion that in their present state Sikes's instrument (which has 9 weights) deserves to be preferred to the rest."[14] In addition they recommended that "Proof Spirit" be redefined as a spirit with a weight 12/13 (of specific gravity 0.923) that of pure water at 52 °F. This gave a quantitative definition of proof spirit that was close enough to the historical one that it had no deleterious effects on manufacturers, consumers, or the nation's revenues.

The report was written by Wollaston, and it is clear from other documents in the hydrometer files that he had taken on a leading role in the evaluation process. For example, on the same day that the Examiners' report was sent to the Board of Excise, Wollaston also delivered a separate letter with a number of his own recommendations.[15] One was for the design of a sliding rule that could, for the most part, be used in place of conversion tables for Sikes's hydrometer. A second was a reminder that the hydrometer chosen should be one that struck an appropriate balance between practical convenience and gauging accuracy, and the third explained a clever idea of his that exploited the newly-proposed definition of proof spirits. He suggested that each hydrometer be accompanied by a weight equal to 1/12 of the hydrometer weight that could be fitted over the top of the uppermost stem. A hydrometer with the weight added would then sink to the same level in water as it would in the newly-defined proof spirit without the weight. This shrewd innovation would allow makers to test and validate the accuracy of each hydrometer produced, and excise officers to verify the continued accuracy of their instrument without recourse to any supplementary information. Such a simple indicator of accuracy would be invaluable for gaining the confidence of distillers and minimizing grounds for appeals of excise decisions in courts of law. The ratio of the test weight to that of the hydrometer itself was, of course, fixed by the composition of proof spirit recommended by the hydrometer-examining committee, so it is hard not to conclude that the proposed redefinition of proof spirit was intimately connected with Wollaston's conception of a self-testing hydrometer.

By early June 1803, Wollaston had completed his design of a sliding rule for Sikes's hydrometer, and he sent a working example to the Board of Excise. However, he alerted the Board to the fact that no sliding rule could give results to the same accuracy as an extensive set of tables, although it was capable of yielding values good enough for excise purposes.[16] In July, the government's Treasury Committee announced that it was prepared to accept Sikes's hydrometer as the official instrument for measuring the strength of spirits, although it wished to be informed if any improvements to the design of the hydrometer were under consideration and what changes to the law regulating the taxation of spirits might be required by its adoption.[17] Thomas Groves, the Inspector of Imports for the Port of London, who had many years of experience with Clarke's hydrometer, worried that Sikes's hydrometer would prove to be too sophisticated for general use, and he began to argue for a reduction in the number of weights that would be necessary for gauging accuracy. The Board of Excise asked Wollaston to consider the possibilities, and, after some initial enthusiasm for the idea, he concluded that any reduction in the number of weights would impair the accuracy of the instrument.[18] Wollaston might well have believed that this decision marked the end of his commitment to the Board of Excise, but his familiarity with the instrument and his comprehension of the theory behind the tables of conversion and alternative sliding rule, together with his access to the device's designer and fabricator, had made him invaluable to the decision-making committees of the government. Therefore, as objections to the introduction of the hydrometer began to surface, he agreed to act as a paid consultant to the Excise and the Treasury. He then became entwined in the slow political process that preceded final parliamentary approval—and payment for work done—a long fifteen years later.

WOLLASTON AS PAID CONSULTANT

Bartholomew Sikes, the designer and fabricator of the favored hydrometer, died in October 1803, prompting the Treasury Committee to gain ownership of all of Sikes's papers relevant to the design and manufacture of his device.[19] Perhaps because of the 1804 change of governments, it was not until early 1806 that the Treasury instructed the Board of Excise to negotiate with Mrs. Sikes for the rights to the hydrometer, on the condition that there be sufficient documentation of Sikes's methods to ensure that the instruments could be made to the necessary accuracy. To satisfy this requirement, the Excise Office asked Wollaston in September 1806 to inspect all of Sikes's notebooks and records to see if there was sufficient detail to merit payment for them. Wollaston agreed to the request, obtained all the relevant materials from Mrs. Sikes, and was able to verify that the documents contained all that was needed to make the desired hydrometers and compile the requisite conversion tables.[20] This was no simple task as the documentation included twenty-seven books of notes and calculations that recorded the results of investigations into the many physical parameters that affected a hydrometer's performance.[21] Soon after obtaining a positive report from Wollaston, the Treasury Board instructed Excise to purchase the rights to the hydrometer, with the directive that Wollaston be contracted to superintend the production of the nation's hydrometers.[22] Obviously, Wollaston had won the confidence of both the Treasury and the Excise officials.

After paying Mrs. Sikes for the rights to her husband's instrument, the Treasury asked Wollaston to again render a decision on the complaint of

Groves that the hydrometer was too complicated for general use. Wollaston replied that the hydrometer could not be simplified in any way which would not be "inconsistent with the accuracy required"; it was designed to give specific gravity measurements accurate to one-fifth of a percent and was more than adequate for the needs of brewers, distillers and the Board of Excise.[23] In short, as he had been telling the relevant government officials for four years, the Sikes's hydrometer was as well designed as it needed to be, and excise agents could quickly learn how to use it properly.

The Treasury and Excise Boards must have agreed with Wollaston's conclusions for by mid 1808, about 2,000 newly constructed Sikes's hydrometers (at a unit cost of £2.2.0) had been distributed throughout the United Kingdom and the British colonies.[24] They did not, however, include the test weight that Wollaston had first recommended in 1803 because the requisite change in the definition of proof spirits had not been passed into law. By 1810, the satisfactory performance of the Sikes's hydrometers provided new impetus to the passage of a bill that would bring consistency to the gauging and taxing of spirits. The Excise then asked Wollaston if he still favored the change in definition that would give the test weight its validity, and if so, what wording he might suggest to the government's Solicitors for the wording of the bill to be presented to parliament. In reply, Wollaston reiterated his conviction that the legal definition of a proof spirit be changed to one that had a weight equal to 12/13 of an equal volume of water at 51°F (a slight alteration from the previously recommended 52°F), and that a test weight be supplied with each instrument.[25] With the technical decisions thus completed, Wollaston urged the government to move quickly on the long-delayed hydrometer bill. However, it was soon discovered that its use would cost the Treasury thousands of pounds in lost revenue.

Clarke's hydrometer, which was scheduled for replacement, did not measure specific gravity as precisely as Sikes's hydrometer. As a result, the stepped increases in measured alcohol content, and the corresponding jumps in taxation rates, were smaller when Sikes's instrument was used. Consequently, the man who first brought the financial concerns to the government's attention (and the one who thought Sikes's hydrometer was too complicated for general use), Thomas Groves, was appointed to join with Wollaston in a comprehensive comparison of the two hydrometers. The two proceeded to carry out comparative trials on a whole range of spirit concentrations in November 1810. Their results confirmed that Sikes's hydrometer gave more accurate measurements of alcohol content that, in nearly every instance, were less than the corresponding values obtained with Clarke's hydrometer.[26] These trials presented the Treasury with a dilemma. Introduction of the more accurate hydrometer would expose the fact that the Excise had, by reliance on an inaccurate instrument, been overtaxing alcoholic spirits for decades, and, even more discomfiting, transition to the better instrument would result in an intolerable loss of revenue. This gave pause for reflection, and so the hydrometer decision was set aside until early 1812 when Vansittart, who had initiated the hydrometer review in 1802, was appointed Chancellor of the Exchequer. He was determined to resolve all outstanding issues before taking the hydrometer bill to parliament and he began by instructing the Board of Excise to collect Wollaston's and Groves's views on the matter. Not surprisingly, each man stuck to their opinions: Wollaston in favor of the new hydrometer, Groves opposed to it.

On January 25, 1813, a letter was sent on Vansittart's behalf informing the Board of Excise that, despite Groves's concerns, the bill for the approval of Sikes's hydrometer was in the final stages of preparation. In fact, a draft of the proposed bill had already been sent to Wollaston, who sent a reply to Excise on the very same day, recommending two minor changes.[27] One day later, Vansittart's assistant, A. Nicholas, wrote to the Board of Excise urging them to accept Wollaston's insertions and to defend them against criticism by Groves or anyone else. In another letter to Excise written another day later, Nicholas relayed Vansittart's position even more bluntly: Wollaston's arguments for the superiority of Sikes's hydrometer and the proposed bill that would mandate its use were not to be delayed any longer by Groves and other enthusiasts of Clarke's inferior hydrometer.[28] Vansittart's convictions, so reliant on Wollaston's expertise, carried the day because the draft wording of the bill was approved by the Board of Excise, which then passed it on for legal rewording. Finally, in July 1816, the first temporary Act establishing Sikes's hydrometer as the official instrument for gauging spirits was approved by parliament, becoming effective in January 1817.[29] Not surprisingly, the Act contained measures to maintain the revenue stream from taxes on alcohol by specifying that higher rates of taxes would in the future be assessed on those few spirits for which Sikes's instrument gave higher values than Clarke's. The time-limited Act of 1816 was soon superseded by the final Spirits (Strength Ascertainment) Act of May 1818, which repeated much of the language of its predecessor but omitted reference to the less precise sliding rules, thus removing them from legal standing.[30] This hydrometer bill, sixteen years in the making, established the regulatory framework for spirit measurement in the United Kingdom that remained in force for over a century and a half.

Wollaston's role in the refinement of Sikes's hydrometer, and the redefinition of proof spirit that made the self-testing feature of the instrument an attractive option, has gone largely unnoticed by historians. There is only one tenuous reference to his work on the hydrometer in Wollaston's extant notebooks, and that is an 1818 entry "Excise Office £525" in his daybook.[31] This must represent payment to him for his service to the government, remitted shortly before the final Hydrometer Bill was passed into law, and many years after he had first been chosen to test replacement hydrometers. His close interaction with the Chancellor of the Exchequer in the perfection of the spirit hydrometer and the construction of the law mandating its use is an illustrative example of how much scientific advice was valued by those who formulated laws regulating British commerce. Wollaston's service on this and other government commissions is also illustrative of an industrial society's shift towards more objective and quantitative measures of commerce in the late eighteenth and early nineteenth centuries, a trend well recognized by economic historians.[32]

SERVICE ON GOVERNMENT COMMITTEES

Wollaston's other government-related activities were not spread out over as many years as the hydrometer project. In fact, only his involvement with the Board of Longitude, which took place over the last eight years of his life, required a long-term commitment. The three other committees on which he served were active for relatively short periods of time before they were able to complete their mandate. They were committees on coal gas safety, weights and measures, and banknote forgery.

In April 1812, the Westminster Gas Light and Coke Company was given a Royal Charter to supply London with coal gas (a mixture mostly of the combustible gases hydrogen, carbon monoxide, and methane) for lighting, marking the incorporation of England's first public gas works. After a small explosion at the works in Peter Street, Westminster, in late 1813 the Royal Society was asked by the Secretary of State for the Home Department to form an expert committee to examine the safety of the works and to report on the likelihood and consequences of future explosions.[33] A committee of eight persons that included Wollaston and Tennant was set up to provide answers. The members inspected the gas works, interrogated both the superintendent and the designer of the storage and delivery systems, and submitted a comprehensive report on February 24. Together with a number of sensible suggestions for the safer construction and operation of such a dangerous facility, the report mentioned some relevant experiments carried out by Wollaston and Tennant, which showed that the flame of a gas lamp could not pass through small diameter supply tubes to the source reservoir. The report was duly presented to the Home Department, and the Gas Light Company went on to prosper as gas lighting spread around the city. The experimental findings contained in the report would likely have been forgotten had they not been brought to life again in a later controversy about the discovery of the miner's safety lamp.

In August 1815, Humphry Davy, then recently returned to London from an extended tour of the continent, set about to design a lamp for coal miners that would not ignite the explosive gas (known as fire-damp, largely methane) that often seeped into mine shafts.[34] By late October, he had discovered through a well-planned and extensive set of experiments that fire-damp combustion could not pass through tubes of diameters smaller than one-sixth of an inch. When these results were published in 1816, Davy added a footnote to the statement of findings that read, "Since these experiments were made, Dr. WOLLASTON has informed me, that he and Mr. TENNANT had observed some time ago, that mixtures of the gas from the distillation of coal and air, would not explode in very small tubes."[35] This made it clear that he carried out his research without knowledge of the work of his colleagues. In subsequent research, Davy found that even a metallic gauze was sufficient to prevent the escape of the lamp flame to the combustible airspace of the mineshaft, and confinement of a flame within such a metallic screen became a key design feature of the resultant, and much-heralded, Davy lamp. However, Davy's claim to priority in the design of a miner's lamp was disputed by a contemporary who had also produced a lamp with many similar features, and, in the debate that followed, some who wished to question Davy's originality hinted that he had not given appropriate credit to the deceased Tennant's observations. Such skepticism was based, of course, on the unfounded assumption that Davy knew of those results, and ignored Davy's published claim to the contrary. The issue of priority became an important part of the debate when awards were to be bestowed on the discoverer of the safety lamp. So, in 1817, Davy enlisted the support of four "first chemists and natural philosophers of the country," one of whom was Wollaston, to set the matter straight.[36] The committee of four issued a joint statement confirming Davy's originality, declaring that he had no "knowledge of the unpublished experiments of the late Mr. Tennant on Flame."[37] This resolution makes it clear that it was Tennant who actually did the coal gas experiments, although it is probable that Wollaston helped in the design of the apparatus and assisted with the combustion tests. One cannot be certain that the statements of the fact-finding committee resolved the issue, but Davy was soon after rewarded for his lamp, as he was awarded the Royal Society's Rumford medal in 1816 and had a baronetcy bestowed upon him in January 1819.

In the same year that he was collaborating with Tennant on the coal gas experiments, Wollaston was called before the House of Commons Select Committee on Weights and Measures. There he was asked to give advice on the best natural standards on which to base units of length, weight, and volume for what was to become the Imperial System of Weights and Measures. Although Britain was on the verge of becoming the world's greatest trading nation, its weights and measures in the early years of the nineteenth century were still a perplexing patchwork of local units for length, weight, and volume.[38] In an effort to bring consistency to mensuration, a parliamentary committee was set up to select a universal standard as the basis for rational units of length, weight, and volume. Its report was presented to parliament and published in 1814.[39]

The committee called on John Playfair, the Edinburgh professor of mathematics and natural philosophy, and Wollaston for expert scientific opinion. Each man understood that the desideratum was a distinctly British standard that could be introduced without wholesale replacement of existing units of measurement. Both recommended the same fundamental and invariable standard of length: a pendulum that beat with a frequency of 60 times a minute at the latitude of London. Such a pendulum had a length of 39.13 inches.[40] With inches thus defined, a cubic foot of pure water ($12 \times 12 \times 12 = 1728$ in^3) at 56.5°F was taken to weigh 1,000 ounces avoirdupois, and 16 such ounces made up one pound. The avoirdupois weights (already in general use) were recommended as a replacement for troy weights in all commerce, except for the quantification of the monetary metals, gold and silver. With these base units established, both experts recommended that standard volumes should be defined by the weight of pure water contained within them, which allowed for greater accuracy than a volume specified by linear dimensions. In addition, Wollaston emphasized one particularly advantageous weight–volume relationship. He pointed out that, if a fluid gallon was defined as the volume that contained exactly 10 lbs. of water at 56.5°F, and such a gallon was made up of 160 fluid ounces (= 8 pints), then the volume of a half-pint (10 oz.) would be exactly equal to 1/100 of the volume of a cubic foot.[41] In their report, the Select Committee heeded the advice of their scientific experts, and a bill containing the recommendations was brought to a vote in 1815, but failed in the Commons after its second reading. One year later, a similar bill passed in the Commons but was defeated in the House of Lords.[42] The following years brought a succession of Royal Commission and select committee reports on weights and measures; Wollaston was a member of most of them.[43] Finally, in 1824, the Weights and Measures Act that established the Imperial System and incorporated most of the recommendations of the 1814 select committee was passed into law. The new units, which remained in use in the United Kingdom for nearly a century and a half, included Wollaston's recommended gallon of 160 fluid ounces with a volume equal to that which held exactly 10 lbs. water.[44]

In 1819, Wollaston agreed to serve as one of the commissioners appointed to investigate means of preventing the forgery of banknotes, likely as a representative of the Royal Society whose president, Joseph Banks, was also a member. The committee presented two reports to the House of Commons, the first with a printing date of January 2, 1819, and the second of February 18, 1820.[45] After digesting all the materials made available, the committee concluded that there could be no foolproof remedy to fraudulent banknote production. Its members recognized that there was sufficient expertise in the English engraving community to produce acceptable copies of Bank of England notes, however technically demanding the engraving might have to be. One recommendation, advanced in the first report, was for the government to make a greater commitment to the apprehension and conviction of the forgers themselves, whom the committee members believed had only been able to ply their trade effectively in concert with corrupted police forces. The second recommendation, included in the second report, was to adopt the (undisclosed) banknote design and printing methods advanced by Applegarth and Cowper, two printing engineers from the London *Times* who had been independently engaged by the Bank to adapt advanced machinery for the printing of notes. There is not enough information to know exactly what Wollaston contributed to the banknote committee's recommendations, but one suspects that his familiarity with both the scientific and engineering aspects of engraving and printing would have been of value to the committee's deliberations.

THE BOARD OF LONGITUDE

In 1714, the British parliament had passed a bill known as the Longitude Act that offered a reward as high as £20,000 for a practical and accurate method of measuring longitude at sea. To assess instruments and methods proposed for a solution the Act established a group of twenty-two commissioners who made up the Board of Longitude and reported to the Admiralty.[46] The Board's awards to Tobias Meyer for his tables of lunar positions and to John Harrison for the marine chronometer later in the eighteenth century fulfilled the original mandate of the Board of Longitude, but the usefulness of the Board's decisions to the vitality of Britain's navy prompted parliament to pass a sustaining Act in 1774. That Act redirected the Board's attention to wide-ranging improvements in marine navigation, including instrumentation, as well as research into astronomy. It even allowed the Board to become active in promoting voyages of discovery and mapping. This broader mandate had less appeal to the naval members of the Board during the Napoleonic wars, allowing the influence of the Royal Society members, under the impetus of its president, Joseph Banks, to grow ever stronger. A further Act of 1818 directed the Board to encourage "Attempts to find a Northern Passage between the Atlantic and Pacific Oceans, and to approach the Northern Pole," and instituted a new prize (never claimed) of £20,000 for reaching the Pacific via an Arctic route.[47] The Act also specified that, in addition to the usual occupational spectrum of commissioners, there were to be three extra commissioners who were permanent residents of London "well versed in the Sciences of Mathematics, Astronomy, or Navigation." The first three "Resident Commissioners" as they became known, were all well-qualified Fellows of the Royal Society. One was the surveyor and cartographer Col. William Mudge, another was a specialist in scientific instrumentation, Capt. Henry Kater, and the third was Wollaston.[48]

From 1818 until its demise in 1828, the Board met four times a year at the Admiralty Offices. Wollaston attended every one of those meetings except for two in 1826, when he was out of the country. He proved to be a very active member. He was quickly chosen to be one of the examiners of the financial accounts of the Board, a position he retained until just before his death, and in 1821 he also became a member of the Committee for Examining Instruments and Proposals. It was in the latter capacity that he, with the assistance of Kater and the instrument maker William Cary, examined the dividing engine made for the Board of Longitude by Jesse Ramsden in the late eighteenth century.[49] This investigation is mentioned here to illustrate how useful Wollaston's talents for careful testing and keen observation were to be for the evaluation of the many instruments presented to the Board, although there is little concrete evidence of his specific contributions to the Board's decisions. Not surprisingly, Wollaston became intensely interested in the polar expeditions newly promoted by the Admiralty and the Board, and saw in them an opportunity to test some of his own optical instruments.

In 1818 the Admiralty decided to outfit two exploratory voyages to the Arctic, the first commanded by John Ross and the second by William

Parry. The polar explorers were to seek out and map a passage through North America's ice-filled arctic waters from Davis Strait on the east to the Bering Strait on the west.[50] Both expeditions were also instructed to carry out a wide variety of scientific investigations and were supplied with a number of specialized instruments designed for those tasks. Instructions for the proper use of the instruments were prepared by the instruments' inventors, printed by the Council of the Royal Society, and given to the ships' commanders. The instructions were also published in the journal of the Royal Institution.[51] Among the instruments described were three of Wollaston's design discussed in Chapter 3. Two of them, the dip sector and the dip micrometer, were meant to give more accurate measurements of the depression, due to refraction, of the visible horizon at sea. The third, the macrometer, measured the distance from the observer to objects located from 70 to 31,000 yards away. The dip sector had found some use after Wollaston had described it in his Bakerian lecture of 1802, but the other two were first used on the Ross and Parry voyages.

Although his instruments did little to improve navigation in the arctic, Wollaston's chemical expertise was put to good use in the analysis of some materials brought back by the explorers. For example, Ross brought back melted samples of crimson-colored snow collected on the shores of Baffin Bay, and he gave some to Wollaston for analysis. After chemical and microscopic analysis, Wollaston concluded that the red coloring matter was of "vegetable origin" and was contained in "minute globules from 1/1000 to 1/3000 of an inch in diameter."[52] The vegetable origin of the red dye was independently identified as the product of a "fungus" by another analyst soon after.[53] Ross also gave Wollaston a sample of iron collected in the far north, which he easily identified as being of meteoric origin because of its characteristic nickel content.[54]

Service on the Board of Longitude added breadth to Wollaston's familiarity with the latest innovations in instrument design. It also involved him in the distribution of government funds for scientific advances and exposed him to the triumphs and tragedies of polar exploration in the 1820s. But a greater reward was the broadening of his circle of friends to include several with connections to the Board's work. One of these was Sir Edward Codrington, the famed naval officer who had commanded ships in the battle of Trafalgar, the defense of Spain, and the North American fleet after the war of 1812. Codrington had a strong interest in science and must have shown enough ability that Wollaston put his name first on the list of supporters of Codrington's certificate of admission to the Royal Society in

November 1822. And it was to Codrington's son that Wollaston gave one of his first rhodium-tipped pens in 1822. The warmth of the relationship between the two men is evident in the many letters they exchanged. They corresponded frequently while Codrington was at sea, and the naval commander even tried to fulfill some of Wollaston's scientific requests on his voyages. One he was unable to execute was for water samples to be drawn from the deepest parts of the current flowing out of the Mediterranean through the Straits of Gibraltar.[55] Wollaston had requested samples of that water because he, like most of his contemporaries, believed that the salinity level of the Mediterranean could only be maintained if the inflow of surface water from the Atlantic was counterbalanced by a deeper outflow of even saltier water. It does not appear that Wollaston ever received the desired water samples from Codrington, but he did manage to obtain some from the captain of a naval survey of the Gibraltar region. In a paper read by another to the Royal Society a few days before he died, Wollaston reported that water taken from a depth of 670 fathoms (approx. 4,000 feet) in the Straits was indeed four times saltier than the surface water flowing in the opposite direction.[56] This example illustrates the close interaction between the Admiralty and several naval officers with the men of science they encountered at the Royal Society and the Board of Longitude, an interaction that continued even during expressly military operations.

In other letters to Wollaston during the Mediterranean posting, Codrington kept his friend abreast of the naval and political developments while adding scientific comments on topics of mutual interest. He mentioned the failure of Davy's cathodic protection of the copper cladding of his ships' hulls, the perverse effects of the dry, dusty sirocco winds originating in North Africa, and an eruption of Mount Etna.[57] But most interesting of all, Codrington sent Wollaston an extensive account of the Battle of Navarino (October 20, 1827) in which he, with French and Russian naval support, had destroyed a combined Ottoman, Egyptian, and Algerian fleet in the Bay of Navarino, a battle that played a crucial role in the Greek war of independence. The letter, written a few days after the event, opened with these words: "That I frequently have wished you here with me, my dear Wollaston, is true, because you would have seen extraordinary sights."[58] Accounts of the battle soon made their way into general circulation, and although Codrington's actions were applauded by the British public and many politicians, the political fallout created such problems for the Admiralty that they elected to relieve him of the Mediterranean command in June 1828. Wollaston, not surprisingly, was among those who thought such a politically-motivated action was indefensible. Such support was much appreciated by Codrington, who mentioned it in a letter to a shared friend

> let our friend Wollaston indulge himself in accusing the Government of injustice; it is only what he thinks just, and he gratifies me by it extremely; for I consider the warm regard of such a friend as he is, ample balance against the hostility of a Secretary of State.[59]

Codrington returned to London in October 1828, just in time to witness the decline of Wollaston's health and to visit with him prior to his death.

THE PRESIDENCY OF THE ROYAL SOCIETY

Joseph Banks was a wealthy landowner who had made himself into a competent naturalist before being elected president of the Royal Society in 1778, at thirty-five years of age. Over the forty-two years of his presidency, he worked tirelessly to integrate the affairs of the Royal Society into the political and economic cycles of Georgian England. And he did so with such success that his yearly re-election became a relatively uncontested event, even though discontentment with his autocratic style simmered among some of the Fellows. Over time, there arose considerable opposition to the type of Royal Society that he had molded and sought to sustain. The wish for reform of the Society spread over disenchanted geologists, mathematical practitioners, astronomers, and Cambridge rebels, among others.[60] The reformers could not amass enough power to act while Banks was in charge, and they could do little more than wait for the end of his presidency to press for changes.

The opportunity for societal reform arose in the spring of 1820, when Banks's health began to fail precipitously. He then directed the Council of the Society to seek out a successor. Not surprisingly, there was much maneuvering, and considerable intrigue, among those who hoped to gain the presidency.[61] Wollaston, who was a Council member fully aware of the situation, knew that he had the support of many for the position, including a great number of the reform-minded. After giving some thought to the matter, however, he decided not to seek the presidency. He explained his decision in a letter to Marcet, then resident in Geneva.

> It was you yourself who first suggested to me standing for the Presidency as not an act of mere presumption. —Others have also named it as attainable —My brother [probably Francis John] pressed it strongly

> as desirable—Blake urged me too, & that repeatedly till at last I felt it right gravely to consider whether if attained it could contribute to my happiness, & fit that I should answer Blake decidedly that he & others might know how to act in case of a demise [of Banks].
>
> I did consider & I did decide & answer in the negative—Blake thought me right & I was satisfied.[62]

One cannot dispute Wollaston's conclusion that the presidency would not contribute to his personal happiness, because he was not one for whom the prestige and responsibilities of Britain's most influential and visible scientific post meant as much to him as his own independence and freedom of movement. Nonetheless, Wollaston certainly had many of the qualities several of his colleagues would have wished for in a president. His breadth of knowledge and experience as an experimental philosopher were unsurpassed by any of his contemporaries, he knew the workings of the Royal Society inside and out, he had strong connections to Cambridge and to the members of parliament with whom he had interacted in various committees, and he had the respect of the naval and military officers he encountered through his work on the Board of Longitude. Of most importance for the scientific members of the Society, although they were still a minority of the total (largely inactive) membership, he was a man of science who could be relied on to advance the scientific agenda of the Society. Furthermore, he shared many of the ideals of the (mostly younger) reform-minded members, who wished the Royal Society to become less hostile to competing scientific organizations and more welcoming to potential members from nontraditional areas of science.[63] In addition, he had sufficient wealth to fund the social gatherings expected of the president. But there were other qualities, mostly personal, that clouded the issue: he had no patience for pretense, no compulsion to be a public ambassador for science and no ego to sustain. We know that Banks was willing to overlook the negatives and hope that Wollaston would be interested in succeeding him, for he told his colleague John Barrow that "he regretted exceedingly to find that Dr. Hyde Wollaston would not consent to be put in nomination for the chair, —'so excellent a man, of such superior talents, and every way fitted for the situation!' "[64] However, Banks's support for Wollaston does not mean that he believed him to be the best choice.

Banks instead held out hope that a successor in his own image could be found, and he believed that man to be Davies Gilbert, a member of parliament from Cornwall with mathematical skills who had promoted the cause of science in many parliamentary committees. In April 1820, Banks informed the three vice-presidents of the Society (one of whom was Gilbert) of his impending resignation and his wish that Gilbert be his successor.[65] Fearing a disorderly succession contest the Council (which included Wollaston) chose not to act on Banks's recommendations and instead urged him to withdraw his resignation and to continue his presidency.[66] This Banks did, but only for the short period that preceded his death one month later. Once word of Banks's declining health had become generally known, a number of candidates interested in his position had made their intentions known. The political situation was succinctly described by Charles Babbage in a letter to his young Cambridge colleague William Whewell:

> all sorts of plans, speculations and schemes are afloat, and all sorts of people proper and improper, are penetrated with the desire of wielding the sceptre of Science. . . . The Society is in a position of unstable equilibrium.[67]

Among those publically seeking "the sceptre of Science," only one, Humphry Davy, had strong credentials in science, and he had such an interest in the presidency that he rushed back to London from Paris to build support for his cause. Ironically, it was his entry into the race that provoked Wollaston into reconsidering his stance. In a letter to Marcet, Wollaston described the intriguing sequence of events that followed his initial decision to stay out of the succession battle:

> Next in the order of events was the return of Davy, professedly to support Lord Spencer against Prince Leopold & D. Gilbert, tho in reality most eager to start himself. Prince L. withdrew. Lord Spencer declined,—so that the names entered for the sweepstakes previous to the death of Sir Jos. were Gilbert, Davy & the Duke of Somerset. When the vacancy actually took place, some of my over zealous friends would try to force me forward & make a stir in spite of my refusal . . . Davy expressed himself as sorely hurt by the activity of my friends which he represented as hostility to himself, totally overlooking the possibility that they might be actuated by regard for me. I must own their earnestness & the very powerful support of which I felt confident did for a few hours make me hesitate, & an expression of Davy's treating their endeavours with some degree of contempt was for a moment very near to determining me to convince him of his mistake; but I did not see that any good end would be answered by holding the office for a year (if attainable, as I firmly

> believe) & I soon decided to withdraw at once from an odious contest, rather than subject myself even for one year to all the ties of a troublesome station. Let me brag once again of having decided rightly & here ends this egotistic detail. —I have taken it for the present but Davy will succeed in November.[68]

A more complete account of Wollaston's conduct that describes the actions of his "over zealous" friends, the entry of Davy into the contest, and his ire at Davy's behavior, can be found in an unpublished manuscript of one of those friends, John Herschel, the son of the famous astronomer William Herschel and an accomplished man of science himself.[69] Soon after learning of Banks's final illness, Davy returned to London about June 5 and began to meet with members of the Society to promote his cause. Davy certainly had many supporters among both the scientific and nonscientific members, but his tendency toward self-promotion rubbed many people the wrong way. Herschel, who did not know Davy especially well, believed him to be "arrogant in the extreme, and impatient of opposition in his scientific views, and likely if power were placed in his hands to oppose rising merit in his own line."[70] On the morning after Banks's death on Monday, June 19, Herschel met with a small group of like-minded Fellows, nearly all of whom were committed to reform of the Society. They came to the unanimous conclusion that Wollaston was the best man to oppose Davy. In the afternoon of the same day, Herschel and Babbage went to Wollaston's house to sound him out on the idea. They left that meeting encouraged by the belief that Wollaston could be persuaded to stand for the presidency if he could be convinced that his candidacy would benefit the Society. Accordingly, Babbage and Herschel quickly initiated their recruiting drive, which included visiting Wollaston's friends, placing notices of his qualities in the newspapers and writing letters to potential supporters. By Thursday morning, Herschel believed he had enough committed votes in hand to justify a return to Wollaston's home and to push him once again to enter the contest.

Herschel appears to have weakened Wollaston's reticence by the complementary tactics of impugning Davy's character and emphasizing the strength of support from Cambridge for his candidacy. But, after Herschel left, Wollaston's doubts returned, and he fretted about the turn of events before attending the dinner of the Royal Society Club that evening, at which an encounter with Davy loomed. Herschel tells us what happened between these two men of such contrasting personalities.

> The Club Dinner at the Crown & Anchor this day was fully attended—Sir H. Davy was briskly in motion, bowing, shaking hands, & ingratiating himself at all points—Wollaston came in late, got seated at the lower corner of table, and presented his usual severe demeanour.[71]

Herschel could not stay for the general meeting of the Society that followed the dinner, but he learned from Babbage and Kater of the emotional exchanges that occurred later that evening.

> Davy when he first heard it suggested that W^{n} might oppose him, made light of it. . . . When D. however found Wollaston's strength (for we had by this time got a formidable list, and had many friends actually at work) he requested some conversat^{n} in private.—Precisely what passed I do not know, but he must have conducted himself in a very overbearing way, as he made W. very angry, & in fact irritated him into a determination to oppose him. D. taxed him with canvassing by his friends. W. assured him it was not by his concurrence . . . and taking Babbage aside, asked him to promise in my name as well as his own, to canvass no more. B. hesitated, on w^{h} W. said "Will you give me this promise if I say I will accept the office."—On this condition (amounting to a formal declaration of himself as candid^{t}) B. gave the promise.[72]

Obviously, it was this contretemps with Davy that led Wollaston, "for a few hours" as he confessed to Marcet, to reconsider making a run for the presidency. Of course, such a decision would have been a disaster for the scientific membership of the Royal Society because it would have split its vote between the two icons of British science, and opened the way for an aristocratic candidate to emerge victorious. Such a consequence must have been obvious to Wollaston, and he deemed the best course of action to be withdrawal of the candidate who did not want the job in favor of one who did. And so Wollaston quit the "odious contest." At the Council meeting of June 29, Wollaston agreed to serve as interim president until the Anniversary Meeting in late November, at which time an election would be held to choose a successor. Humphry Davy won that election easily and remained president until ill health forced him to resign in 1827.

It is possible that Wollaston could well have triumphed over Davy in an open election, but his decision to avoid a showdown was a wise one. And, to both men's credit, they each worked hard to put the hard feelings behind them. After Wollaston wrote to Davy to confirm his withdrawal Davy wrote in reply

> I certainly felt severely wounded when I became acquainted with the active & extensive canvass of many of your friends against me, at a time when I was almost reposing on the hopes of your support. This wound however you have healed & the manner in which you have done so has increased my admiration of your character.[73]

Shortly after becoming president in November 1820, Davy made a more tangible signal of respect for his colleague by naming him deputy president. That meant that Wollaston was to perform all the duties of the president whenever Davy could not, such as during periods of illness or travel.[74]

In the five months that he was president, Wollaston performed his duties diligently and did some things that the reformers would have viewed favorably. At the first meeting (June 29) over which he presided, eight papers were read, thus beginning the practice of clearing the backlog of submitted papers before the weekly meetings adjourned for the summer. At a subsequent Council meeting in July, stiffer penalties were introduced for members who were in arrears with their membership fees. Later, on the resumption of meetings in November the Council resolved to ask the Treasury (whose officials Wollaston knew very well) for more space in Somerset House to store books and instruments, appointed a committee to seek a reliable method of measuring a ship's tonnage, and produced an index to the post 1780 *Philosophical Transactions.*[75] There was one official event, however, that he would have liked to avoid, had it been possible. That was the ostentatious coronation ceremony of King George IV on July 19, to which he was invited as president of the Royal Society. His daybook reveals that he rented obligatory court dress for the ceremony and official introduction to the new king; it does not mention whether or not he enjoyed the day.

The reforms Wollaston promoted in the few months he was president clearly reveal that he viewed himself as more than a caretaker officer and that he was keen to advance many of the ideals of the reform-minded members. But there is no doubt that his happiest, and final, duty as president was to preside over the Council meeting that elected Humphry Davy as his successor. The short time he had served in the post was enough to convince Wollaston that all his reticence in seeking the position was justified, and he admitted as much in letters written to close friends a few days after he had regained his freedom. One was to Marcet, one of his strongest supporters

> [In your letter of Nov. 28] it was really gratifying to me to find my only remaining friend whose opinion I valued converted to the true faith

> & confessing himself satisfied that I have decided wisely & to receive his recantation previous to the day of final decision, so that I was light hearted & thoroughly happy in the certainty of release from responsibility & stately bondage which I would on no consideration bear for 12 months together. I now can go where I please & when I please & am now setting off for the first battue at Beechwood, which has been deferred to this time on account of my Presidency.[76]

So, as Wollaston stated, had he not served as president for a transitional period, he would never have consented to hold the post for more than a one-year term, a period of leadership that would not have satisfied even the staunchest of his supporters. Davy was the right choice for those who wanted a scientific president with a mandate for reform and, after the drama of the Banks's succession played out, even the majority of the Davy skeptics came to realize that.

CHAPTER 13

A Diversity of Interests 1815–1824

Dr. Wollaston, from the time of his being a boy at school, was subject to this peculiarity of vision, that occasionally on looking at an object, he saw only one half of it, it appearing to be divided in a vertical direction.[1]

The Treaty of Paris, signed in November 1815, finally brought a long-lasting peace and expanded trade and the accompanying prosperity to Europe. Early in the following year, Britain's parliament repealed the income tax that had been re-instituted in 1803 to finance the renewed war against Napoleon. Wollaston was one of those happy to see the demise of the tax, and he celebrated with his friends at a Chemistry Club meeting on the day following the parliamentary vote. One of the co-celebrants, Marcet, wrote to the Scottish geologist Leonard Horner (brother of the Scottish MP Francis Horner) to say:

> As for *us* philosophers, chemists, and scientific amateurs, we celebrated this event by generous libations of champagne at the Chemical Club. The two members who showed the most spirit upon that occasion were Dr. Wollaston (the Pope!) and myself. Being rather out of practice we both exceeded a little the capacity of our brains, and as we rose from table, we discovered by certain vulgar symptoms, the fragility of our nerves.[2]

Other glimpses of this decidedly unreserved Wollaston surface occasionally in the letters of Marcet and Hasted, but the most valuable insights into Wollaston's unguarded social persona are presented in two personal accounts written after his death by a young woman who developed a strong affection for him, Julia Hankey.

FRIENDSHIP WITH JULIA HANKEY

Julia, born in 1798, was the only daughter of the London merchant and banker John Hankey and his wife Isabella (née Alexander).[3] After John Hankey's death in 1807, Isabella and her children moved in with her brother William Alexander, who was a city barrister. The relatives divided their time between William's London home and a country estate at Finchley, a few miles northwest of the city. How Wollaston became a friend of the Alexanders is unknown, but he was a frequent visitor and guest of theirs from the second decade of the 1800s until the last weeks of his life. We learn most of the details of Wollaston's interactions with the family from Julia's remembrances of him written in response to a request for such from Henry Warburton in 1829.[4] Julia's first meaningful interaction with Wollaston occurred shortly after the rebuilt Drury Lane Theater opened in 1812. When the workings of the theater's novel lighting system came into question at dinner at the Alexanders one evening, Wollaston was called upon to supply the answer. The description made a lasting impression on the young Julia.

> I have the most perfect recollection of Dr. Wollaston's figure as he stooped down with a letter & his pencil in his hand, making a diagram rather than a drawing of the internal construction of the lamp, my Uncle standing behind me with a hand on each shoulder, of myself listening with the most fixed attention & now & then raising my eyes to his face & when the explanation was finished, never could I forget the dismay with which I heard him desire me to repeat all Dr. Wollaston had said to me as a proof that I had understood. I suppose notwithstanding my alarm that I succeeded as well as they expected, for I received some praise for the attention I had paid & from that time Dr. Wollaston never came to visit us without bringing something amusing to show me & my brothers. We soon learnt to consider him as our property and as soon as he appeared I regularly took my station by his side. I love to think of the commencement of a friendship to which I owe so much of the happiness of my life.[5]

This passage reveals that Julia was an intelligent and curious young woman, just the type of person whose company Wollaston enjoyed, and he frequently spent weekends with the Alexanders at Finchley.

The warm relationship the group enjoyed was exhibited most clearly by the invitation extended by Isabella and William for Wollaston to accompany them and Julia on a tour of the continent in the fall of 1816. The timing was convenient for Wollaston; Julia recounted the details of his acceptance.

> In 1816 my Mother & Uncle determined to make a tour through parts of Switzerland & invited him to take a place in the carriage this gave me great pleasure & I perfectly remember the day he came to Finchley to talk over our plans. I remember his countenance, I remember our sitting in the little breakfast room with the large map book on the table before us, I remember his hesitating & looking first at one & then at another of the party doubting whether it would be agreeable to all that he should accept & well do I remember [him] at last exclaiming "well it is too agreeable to be resisted so you must take the consequences."[6]

Nearly all the available information about the trip is contained in Julia's remembrances, which are even more pertinent for the fresh characterization of Wollaston composed by the young admirer who was to become a lifelong friend.

The travelers moved in a roughly clockwise direction from Calais through Belgium to Germany, where they followed the Rhine south to Switzerland and continued thereafter to Paris and finally back to Calais for the return trip to London in November, an overall excursion of eighty days. Julia's stories of Wollaston's actions on the trip show that, even at fifty years of age, he had lost little of his youthful enthusiasm, delight in new discoveries, or zest for inspiring others.

> The instant the carriage stopt, if but to change horses, Dr. Wollaston got out, examined the horses, the people, their occupations, the tools they were using, and by signs and the assistance of his pencil and the back of a letter, & the few words of their different dialects which he picked up, he contrived to ask questions & to receive a variety of information. . . . He walked up every hill to examine the vegetable & mineral productions by the roadside & he frequently clambered down precipices to obtain a plant with which he was unacquainted. Wherever he found a person with any pretensions to science his name was sufficient to ensure their attention and to procure admittance for all the party to every collection of natural history or of works of art.[7]

On their way through Belgium, they visited the battlegrounds of Waterloo, and there collected grass seed and houseleek which the Alexanders replanted at Finchley. After passing into Germany, the party joined the

Rhine at Cologne and visited the famous Catholic cathedral, where they stayed to hear "a remarkably fine Mass performed, which delighted Dr. W particularly."[8] After spending a few days in Frankfurt, the party traveled through the Black Forest on their way to Geneva, from where they caught their first sight of the Alps. Unfortunately, Wollaston found that he was unable to sketch the heights with the camera lucida he had used extensively up to that point because the scale of the mountain scenery was too grandiose to be captured adequately by the sketching device. In Geneva, Wollaston met up with Sebright and some of his daughters, but his enjoyment of that city was shattered when he learned, by a letter from Warburton, of the death on September 27 of his close friend and Chemical Club regular, the London chemist and sugar refiner, Edward Howard.

Howard's health had been in a state of decline even before Wollaston's departure, and it had grown progressively worse thereafter. As Howard grew increasingly frail, Warburton anguished over whether or not to tell Wollaston, who apparently had acted as Howard's physician, of the worrisome developments.[9] Warburton's decision was made for him when Howard requested, the day before he died, that his remembrances be conveyed to Wollaston. Warburton subsequently sent letters addressed to Wollaston to contacts in Paris and Geneva, informing him of Howard's death. Wollaston's grief upon learning of the news was remembered by Julia.

> At Geneva Dr. Wollaston received letters from England announcing the death of his friend Mr. Howard, this had a violent effect on him he retired to his room & threw himself on his bed in an agony of grief. My Uncle found him in this state some hours after and through the thin wooden partition which divides the Swiss apartments we heard him sobbing almost all night. The next morning his eyes were swollen & red with weeping, but he was perfectly composed, and when he arrived at Paris [having traveled there separately from the Alexanders] he had recovered his spirits so as to join with his usual cheerfulness in our amusements, altho' when he spoke of Mr. Howard a cloud seemed to pass over his brow & tears started to his eyes.[10]

This passage makes it clear how emotionally attached Wollaston was to his closest friends, and how perceptive it was of Warburton not to inform Wollaston of Howard's decline. Wollaston would almost certainly have abandoned his continental tour to return to his friend's (and patient's) bedside, even though such could have done nothing to alter the outcome.

The Alexanders left Geneva for Paris before Wollaston, who elected to remain a few days longer before joining the Sebright's carriage for the trip to the French capital. Once there, he moved into a small room adjoining the Alexanders' lodgings. A few of his activities in the city are again captured in Julia's memorials:

> After breakfast my Uncle always went out & my Mother into her own room to write & settle her accounts. I used to sit at work in the hall with Dr. W and was much diverted with the strange faces of the learned people who came to see him, particularly with a M. Biot the Great Astronomer. . . .
>
> One evening he was invited to dine with Mad. Lavoisier in Ruenfond [*sic*]. She wrote him a most polite note to which he wanted to write an answer in french and to my great diversion he sat with his pen in his hand half the morning doubting whether he should write & what he should write. I told him I could think of no french letter but one which I had composed on a former occasion which began "Cher & adorable Jeanie." This Commencement did not I believe prove of much use for his note to the Comtesse but he afterwards wrote me a note beginning "Chere et adorable Cousine" which showed me he had not forgotten the lesson.[11]

Julia obviously enjoyed Wollaston's company and he was equally delighted to have such a young, vivacious companion for his second visit to Paris. Together they explored places of interest such as the mint, the Jardin des Plantes, and the Musée des Arts et Métiers. Wollaston was, as usual, constantly on the alert for items of novelty, and he did his best to keep Julia engaged by methods she well remembered:

> he took the greatest pains to explain to me what he thought would interest me and . . . he adopted a particular sign to make me remark what he wished to speak of afterwards, he used to put his finger & thumb to his lips & then point his forefinger very quickly to what I was to observe & when we got home, he said You remember what I pointed to on the right hand as we went in, that was so & so [and] on the left in the second room you saw such & such a thing & I generally contrived to remember all the different objects so as to comprehend what he had to say about them.[12]

At the end of October, Wollaston and the Alexanders departed Paris and returned to London, arriving there on November 5, 1816.

Wollaston remained close to the Alexanders after their European tour, and although the family usually wintered in the warmer climates of the Mediterranean, he frequently visited them during the parts of the year when they were at Finchley. When there, he often accompanied Julia on her regular horseback rides. When she ventured into the city, Julia often joined Wollaston and some of his sisters for lectures at the Royal Institution or orchestral performances at the Ancient Concerts (a concert series devoted to the music of the classical composers that Wollaston regularly attended). Julia mentions in her memorials that she frequently exchanged letters with Wollaston, although few of them are extant. The age difference between the two, and the determination of Wollaston to remain independent, suggest that there was no romantic attachment between them. It is possible, of course, that the Alexanders might have imagined a marriage to be a potentially beneficial outcome, but it seems more likely that Julia viewed Wollaston as a surrogate father or an honorary uncle, and he was happy to enjoy her company in reciprocal fashion.

MORE LEISURE TIME

Several months after the trip with the Alexanders, while suffering from another attack of lumbago, Wollaston wrote to Hasted with more information about the tour of Switzerland and France. He commented on his exposure to European Catholicism and his irritation at some of the necessities of international travel, especially the requisite surveillance of his movements. But he made it clear that the benefits of such experiences outweighed the negatives, and that he was planning on crossing the Channel at least one more time. In his words:

> If you ask what most exceeded expectation I should say the immensity of the snowy mountains of Switzerland. If you would know what of human concerns recurs most frequently to my recollection I should say the idolatrous follies of our brother Christians in the Catholic Countries we passed— . . . the recollection is melancholy as proof of the extreme weakness of human intellect upon such subjects.
>
> It is more gratifying to call to mind the discomfitures of bad accommodations or even the nausea excited by the nastiness of many, & in so doing enjoy the comparative neatness to be found at home, but above all things to reflect upon the petty watchfulness of suspicious governments & interrogatories of their officious police. . . . Yet tho' a foreign trip so much resembles descent into a mine that half the pleasure arises from

> the return to daylight, it is an experiment that will better bear repetition & I should certainly prefer it to any other disposal of my time.[13]

From his comments on the "idolatrous follies" of Catholics, one suspects that Wollaston tended toward the Latitudinarian version of Anglicanism, the less doctrinaire version of the Church of England, which had taken root in the eighteenth century and was much favored by many, like his father, who believed there was a role for human reason in matters of faith. Wollaston's religious leanings have to be guessed at, for there is no evidence of any sustained religious commitment in his life. He was also happy to be an Englishman, an easy enough attitude for one who was able to achieve such success under that country's laws, but it is also clear that he valued the freedom from governmental interference in personal affairs (income tax excepted) so characteristic of Georgian England.

In August 1817, Wollaston set out with Warburton for a four-week tour of Belgium and the Netherlands, spending ten days in Amsterdam.[14] There is little information available about this trip, other than a single notebook entry detailing the wide variation in the rental rates for horses and carriages in the various locales.[15] The travelers returned to London in time for Wollaston to head to Beechwood for the first of his three visits there in the fall of 1817. The number would have been four except for an attack of rheumatism in his right shoulder which prevented him from joining Marcet for another visit there in December.

One significant decision Wollaston made shortly after his return from the continent in 1816 was to resign as one of the two secretaries to the Royal Society. This freed him from many of the routine tasks and responsibilities that were delegated to the secretaries, but it did not end his active involvement in the Society's affairs. He was immediately elected in 1816 to the Council of the Society, a group of twenty Fellows who met regularly to assist the president with governance decisions. As a councilor, Wollaston was able to remain as a member of the Society's dining club, and he continued to be a regular at the weekly meals, often bringing with him one or two guests. He also stayed on as a member of the Committee of Papers, which decided which papers read to the Society would be printed in the *Philosophical Transactions*. In short, he retained many of the more intellectually and socially rewarding aspects of Royal Society engagement while freeing himself from the more tedious duties of a secretary.

This partial retreat from administrative commitments allowed Wollaston to spend more days at different times of the year away from London enjoying hunting and fishing sorties. In fact, the first week he was back from his 1816 continental trip, he made only one appearance at the Royal Society Club before leaving the city to spend the weekend with the Sebrights at Beechwood.[16] In addition, through his friendship with Sebright, Wollaston became acquainted with Henry Lascelles, a member of parliament for Northallerton, North Yorkshire, who was to become the second earl of Harewood in 1820. Lascelles was nearly the same age as Wollaston and had married John Sebright's sister, Henrietta, in 1794; together they had eleven children. The family residence was the palatial Harewood estate a few miles north of Leeds (the house is currently one of the historic Treasure Houses of England). Like his brother-in-law, Lascelles was very fond of shooting game and riding with the hounds, and regularly invited guests to Harewood at the peak sporting seasons. Wollaston became one of those guests, and from 1818 to 1825 he spent two to four weeks there each summer visiting with the large family and grouse shooting on the moors, often as part of a larger tour of northern England and southern Scotland.

Wollaston's excursions into the country became ever more frequent in the 1820s, but instead of the type of sightseeing tours that characterized his earlier holidays, they usually were structured around visits to the country estates of the wealthier men and members of parliament with whom he began to associate after the deaths of his father and Tennant. It is not unlikely that his own financial well-being and scientific reputation eased his entry into a more privileged circle of friends, but interacting with large families and engaging in outdoor activities appear to have been the principal factors that determined the associations.

In what appears to be a new initiative in his interpersonal relationships, Wollaston was elected in April 1818 to the exclusive Literary Club of London, known to its members simply as The Club. The Club's membership consisted mostly of parliamentarians, but other notables in the arts and sciences could be elected to one of the forty positions. The members met every second Tuesday for dinner at 6 o'clock during the sitting of parliament. Lord Glenbervie (Sylvester Douglas), a member who was elected a few months prior to Wollaston, listed good breeding as the first of the characteristics shared by the clubbers.[17] Aristocrats with philosophical interests did indeed form the core of the membership, but commoners of sufficient professional distinction could become members. This does not seem to be the type of social group in which Wollaston would have felt comfortable, but his work on several government committees must have gained him sufficient support from the parliamentarians to secure the requisite unanimous election vote.[18] Although Wollaston never became a frequent attendee at the fortnightly meetings, he did attend several times a year right up until a few months before his death. The Club was the only formal association that Wollaston is known to have joined that did not have a close connection to the sciences.

CONTINUING SCIENTIFIC WORK AND THE END OF THE PLATINUM BUSINESS

Not surprisingly, Wollaston's broader interactions with England's privileged classes and his renewed zest for the outdoors encroached somewhat on his scientific productivity. No longer did it seem as though each new issue of *Philosophical Transactions* contained something new and perceptive from him. And his colleagues noticed it. For example, Marcet wrote to Berzelius in 1819 and commented, "Our Wollaston is doing a thousand ingenious and useful little things, but for quite a while he has not undertaken anything glorious."[19] Nonetheless, Wollaston never entirely stopped pursuing his scientific interests.

Some of the "ingenious and useful" things that he did publish included an improved method of extracting iodine from kelp, one year after the discovery of the new chemical element (1813) by French chemists.[20] Another was the publication in 1816 of his study of the crystal edges of natural diamond, in which he explained the characteristics of the cutting edge that made diamond points so useful for scoring glass.[21] Also in 1816 he published an analysis of a sample of meteoric iron collected from the Bendigo iron mass in Brazil. Wollaston detected nickel in a one milligram sample of the iron by one of his unique micro methods and then used traditional small-scale methods to determine that nickel made up 4 percent of the iron sample.[22] These three papers, together with others (previously discussed) published about the same time, are again indicative of the breadth of his interests. None of them, after the 1814 publication on the scale of chemical equivalents, can be described as glorious achievements. But we should not forget that the bulk of Wollaston's laboratory time at this period of his life was still devoted to the purification, consolidation, and sale of malleable platinum.

As mentioned in Chapter 7, more than half of the 47,000 oz. of platina purchased by him was acquired from 1815 to 1819, and sales of malleable metal over that same period amounted to nearly 20,000 oz. It seemed as if the platinum business would thrive for years to come. Demand for the metal by gun makers remained strong, and requests for platinum boilers by sulfuric acid producers outstripped Wollaston's ability to supply them. However, after the Treaty of Paris brought peace to Europe in 1815, Spain tried to reassert control over New Granada, but the South American colony resisted the re-establishment of Spanish domination and accelerated its battle for independence. Finally, Venezuela and New Granada joined together as the Republic of Colombia and declared independence from Spain in December 1819. Naturally, the new country wished to keep its natural resources to itself and quickly shut down the contraband trade in platina. Wollaston's last payment to Johnson for platinum ore had occurred two years earlier, on September 26, 1817, and in his account book he wrote that the payment signaled "closing of the account with interest."[23]

Not surprisingly Wollaston explored every opportunity to obtain other sources of crude platina, and his notebooks contain several entries of unsuccessful attempts to obtain large amounts of the ore after 1819.[24] Finally, he collected all the platinum scraps he had bought back from purchasers of the metal and began his last series of purification batches on these remnants in November 1820. In July of the following year, he purified his last sample of platinum powder and in October 1821 delivered the last of the platinum ingots made in the normal way to William Cary for sale.[25] Only a few more small ingots of platinum made from recycled materials were intermittently supplied to Cary over the following four years, and they were sufficient only for a few specialized applications. Thus Wollaston's lucrative and technically-advanced platinum business came to a close in the early 1820s. Nonetheless, he still held the details of its production secret, perhaps hopeful that he could start up again should he ever be able to access new sources of the ore. Unfortunately, no such opportunity was to present itself.

ELECTROMAGNETIC ROTATION AND THE FARADAY INCIDENT

In early September 1821, Michael Faraday, who had begun his scientific career eight years earlier as a laboratory assistant to Humphry Davy, made the first major scientific discovery of his glittering career. He used an ingenious device of his own design to demonstrate that a metal wire with an electrical current flowing through it could be made to rotate continuously around a bar magnet, thus showing for the first time that the forces contained in electrical and magnetic fields could be made to do mechanical work.[26] This demonstration of electromagnetic rotation was quickly published and brought instant fame to its humble discoverer. Oddly, however, the scientific breakthrough brought tension to Faraday's relationship with both his mentor, Humphry Davy, and Wollaston. The unhappy personal and political aftermath, resulting from a jumble of miscommunication, unfounded rumor and intergenerational conflict, caused Faraday much grief. He even, for a time, thought that the consequences would be severe enough to jeopardize his chances of gaining admission to the Royal Society.

The discovery that launched a flurry of new experimentation into the interactions between electric and magnetic fields was Hans Christian Oersted's observation in 1820 that a current-carrying wire (a platinum wire was the first to be used) generated a magnetic field around its circumference.[27] Oersted's experiments came to Wollaston's attention during his term as president of the Royal Society, and he discussed their results at a Chemical Club meeting in August or early September.[28] After Davy learned of the discovery, he proceeded to replicate the crucial experiments with Faraday's assistance, and to try some new ones, at both the Royal and London Institutions in October and early November.[29] One novel experiment, designed to "put bodies magnetized by electricity in motion" was prompted, as Davy noted, by a conversation with Wollaston. In the experiment, wires of platinum, silver, or copper were laid in turn across two knife edges of platinum that were connected to the two ends of a voltaic battery. When one pole of a bar magnet was brought close to the current-carrying wire, it could be made to roll on the knife edges towards the magnet. The opposite pole of the magnet caused the wire to roll away from it. Although Davy said nothing in his paper about the nature or spatial orientation of the forces that caused the phenomena, it is obvious that both he and Wollaston were trying to devise experimental techniques to learn more about the action of those magnetic forces. But none of the three participants in these early experiments did much more on the topic for several months. Faraday was involved with other projects at the Royal Institution, while Wollaston's and Davy's attention turned more to the succession of one for the other as president of the Royal Society.

Nonetheless, Wollaston continued to think about electromagnetic interactions, and he speculated that a circumferential "electromagnetic current" surrounding a wire could be generated by a "helical electrical flow through it."[30] He then deduced that an external magnet, when brought close to a current-carrying wire, might interact with the magnetic field of the wire in such a way that the wire could be made to rotate continuously around its longitudinal axis. The challenge then became the design and construction of an apparatus that could demonstrate the predicted effect. Wollaston did not have a strong enough voltaic battery himself, so he continued to work with Davy at the Royal Institution to execute the necessary experiments. Their working relationship must not have been damaged much by their presidential skirmishing.

In April 1821, Wollaston brought an apparatus of his own (unknown) design to the Royal Institution, where he and Davy tried without success to make a current-carrying wire rotate on its own axis. Faraday was not present at the failed experiments, nor did he see the apparatus, but he did enter the room shortly thereafter and overheard the two men discussing their results. He then joined with them to carry out some more experiments on the magnetically-induced rolling of wires on knife edges, and he learned of Wollaston's desire to effect electromagnetic rotation.[31] Shortly after, Faraday agreed to write a review article on electromagnetism, and in preparation he repeated many of the novel experiments that had been done since Oersted's discovery. Then, in early September, he too tried to make a wire rotate on its own axis, with the same negative results that Wollaston and Davy had obtained. But, in the course of the experiment, he realized that a current-carrying wire consistently reacted to a magnet by moving sideways. Acting on this new insight, Faraday successfully designed an ingenious apparatus in which a current-carrying wire could be made to rotate around a bar magnet and vice versa.

Naturally, Faraday was eager to get his discovery into print as soon as possible. But he wanted first to get permission to acknowledge Wollaston's speculative ideas on electromagnetic rotation. Unfortunately, Wollaston was out of town when Faraday went to his house to get the desired approval, and so, wishing to publish his discovery before he too left London, Faraday published his paper on October 1 without the desired references to Wollaston's suggestive ideas and experiments.[32] Faraday soon came to regret this decision because, when he returned to the city, he was shocked to find that both his originality and character had been called into question by some from whom he had expected to receive compliments. In anguish, he wrote to James Stodart:

> If I understand aright, I am charged, (1) with not acknowledging the information I received in assisting Sir H. Davy in his experiments on this subject; (2) with concealing the theory and views of Dr. Wollaston; (3) with taking the subject whilst Dr. Wollaston was at work on it; and (4) with dishonourably taking Dr. Wollaston's thoughts, and pursuing them, without acknowledgment, to the results I have brought out.[33]

Faraday wrote to Stodart because he wished him to arrange a meeting with Wollaston (whom Stodart knew well) so that Faraday could explain his actions and refute the damaging comments about the genesis of his ideas. Faraday mentioned that he had already resolved the situation with Davy, but the rumored infringement on Wollaston's intellectual territory was especially troubling, because he knew that the prestige imbalance would inevitably work against him. After three weeks elapsed with no progress, Faraday took the matter into his own hands by writing to Wollaston to arrange a meeting to explain his actions and to give his older colleague a simplified version of the electromagnetic rotation apparatus "as a mark of strong and sincere respect."[34] Wollaston replied immediately, but only in the standoffish way he often used with people who had yet to gain his trust.

> Sir, —You seem to me to labour under some misapprehension of the strength of my feelings upon the subject to which you allude.
>
> As to the opinions which others may have of your conduct, that is your concern, not mine; and if you fully acquit yourself of making any incorrect use of the suggestions of others, it seems to me that you have no occasion to concern yourself much about the matter. But if you are desirous of any conversation with me, and could with convenience call tomorrow morning, between ten o'clock and half-past ten, you will be sure to find me.[35]

Faraday's predicament should have concerned Wollaston, and it should have prompted him to discredit the opinions of those who were expressing disapproval of the younger man's conduct. He might even have seized the opportunity to compliment Faraday on the originality of his discovery. Nonetheless, he did at least agree to a meeting. It is not known if that meeting materialized, but it is unlikely that Faraday would have passed up the opportunity to explain his side of the story to one whose support he so ardently sought.

Faraday continued to pursue his electromagnetical investigations, and, late in 1821, he was able to show that a current-carrying wire could be deflected by the earth's magnetic field, a phenomenon that was expected but had proven difficult to demonstrate. Pleased with his newest discovery, Faraday invited Wollaston to witness the demonstration device at the Royal Institution, and this Wollaston did on two or three occasions. At one of those encounters, Faraday tried again to gain Wollaston's permission to incorporate his ideas in the paper being written on the latest results. But, stubbornly, Wollaston told Faraday that he did not think it necessary to acknowledge his work, and Faraday did not. Unfortunately, heeding Wollaston's advice did little to silence Faraday's detractors, who saw the second paper as another missed opportunity to set the record straight.

Finally, in early 1823, Faraday felt obliged to clarify the misunderstandings by publishing a short historical summary of his and Wollaston's roles in the electromagnetic rotation discovery.[36] He showed Wollaston the account before publication, and Wollaston, after making some minor corrections, declared it to be "perfectly satisfactory."[37] In that historical summary, Faraday reported his conviction that Wollaston did not wish to have his preliminary ideas included in any of Faraday's accounts of his discoveries, and so he put the record straight by stating:

> It may have been about August 1820, that Dr. Wollaston first conceived the possibility of making a wire in the voltaic circuit revolve on its own axis . . . [and] it was at the beginning of the following year that Dr. Wollaston . . . came to the Institution with Sir Humphry Davy, to make an experiment of this kind. I was not present at the experiment, nor did I see the apparatus, but . . . I heard Dr. Wollaston's conversation at the time, and his expectation of making a wire revolve on its own axis; . . .
>
> It has been said I took my views from Dr. Wollaston. That I deny.[38]

For two reasons, Faraday believed it was critical to make a public statement at this time on the connection of his researches to Wollaston's. One was to correct a new and possibly damaging statement by Davy on the topic, and the second was to prevent his candidacy for Royal Society membership from being negatively affected.

On March 6, 1823, Davy read a paper to the Royal Society describing how he had been able to make liquid mercury rotate around a current-carrying wire by the action of a magnetic field.[39] In the concluding paragraph Davy went off topic to state that "we owe to the sagacity of Dr. Wollaston, the first idea of the possibility of the rotations of the electro-magnetic wire round its axis, by the approach of a magnet."[40] On the surface, Davy's comment was simply an appropriate reference to an influential suggestion, but its disconnection with the experimental content of the paper could easily lead one to question why Faraday had not made a similar acknowledgment in his publications. Further confounding the matter was a summary of Davy's oral presentation that was prepared for a report in the *Annals of Philosophy*, which appeared prior to publication of the full paper in the *Phil. Trans.* It was there that a controversial claim was added to the text by the report's author, Mr. Edward Brayley:

> Had not an experiment on the subject made by Dr. W[ollaston] in the laboratory of the Royal Institution, and witnessed by Sir Humphry, failed *merely through an accident* which happened to the apparatus, *he would have been the discoverer of that phenomenon.*[41]

Davy immediately thereafter denied making such a claim, and a retraction acceptable to Faraday was published in the subsequent issue of the reporting journal. Nonetheless, Brayley's baseless statement fueled the suspicions of those who continued to suspect that Faraday had benefitted from Wollaston's ideas. For Faraday, the misleading statement could not have appeared at a worse time, for on May 1, 1823, his certificate of admission to the Royal Society was first placed on display for the required ten-meeting period that preceded balloting for admission. Although Wollaston had done little to this point to quell the rumors that were harming Faraday's integrity, he finally acted honorably by placing his name first on Faraday's certificate of admission. Unfortunately, even that gesture was insufficient to placate everyone.

In late May, a few weeks after Faraday's certificate had been placed on display, Davy angrily confronted him to say that he wished the application to be withdrawn and intended to use his power as president to do so if Faraday did not act. Faraday replied that only his nominators could withdraw his admission certificate.[42] With Davy's intentions so ominously stated, Faraday needed his primary nominator's support more than ever, for Wollaston's status in the Society was nearly equal to that of Davy's. But Wollaston was not the type to campaign actively for Faraday's cause, and his more politically active friend Henry Warburton was one of those who believed Faraday had acted improperly. In fact, Warburton had even decided to speak out in opposition to Faraday's candidacy. Such an action from one who claimed to represent the feelings of other Wollaston supporters could have been a fatal blow to Faraday's chances, so he wrote to Warburton on May 30 to request a meeting of reconciliation.[43] Warburton did meet with Faraday on June 5, and that meeting, together with another with Wollaston on June 14, likely influenced the construction of Faraday's historical statement on electromagnetism, which was published on July 1.[44] Certainly the published statement mollified Warburton, who thereafter informed Faraday that he had decided to support his election.[45]

The voting for Faraday's membership, conducted on Jan. 8, 1824, was successful; only one negative vote, by an unknown member, was cast. Davy elected not to attend, and Wollaston was away at the Althorpe estate of Lord Spencer at the time of the balloting, but both were present the following week when Faraday paid his admission fee and formally became a member.[46] As the course of events suggests, widespread support for Faraday's election depended largely on resolving the electromagnetic priority issue. That support could then overwhelm whatever negative sentiment Davy's opposition had generated. Wollaston's approval of the historical statement could well have been crucial to the successful outcome.

MORE NOVEL OBSERVATIONS

Although the period from 1815 to 1820 was a comparatively fallow one for the type of scientific discovery that had become synonymous with his name, Wollaston's creativity re-emerged about 1820 with the publication of a few chemical and optical works together with two pioneering physiological studies. In 1820, he published directions for the construction of a two-component quartz prism capable of splitting a ray of light into two separate beams (now known to be rays of orthogonal linearly polarized light).[47] A similar prism had earlier been reported by Abbé A. Marie Rochon, but without details of the prism's design. Wollaston was able to duplicate and improve the design by eliminating the effects of chromatic dispersion and specifying an arrangement of the two glued-together triangular segments that doubled the normal angle between the emergent light rays. The design, now known as a "Wollaston prism," continues to be widely used in modern optical devices such as beam splitters and polarizers. Two years later, Wollaston described a method, based on a careful analysis of a Dollond-made triple object-glass in the telescope passed on to him by his father, of aligning the three lenses of the object lens to obtain the sharpest possible image.[48]

In 1823, in addition to his paper on the finite extent of the atmosphere, he published two analyses of small metallic-looking cubical particles taken from the slagheap of a Welsh ironworks. He mistakenly concluded they were made of titanium, an element that had yet to be produced in its metallic form by chemical means.[49] However, it was shown two decades later that the particles were a composite of titanium nitride and titanium cyanide. In a second paper on the same material, Wollaston correctly concluded that the weak magnetism detectable in the cubical "titanium" crystals was due, not to the purported metal itself, but to small amounts of iron impurities.[50]

The third paper published the same year was a description of Wollaston's method of identifying magnesia in chemical compounds by application of one of the micro methods that so impressed his contemporaries.[51] Whereas other analysts typically used flasks and crucibles with reagent weights of ounces or more, Wollaston customarily worked on a scale hundreds of times smaller, generally using only grains of substances. An excerpt from the magnesium procedure provides a nice example of his unique and impressive techniques.

> Dissolve in a watch glass, at a gentle heat, a minute fragment of the mineral suspected to contain magnesia . . . in a few drops of dilute muriatic acid; to this solution, add oxalic acid . . . [and] a few drops of a solution of

> phosphate of ammonia or soda. Allow the precipitate to settle for a few seconds, and decant a drop or two of the supernatant clear liquid on a slip of window-glass; on mixing with this liquid two or three drops of a solution of the scentless carbonate of ammonia, an effervescence takes place; draw off to one side with a glass rod, a little of the clear solution, and trace across it, with the pressure of a point of glass or platina, any lines or letters on the glass plane; on exposing this to the gentlest possible heat . . . white traces will be perceived wherever the point was applied. These consist of the triple phosphate of ammonia and magnesia.[52]

When showing this analysis to young people Wollaston would trace out their name in the penultimate step, so that the precipitate spelled out their names. By such engaging methods, Wollaston was able to instill a love of chemistry into the offspring of many of his closest friends.

PIONEERING PHYSIOLOGICAL RESEARCH

At the meeting of the Royal Society at which he was elected president on June 29, 1820, Wollaston delivered a paper on the inability of some people with otherwise normal hearing to discern sounds of very low or very high frequencies.[53] He had noted years earlier, while still practicing medicine, that persons with partial hearing loss were especially insensitive to low-frequency sounds. He discovered that he could mimic the condition by holding his mouth and nose closed while forcefully attempting to inhale. This had the effect of lowering the pressure on the inner side of the eardrum (via the Eustachian tube) such that the greater relative pressure on the outer eardrum rendered it less responsive to low-frequency sounds. By repeated experiment, Wollaston found that, in such a state, his ear could no longer hear sounds below the musical tone denoted by base clef F (ca.175 Hz, in modern units), significantly above the normal low-frequency cutoff for humans of 20–30 Hz. However, he could not better quantify low-frequency abnormalities because of the lack of suitable tone generators.

Wollaston's investigation of insensitivity to high-frequency sounds produced a much more interesting result, as he was the first to discover wide variability in an individual's ability to hear high-pitched sounds, as well as the general loss of high-frequency hearing with age. He was first drawn to the subject by observation of a friend's inability to hear the F note (5588 Hz) four octaves above middle E of a pianoforte, although the friend could discern the E note immediately below it. Wollaston judged his own high-frequency hearing limit to be that of a sound from a small pipe one-quarter of an inch in length, which he equated to six octaves above middle E (ca. 22,300 Hz, near the upper limit of normal human hearing). From conversations with children, who could hear the sounds of some insects which were imperceptible to himself, Wollaston concluded that those sounds might be an octave higher than his own upper limit. Consequently, he predicted what we now know to be true, that some species of animals can communicate via sounds far beyond the capability of humans to hear them.

In addition, Wollaston determined that the range of human hearing extended over nine octaves, from the lowest-frequency organ notes to the highest frequency sounds of insects. Moreover, the transition from audibility to non-audibility at high frequencies occurred suddenly, often from one musical note to another. One of those subjected to Wollaston's hearing experiments was Sir Henry Bunbury, a longtime friend who frequently hosted him in the 1820s at the family estate near Bury St. Edmunds. Bunbury recounted the details several years later.

> Whenever he came to Barton, or to Mildenhall, he was sure to have some new object of inquiry in his mind, or some new discovery to communicate. One year he would pretend to be examining a book in a distant corner, when there was a large party in the library; then would he sound an extremely acute note on his little pipe, and glance round to observe who caught the sound, and who were unconscious of it.[54]

Wollaston's investigation of hearing abnormalities was novel chiefly because it focused on hearing loss dependent on the frequency of sounds, rather than their loudness. In addition, it directed the attention of later investigators to the sudden cutoff of sensibility to high-frequency sounds, the individual variability of the point of that cutoff, and the general decline of high-frequency response in adults. It was a fine paper to deliver to the Royal Society on the evening he became its president.

Wollaston's second innovative physiological study in these years was an investigation into the structure of optic nerves, deduced from visual aberrations he had himself experienced.[55] The study was seen at the time as an important insight into the possible microstructure of the optical nerves, but has lasting value for its connection of visual disturbances with what came to be known as migraine headache. The investigation was sparked by two occasions on which Wollaston lost for a short period of time half of his visual field (an optical affliction now known as hemianopsia), as he describes in the paper.

> It is now more than twenty years since I was first affected with the peculiar state of vision, to which I allude, in consequence of violent exercise I had taken for two or three hours before. I suddenly found that I could see but half the face of a man whom I met; and it was the same with respect to every object I looked at. In attempting to read the name JOHNSON, over a door, I saw only SON; . . . and was the same whether I looked with the right eye or the left. . . . The complaint was of short duration, and in about a quarter of an hour might be said to be wholly gone, . . .[56]

Wollaston dismissed this first visual impairment as a temporary, and nonworrisome, consequence of overfatigue. But his curiosity about the phenomenon was increased when it occurred again two decades later. In this second instance, he lost the right half of his field of vision. His explanation, unaccompanied by any illustrations, can be understood by reference to an eighteenth-century sketch of the optic nerves published by John Taylor, shown in Figure 13.1.[57] In Taylor's sketch, probably unknown to Wollaston, the optic nerves are greatly expanded to illustrate the postulated structure of their crossing (known as decussation) in the nerve bundle (the optic chiasm) just before entering the two halves of the brain (not shown, but at the very bottom of the diagram). Wollaston had a similar construction in mind when he speculated that objects in the right half of human vision, such as the arrow head *A*, impinged on the left half of the retina of each eyeball. The nerve signals generated there passed through the left optic nerves (shown unshaded), to the optic chiasm, where the nerve fibers from the left hemisphere of the right eye crossed to the left to join up with those from the left eye and proceed on to the left half of the brain. A similar routing existed for the transmission of objects in the left field of vision, such as the arrow tail *B*, through the right half of each eyeball's retina, shown shaded, to the right half of the brain. In the brain, the nerve signals from the two eyes were integrated into a single image. Since only half of the optic nerves from each eye had to cross over in the optic chiasm, Wollaston dubbed the postulated structure "semi-decussation of the optic nerves." From such an anatomical supposition, Wollaston was led to conclude that, for example, an "injury to the left thalamus [of the brain] would occasion blindness . . . to all objects situated to our right."[58] Unfortunately, in the early nineteenth century, the course of optic nerves through the optical chiasm could not be verified or refuted by dissection, which was incapable of revealing the pathway of specific nerve bundles. Moreover, the suggested semi-decussation of the optic nerves was not a novel one, for others had previously invoked the same type

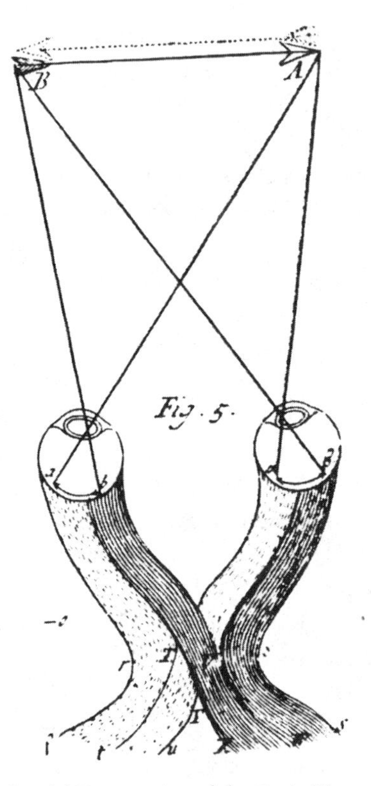

Fig. 13.1. Partial Decussation of the Optic Nerves, 1738.

of structure as a requirement for binocular vision or for an explanation of half-vision loss.[59] However, there is no evidence that Wollaston was aware of such prior thought, because he made no reference to it in his paper.

The loss of half of the field of vision in both eyes, he was led to believe, resulted from some sort of disease in the one half of the brain that received the relevant optical nerves from the optical chiasm. From other observed cases, Wollaston concluded that half vision could be either permanent or, as in his own case, temporary. In the gravest case he cited, a friend was afflicted with severe pain behind his left eye, which was accompanied by a permanent loss of vision to his right. Wollaston attributed the affliction to a permanent compression on the left thalamus, a brain tumor in other words. Not surprisingly, Wollaston thought the chances of recovery were "very doubtful."[60] Two other cases of hemianopsia, brought to Wollaston's attention after the bulk of his paper had been written, were temporary like his own, but both reoccurred frequently over a period of years. In one case, the visual abnormalities were usually accompanied by stomach distress and in the other by headache. Wollaston reported neither of these accompanying symptoms for his own two events, which were unusual because they differed in the half-field of vision lost. Consequently, he did not think there was a relation between them nor that the symptoms were cause for concern.

Wollaston's paper was widely reprinted in continental journals and did much to spark interest in the structure of the optic chiasm. The German physiologist Johannes Müller did much to popularize Wollaston's ideas on semi-decussation, and ultimately the proposed structure was demonstrated to be correct. Unfortunately, Wollaston failed to make a connection suggested by the symptoms experienced by his last two examples of hemianopsia. In those cases, and perhaps his own, visual disturbances, upset stomach, and headache could all have been symptomatic of migraine, an affliction which only became satisfactorily characterized much later in the nineteenth century, and more completely understood in the twentieth. It is now known that some migrainous attacks can occur with no other symptoms than visual disturbances such as hemianopsia. Wollaston was ultimately to die of a brain tumor situated in his right thalamus, and most of those who have commented on Wollaston's episodes of half vision have attributed them to the tumor, but there is good reason to believe that his hemianopsia was instead a manifestation of migraine.

In the last year of his life, Wollaston experienced two instances of a new type of visual disturbance, later to be described as "fortification spectra" and connected with migraine.[61] The second of these occurred while on a visit to the Isle of Wight in August 1828, and Wollaston made a sketch, never published, of the zigzag line that appeared in his visual field, as shown in Figure 13.2.[62] Such fortification spectra were later described and related to migraine attacks by the Astronomer Royal, George Airy, who had been drawn to the topic in part by Wollaston's paper.[63] Although it is impossible to rule out a brain tumor as a cause of some of Wollaston's later visual irregularities, its presence in the right thalamus would be expected to cause left-side vision loss, such as Wollaston's first reported incident. But it is extremely unlikely that such a tumor, if it existed at that time, would fail to cause problems for over twenty years. More probably, the first occurrence of half vision, the subsequent change in the direction of vision loss in the second incident, and the incidences of fortification spectra, are all symptomatic of migraine.[64] Such a conclusion is strengthened by a statement made by the surgeon Benjamin Brodie, who attended Wollaston in his final days and claimed that Wollaston had told him he had experienced episodes of half vision since his childhood.[65] Another colleague, William Brande, mentioned that Wollaston was subject to frequent and violent headaches,

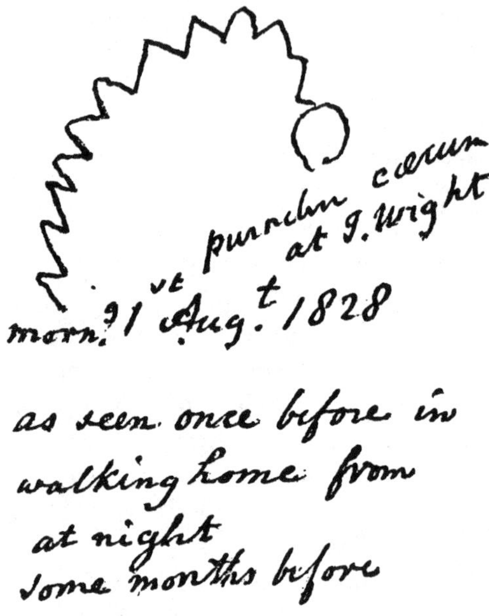

Fig. 13.2. Wollaston's Fortification Spectrum, 1828. Reproduced by kind permission of the Syndics of Cambridge University Library.

which he endured by taking a dose of powdered colomba root for his upset stomach and then sitting perfectly quiet for half an hour, after which he would declare that he was "himself again."[66]

Of course, Wollaston could not have known that his visual disturbances, headaches, and upset stomach were symptoms of the ailment later to be characterized as migraine. Instead, the possibility that his instances of half vision might have been caused by some sort of brain disease must have troubled him in the last decade of his life. And Wollaston's worst fears were to be realized a few years later, as unmistakable signs of brain damage began to manifest themselves. But, as we will see in Chapter 14, Wollaston remained very active until a month or two before his death.

THREE REMARKABLE WOMEN

Another association that blossomed during the early years of Wollaston's term on the Board of Longitude was that with his fellow resident commissioner, Henry Kater, and his family. Kater, a fellow member of the Royal Society, served on the royal commission that sought to regularize the imperial standards of weights and measures and designed several instruments for use by the polar expeditions. So he and Wollaston would have known each other well by the time they were appointed to the Board of Longitude in 1818. The Kater house, together with the Beechwood estate of the Sebrights, became the venues at which Wollaston interacted most frequently with three of Britain's most accomplished female writers of the early nineteenth century, Jane Marcet, Mary Somerville, and Maria Edgeworth.

We learn much about this social group from letters written in 1822 by Maria Edgeworth, the great Irish novelist. Maria and her much-younger half sisters Harriet and Fanny visited Sebright's Beechwood estate in January 1822, where they were welcomed by Mrs. Marcet, Dr. Wollaston, and the famous Mrs. Somerville, as Maria referred to them.[67] Jane Marcet, the London-born daughter of the banker Anthony Haldimand, had married Alexander Marcet in 1799 and became mother to four children. With the encouragement of her husband she wrote (at first anonymously) the very popular instructional book *Conversations on Chemistry*, first issued in 1805. This was followed by *Conversations on Natural Philosophy* (1819), and, two years after the Beechwood meeting with the Edgeworths, *Conversations on Political Economy* (1824). It was she who initiated a lifelong friendship with Mary Somerville shortly after the Somervilles took up London residence in 1816 a short distance from her own home, then in Russell Square.[68]

Mary Somerville (neé Fairfax) was born in Jedburgh, Scotland, and had two children with her first husband, who died in 1807. In 1812 she married the London doctor and Fellow of the Royal Society William Somerville, with whom she had four more children. While raising her family, she continued the careful study of the physical sciences she had begun as a young girl, often by study of French and Latin texts, and she soon became well-known for her mathematical and astronomical prowess. She was later to publish an informed exposition of Laplace's complex celestial mechanics, which she entitled *The Mechanism of the Heavens* (1831). This work established her reputation as a formidable intellect and reliable science writer. She was to follow up that work with the oft-republished *On the Connexion of the Physical Sciences* (1834), as well as *Physical Geography* (1848) and *On Molecular and Microscopic Science* (1869).[69]

Shortly after the Somervilles and their young family returned to London in 1818 after a year abroad, they began to associate regularly with

Wollaston, who was happy to provide scientific expertise whenever asked. Their association was described by Mary in her autobiography

> At this time [1818] we formed an acquaintance with Dr. Wollaston, which soon became a lasting friendship. He was gentlemanly, a cheerful companion, and a philosopher. . . . We bought a goniometer, and Dr. Wollaston, who often dined with us, taught Somerville and me how to use it, by measuring the angles of many of our crystals during the evening. I learnt a great deal on a variety of subjects besides crystallography from Dr. Wollaston, who, at his death, left me a collection of models of the forms of all the natural crystals then known.[70]

Mary Somerville also provides information on the London group centered on the Katers, which provided her with social and intellectual support.

> Somerville and I used frequently to spend the evening with Captain and Mrs. Kater. Dr. Wollaston, Dr. Young, and others were generally of the party; sometimes we had music, for Captain and Mrs. Kater sang very prettily. All kinds of scientific subjects were discussed, experiments tried and astronomical observations made in a little garden in front of the house.[71]

It was Wollaston's close relationship with the Somervilles that finally made him relent in his persistent refusal to sit for a portrait. Mary was so insistent in her desire for one that the two finally agreed that John Jackson be chosen as the artist.[72] The preliminary pencil sketch of Jackson's subject is shown in Figure 13.3,[73] and the oil portrait made from it about 1823 remained in the Somerville's possession for many years before it was given to a member of the Wollaston family. After Wollaston's death, William Somerville engaged Jackson to make a life-size copy of the original, which was thereafter given by Wollaston's brother and executor, George, to the Royal Society, where it now hangs.

In contrast to Jane Marcet and Mary Somerville, Maria Edgeworth was exclusively a literary novelist. She was the eldest daughter of the Irish landowner and educational writer, Richard Lovell Edgeworth, who was to have a total of twenty-two children with four wives. Although born in England, Maria moved to Ireland in 1782 with her father to help him manage the family holdings near Edgeworthstown and to superintend the education of her many half-siblings. She began to make a name for herself by promoting female writing in *Letters for Literary Ladies* (1795) and attracted much

Fig. 13.3. John Jackson's Sketch of Wollaston, ca. 1823.
© National Portrait Gallery, London.

attention with her first historically-based novel *Castle Rackrent* (1800). Thereafter she continued to publish a succession of novels and educational books, becoming "the most commercially successful novelist of her age."[74] She was also a compulsive letter writer, and several of her letters mention Wollaston and his circle of friends.

While in the presence of her more scientifically knowledgeable companions at Beechwood in early 1822, Maria Edgeworth took advantage of the opportunities that arose to learn what she could of scientific matters, much of it from Wollaston. Later, during the few months they stayed in London in the spring of 1822, Maria and her sisters interacted with "the best scientific and literary society" that frequented the homes of the Marcets, the Somervilles, and, especially, the Katers—the last becoming their favorites.[75] And Wollaston, not so tied to his platinum process as he had been in earlier years, was a frequent guest at all three domiciles. On one occasion, the three Edgeworth sisters were guided by Wollaston and Kater on a tour of Greenwich Observatory, and in the carriage the sisters benefitted from the men's "delightful conversation all the way there and back again."[76] Of course, science was not the only unifying force in this close-knit social group, and in another letter Maria describes an evening of music and Wollaston's total immersion in its effects.

> There was no party—only Somervilles Katers and Wollaston happy together. . . . It was worth while to bring Sir W. Pepys to hear her [Mrs. Kater] singing he really admires and enjoys it so much. So does Wollaston who sits mute as a mouse and still as a statue of a philosopher charmed while she is singing. He has a great deal of feeling.[77]

At the beginning of June 1822, the Edgeworths left London and set out on a rambling return to Ireland. On the way, they stopped for a few days at the country home of Thomas Carr of Frognel, Hampstead, just to the north of the city, where a large group of friends and acquaintances awaited them. Wollaston arrived on their second day there, and Maria described for her stepmother some uncustomary behavior on his part.

> Second day . . . Wollaston particularly pleased with Harriet who sat beside him at dinner and to whom he talked unusually veiling for her the terrors of his beak and lightning of his eye. He gathered a large Rhododendron and put it in her sash—all the rest of the young ladies having one in their heads.[78]

It does not appear as if Harriet herself attached any significance to the gesture, but she did mention another encounter with him a few weeks later at Beechwood in a letter to her aunt. She, too, was struck by his willingness to instruct, and the penetrating effect of his gaze.

> [At Beechwood I again met] Dr. Wollaston—his countenance appears to me when he is silent to be more melancholy than striking—but when he speaks his eyes have a fixed and brilliant light different from any others I ever watched. He is ready to communicate whatever he knows to any person who asks & seems rather to shun display or exertion in general conversation.[79]

The Katers, the Somervilles, the Marcets, and the Sebrights were all vibrant, scientifically-literate families with a broad range of shared interests who were happy to include a middle-aged bachelor in their activities. It is clear from the stories recounted in the letters and memoirs of some of the group that Wollaston enjoyed his social, intellectual, and sporting experiences with them, and that their companionship had largely replaced that of his own family members by the 1820s.

CHAPTER 14

The Last Years 1824–1828

> Die, however, of his present attack, I sadly fear he must . . . I say *fear* when one might rather be expected to say *hope*, considering the physical dependence of one who has ever been the most physically as well as mentally independent of any man I ever knew.[1]

Wollaston remained in generally good health as he approached the age of sixty in the mid 1820s, and his trips to the countryside estates of his many sporting friends increased in frequency. He even became an avid fisherman and, after his short stint as president of the Royal Society, he was never again to commit himself to anything that would infringe on his freedom of movement. Of course, the termination of the platinum business in 1820 caused his yearly revenues to fall precipitously, but the return on his investments in government bonds and dividend-bearing stocks ensured that he could live comfortably for the rest of his life.

FAMILY, FRIENDS, AND FISHING

As discussed earlier, when Francis Wollaston died, he distributed his estate among all the surviving Wollaston children except for the youngest boy, Henry, whom he had previously helped toward a career in banking. But Henry struggled to find success, and there may have been some sort of falling out between him and his father. In early 1823, Henry asked William to intercede on his behalf with the prime minister, Lord Liverpool, for some position in the government, but that was something his elder brother could not bring himself to do. William understood that gentlemen were expected to repay political favors in some way at some time, and he was not willing to take that burden upon himself, even for the benefit of a brother. So, instead, William opted to give Henry an outright gift of £10,000 of 3 percent government annuities.[2] This was a significant sum, amounting to about one-quarter of the amount William held in his Bank of England accounts at the time. Although William informed his brother that he did not believe he would live long enough to have need of the transferred money, there is no evidence that he expected to die anytime soon. But he could not have been blind to his own mortality, for two close friends, Edward Daniel Clarke of Cambridge and Alexander Marcet, both died in 1822. In fact, Wollaston made no concessions to his age and remained active, spending ever more time each year away from London.

Many of Wollaston's excursions to the countryside for the shooting of game have already been noted, and they became supplemented in the 1820s by frequent fishing trips. He had taken up fishing about the same time he began visiting John Sebright at Beechwood, and his daybook contains a record of the purchase of fishing tackle in 1813, followed shortly thereafter by one for the printing of fish identification cards.[3] In the following years, he customarily spent a few days each year visiting prime fishing locations and soon became an adept fly fisherman. When possible, he combined his two favorite sporting pastimes, as in a carefully planned trip to the north of England and Scotland in 1822. While visiting the Somervilles near Jedburgh, Wollaston was invited to join in a day of coursing after rabbits at the nearby Abbotsford estate of the popular novelist and poet Walter Scott, a potentially risky adventure for one whose horseback riding days were well behind him. There he met up with Humphry Davy, also an avid angler, who was in the area fishing with a companion. The contrasting styles of the two recent competitors for the presidency of the Royal Society is well captured in a published account of the event:

> It was a clear, bright September morning . . . , and all was in readiness for a grand coursing match on Newark Hill . . . among a dozen frolicsome youths and maidens . . . [were] Sir Humphry Davy [and] Dr. Wollaston. . . . But the most picturesque figure was the illustrious inventor of the safety lamp [Davy]. . . . Dr. Wollaston was in black, and with his noble serene dignity of countenance might have passed for a sporting archbishop.[4]

Despite their quite different personalities, and occasional clashes, Davy's and Wollaston's mutual interests in science and outdoor pursuits drew them closer in the 1820s. In 1823, they even traveled together on a six-week tour of Ireland that included a visit to the Edgeworths and, one assumes, some shooting and fishing.[5] When Davy, shortly after Wollaston's death, published a conversational book on angling, he made mention of his creative counterpart's fishing interests.

> There was . . . an illustrious philosopher, who was nearly of the age of fifty before he made angling a pursuit, yet he became a distinguished fly-fisher, and the amusement occupied many of his leisure hours during the last twelve years of his life. He, indeed, applied his pre-eminent acuteness, his science, and his philosophy to aid the resources, and exalt the pleasures of this amusement.[6]

Although Wollaston and Davy were frequently in each other's company, their personalities were so dissimilar that it is unlikely they could ever have become close friends. Davy did, however, use Wollaston's character as the basis for one of the individuals in his fishing-themed book, *Salmonia*.[7]

Wollaston's enjoyment of fishing also enhanced his friendship with Francis Chantrey, England's most acclaimed sculptor. Other than his choice of profession, Chantrey had much in common with Wollaston. Both were raised in the country and moved on to build prosperous and financially rewarding careers in London, and both shared a talent for careful observation of the world around them. Wollaston probably first became acquainted with Chantrey when the artist began using a camera lucida to capture the outlines of a subject's head prior to sculpting a marble bust, as discussed in Chapter 6. Subsequently, after Chantrey became a Fellow of the Royal Society (in 1818), he occasionally attended the Royal Society dining club as a guest of Wollaston, and in the early 1820s, if not before, the two regularly fished together. Both men became members of the exclusive Houghton Fishing Club in 1824, most likely through their association with Henry Warburton, one of the expert anglers who founded the club in 1822.[8] The Club had fishing rights to a stretch of the river Test near the village of Houghton in Hampshire, a chalk river widely recognized as the birthplace of modern fly fishing. Chantrey and Wollaston often traveled to the river together for the prime fishing seasons, and Wollaston quickly became one of the most successful anglers of the Houghton Club, as totals of yearly catches in the Club's record books reveal.[9]

It is not surprising, then, that when Julia Hankey wished to commission a bust of Wollaston she selected Chantrey to be the sculptor, and sometime in the late 1820s he finished the work. Julia, who married Thomas Bathurst in 1829, retained possession of the bust until her death in 1877,

Fig. 14.1. Sir Francis Leggatt Chantrey's Bust of Wollaston, late 1820s.

after which it was given by her son the sixth Earl Bathurst to the Royal Institution, where it is now prominently displayed.[10] Chantrey's preliminary plaster cast of the bust (shown in Figure 14.1), is in the collection of the Ashmolean Museum.[11] Chantrey prided himself on capturing the essential character of his subjects, and he drew attention in the bust to the penetrating gaze of his scientific friend. The set of those eyes also impressed the famed portraitist, Thomas Lawrence, who drew them for an illustration in a paper of Wollaston's on the physiognomy of portraits.

Wollaston's interest in art prompted him to seek an explanation for the perception that the eyes of a person in a portrait appeared to retain their direction of vision regardless of the angle from which the portrait was viewed.[12] Wollaston began by confirming that, when we look directly at a person, the most obvious signal of the direction of the eyes relative to the face is the whites of the eyes. Nonetheless, he recognized that such a measure of eye-turn was perceptually overwhelmed by the appearance of the more prominent features of the face, especially the nose, an early foray into the psychology of vision.[13] When he showed his friends simple sketches of faces that had identical pairs of eyes placed in faces aligned in different directions, they were so certain that the representations of the eyes in the different faces had been altered to agree with facial direction that he had trouble convincing them that there was not some visual trickery involved. To prove his point, Wollaston included a number of facial sketches in the paper, showing how an artist could present a variety of facial directions with no changes to details of the eyes themselves. To make the sketches, Wollaston engaged Thomas Lawrence to draw a pair of eyes of a person looking directly forward. After learning of the qualities desired in the sketches, the artist quickly retorted, "'I know the very eyes you require—sit down, for you are the possessor of them.' And Wollaston, though sorely against his inclination, sat to the great painter, and his eyes are those introduced in the engraving."[14]

After concluding that the direction of eyes in a portrait is a consequence of the general position of the face and the turn of the eyes from that position, Wollaston moved on to an explanation of why the gaze of such eyes remains unchanged when a portrait is viewed from different directions. This was simply explained by a principle of perspective drawing, in which one object directly above another must lie in the same vertical plane, which remains the case when viewed from any oblique direction. Although the portrait paper contained no new science, its novel conclusions are illustrative of Wollaston's near universal interests and aptitude, undiminished by age or success, for penetrating observation and rational deduction.

THE END OF SCIENTIFIC WORK

In the autumn of 1823, while preparing to recharge his gun on a shooting expedition at Beechwood in September, Wollaston's gunpowder flask exploded in his right hand, injuring it badly. The explosion drove the nozzle of the flask into the thumb and forefinger of his left hand, cutting and bruising it also. One month later, as Wollaston described in a letter to Hasted, his right hand had healed sufficiently to allow him to write, but the digits of his left hand remained swollen, sore, and slow to heal.[15] To compound his discomfort, a week after writing to Hasted Wollaston learned that his older brother Francis, who had become archdeacon of Essex, had died on October 12 of what was believed to be a stroke. The death of a brother only four years older from the same type of affliction that had taken the life of his father must have made Wollaston fear some familial predisposition to paralytic seizures. Perhaps he even began to wonder whether his instances of half vision were as benign a condition as he had assumed them to be.

In spite of his desire to leave the practice of medicine and all its unhappy connections behind him in 1800, Wollaston acknowledged the benefits he had accrued through membership in the Royal College of Physicians by donating £50 to that organization's building fund in 1822.[16] Encouraged perhaps by that generosity, or by the opportunity to have a past president of the Royal Society play a more active role in their organization, the governing group of the Royal College of Physicians, the Elects, voted unanimously in February 1824 for Wollaston to fill a vacancy caused by the death of one of their number.[17] This was a puzzling decision because Wollaston had not attended a meeting of the College since 1800, and he had no interest in engaging himself in its governance. Consequently, he resigned the position one year later, after having attended only one meeting. However, participation in scientific initiatives still had some appeal, and when an opportunity arose to take part in a new government-sponsored investigation, he agreed to become involved.

On April 1, 1824, the Board of Longitude recommended a committee be formed to oversee a program of research designed to improve the quality of highly refractive optical glass, since it was believed that England's pre-eminence in optical lens construction was being eclipsed by continental glass makers.[18] The following month, the Council of the Royal Society acted on the recommendation by naming several qualified persons, including Wollaston, to a Joint Committee for the improvement of optical glass.[19] That committee one year later created a three-person subcommittee to carry the project to completion. Each of the subcommittee members had specific responsibilities: Michael Faraday was to formulate and prepare improved glasses, George Dolland was to grind them into shape, and John Herschel was to test them.[20] The most challenging part of the study was the production of homogeneous glass blanks that were free from defects such as density variations and striations. From entries in his notebooks, we learn that Wollaston provided some of the platinum used by Faraday for things such as containers in which the glass components were melted together, stirrers and ladles for the molten mixtures, and powder to promote the dissipation of bubbles.[21] Wollaston by this time had only a few small ingots of platinum still in his possession, and he used one of them to prepare platinum sheeting for Faraday's use.[22] The project generated nothing radically new, however, and Faraday brought the investigation to an end in 1829.

By 1824, it became evident to Wollaston that there was no prospect of resuming the production and sale of malleable platinum, so there was no need for him to keep the large house in Buckingham St. and its adjacent rented laboratory. Consequently, in September he moved to 1 Dorset St., Marylebone, taking his three reliable servants with him. The new house had no room for large scale laboratory work, so Wollaston focused instead on small-scale chemistry, mineralogy, optics, and some astronomy. His studies in the new residence resulted in three publications that were rushed into print within a few months of his death in late 1828.

Wollaston had always been interested in astronomy, and after he settled in at Dorset St., he set out to complete a study he had begun in his last years as a physician. He wanted to compare the brightness of the sun to that of two bright stars. He did this by first making equal the image of a candle reflected from a small thermometer bulb as seen with one eye and the image of the sun reflected from a distant thermometer bulb and viewed through a telescope by the other eye.[23] He then compared a reflected candle image with a telescopic image of the stars Sirius and Lyra (a constellation whose principal star is Vega), and after accounting as best he could for atmospheric effects, image size, and losses due to reflection off a curved surface, he concluded that Sirius produced as much light as 13.8 suns and Lyra was equal to about 1.5 suns.[24] Wollaston understood the inherent difficulties of his methods and hoped to refine the observations by obtaining more measurements over a long period of time. However, as his health deteriorated, he arranged for his preliminary results, which he asserted were more valuable for their methodology than for their results, to be presented to the Royal Society a few days before his death.

At the other end of the observational spectrum, Wollaston improved the performance of simple microscopes by making two alterations to traditional designs, which he described in a paper delivered in November 1828.[25] The first improvement was to deliver light that had been reflected off a plane mirror and focused by a plano-convex lens (ET) onto the object to be examined (a), as shown in Figure 14.2. The second was to magnify the object with two plano-convex lenses (M), whose plane surfaces were mounted facing the object. The best image was formed when the focal lengths of the two lenses were in a 3-to-1 ratio and the distance between them, which

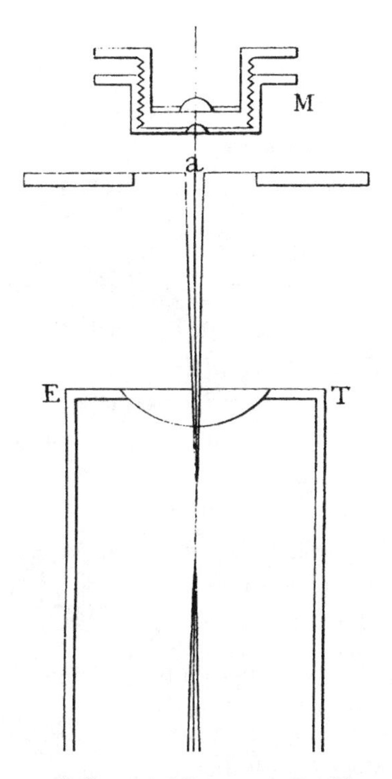

Fig. 14.2. Wollaston's Microscopic Doublet, 1829.

could be adjusted by a screw mechanism, was made equal to 1.4 times the length of the shorter focus.[26] Wollaston reported that with a microscope of this design, adapted to fit in a box four inches square, he was able to obtain images such as "the scales upon a gnat's wing, with a degree of delicate perspicuity which I have in vain sought in any other microscope with which I am acquainted."[27]

Wollaston's incorporation of two plano-convex lenses into simple microscopes was widely hailed as a significant innovation, particularly in France. Charles Chevalier, a leading maker of achromatic lenses and optical instruments in Paris, claimed that Wollaston "was the first to employ multiple lenses to remedy the deficiencies of a simple microscope" and that several microscopist friends of his who viewed specimens with an instrument made to his specifications "all recognized that . . . his instrument produced remarkable results."[28] Wollaston's design went through several modifications in the hands of others, and his combination of two plano-convex lenses, still widely referred to as "Wollaston doublets," became a feature of relatively low-cost microscopes in the early nineteenth century.[29] The general design was still in wide use in mid century but, as the technology needed to produce the curved faces of compound lenses improved and their prices decreased, they gradually replaced Wollaston doublets in most good microscopes.

The third paper Wollaston worked on at Dorset St. was the last one published under his name, communicated posthumously to the Royal Society by Warburton. It described the principle behind, and construction of, a device called a "differential barometer" that employed two fluids to measure small differences in air pressure.[30] He believed his device could be adapted for applications that involved the measurement of small pressure differentials, such as a wind gauges. It does not appear, however, that such a two-fluid barometer ever advanced much beyond the discovery stage or that its principles were incorporated into similar devices. Nonetheless, as the last of Wollaston's instruments to be presented to the world, it stands as an appropriate symbol of his lifelong impetus to measure and quantify natural phenomena. He was the consummate natural philosopher.

THE ONSET OF ILLNESS

In 1826, Wollaston supplemented his many trips into the English countryside with an extended tour of Italy and France from June 26 to October 17.[31] Most of the time was spent in Italy, a place he once feared he would never have the opportunity of visiting. Unfortunately, there is little known about this trip, not even with whom he traveled. Although a slip of paper in the Cambridge collection suggests that Wollaston filled three notebooks with details of the trip, those records have disappeared and all that remains is a short listing of the major Italian cities and towns visited. He began in Turin and then proceeded via Pisa and Florence to Rome, where he remained for a total of six weeks before passing through Bologna, Padua, Venice, Milan, and Geneva before reaching Paris sometime in late September. In early October, he met up with Herschel who was visiting the French capital at the same time, and the two dined at the home of the famed naturalist, Georges Cuvier, on October 7.[32] In mid October, Wollaston returned to London in time for the November start of the Royal Society meetings, where once again he butted heads with Davy.

The two young Cambridge graduates who had mounted the campaign to get Wollaston elected to the presidency of the Royal Society in 1820, John Herschel and Charles Babbage, had both become active thereafter in the affairs of the Royal Society. Herschel had been elected as one of the secretaries of the Society in 1824, and when the second position became vacant in 1826, Babbage sought to fill that post. One of his most important supporters was Wollaston, who had become something of a mentor to the irascible mathematician. Not surprisingly, Davy was not keen on having both men who had opposed his run for the presidency in such positions of importance in the Society. According to Babbage the Council members supported his election, but, at a meeting called by Davy to choose the new secretary, Davy declared that he had decided to fill the vacant position with one of his own trusted friends, the capable and well-regarded chemist John Children. Wollaston asked if Davy was invoking "a right of the President" to bypass the normal voting procedure, to which Davy answered in the affirmative.[33] Although Babbage took umbrage at the decision and castigated the Council members for their perceived lack of moral courage, he failed to appreciate that it was long-established tradition for the president of the Society to make autocratic decisions when it suited his purpose. Once Wollaston had elicited from Davy confirmation that Children's appointment was made by direction of the president, then a contrary vote by Council members would have thrown the governance of the Royal Society into disarray. On balance, it is probable, given Babbage's confrontational nature, that Davy's choice was a better one for the Society.

In early 1827, Wollaston supported Babbage's unsuccessful attempt to secure a professorship at Oxford and a year later was much pleased to learn of Babbage's appointment as Lucasian professor of mathematics at Cambridge. The appointment was especially timely because Babbage had suffered a series of tragedies in 1827. In that one year, his wife, father, and two of his children all died. We can understand then why Wollaston informed Hasted: "I do not know when I have heard of an appointment that has given me more sincere pleasure."[34]

Babbage was only one of the next generation of British men of science who benefitted from Wollaston's knowledge, open-mindedness, and connections. His interest and expertise in the subjects promoted by new scientific societies, such as the Geological (founded 1807) and the Astronomical (founded 1820) made him a valuable conduit to Royal Society affairs. His experience in commercial chemistry, instrument development, and government service prompted many to seek his advice. Much of the information about Wollaston's significant role as a nurturer of scientific talent has to be gleaned from brief entries in his daybook, or from the list of guests at the Royal Society Club, or from snippets of information in letters and autobiographies. A typical example is Roderick Murchison, one of the new breed of geologists emerging in early nineteenth-century Britain, who commented on the advice he received from Wollaston at the outset of his career.

> Among my scientific friends I was of course most proud to reckon Dr. Wollaston, who then and in subsequent years invariably took pains to make me understand the true method of searching after new facts, and often corrected my slips and mistakes.[35]

Another example was the naturalist Leonard Jenyns, who was the son of one of Wollaston's cousins, and the man who declined to be the naturalist on board the Beagle before Charles Darwin was offered the position. In the late 1820s, Wollaston invited the young man to witness some chemical experiments at his house in Dorset St. This Jenyns was delighted to do, and he watched the renowned analyst employ the techniques for which he had become so famous. His account has been taken up as the definitive portrait of Wollaston the chemist.

> On entering the house at the appointed time, he soon appeared, and I asked him if I was going into his laboratory; —upon which he very decidedly answered—"No, you do not go into my laboratory; I bring my laboratory to you." . . . He then showed me into an adjoining room, and after a few minutes he re-appeared with a small tray in his hands, fitted up in the simplest way imaginable; a spirit lamp and blow-pipe, a few reagents in small bottles, watch glasses, and a few plain slips of glass, like the slides of a microscope, and some glass tubes; these things made up, as far as I remember, the whole of his humble apparatus, with which he showed off some of the most striking experiments. . . .[36]

In 1827, Wollaston continued his regular attendance at the Chemical Club and the Royal Society Club, together with meetings of the Royal Society, the Board of Longitude, and the Glass Committee. Fitted around his commitments in the city were several visits to the Houghton Fishing Club and country estates for hunting and fishing, as well as an August tour of Manchester and Wales with his old friend William Blake. In mid October, he made a brief trip to the Cassiobury estate of the earl of Essex, where he noted a worrying, though short-lived, numbness in his left arm, which he was not able to dismiss as simple rheumatism. In a letter to Hasted written shortly after his return to London, Wollaston wrote in words that hinted of his mortality:

> Too much time has passed since [their tour together in 1800], & too many circumstances remind one of the lapse of time; but it is gratifying to reflect on the line I have taken in converting lingering years of misery into a series of enjoyment—to feel that I have not been cut off prematurely, but as a sportsman have fairly bagged seven & twenty good years, & tho hope is not now so active in suggesting prospects of future happiness I have as yet no reason to wish the curtain to drop however quietly.[37]

These sound like the words of someone who is coming to grips with the fact that his best years were behind him, even though he had no intention of limiting his active and productive lifestyle. As usual, Wollaston took an optimistic view of his future, in spite of his certain knowledge that limb numbness was generally a harbinger of worse things to follow.

ANOTHER CHANCE AT THE PRESIDENCY OF THE ROYAL SOCIETY

In late 1826, Humphry Davy's health took a turn for the worse, and he suffered a stroke in December that left him partially paralyzed on his right side.[38] By January, he had recovered sufficiently to travel to Italy in hopes of restoring his health, but while there he decided he could no longer function as president of the Royal Society and would retain the position only until the Anniversary meeting in November. Cognizant of the tensions that existed within the society between aristocratic patrons of science and those "reformers" who actually did science, Davy favored an aristocratic successor and recommended Sir Robert Peel, parliamentarian and future prime minister. Such a suggestion was quite unpalatable to the members of a committee of the Society that had been struck to increase the influence of men of science within their society.[39] Wollaston was an influential member of that committee, as also were Babbage and Herschel. The reformers were opposed by a group of Fellows who preferred that the Royal Society recapture the style of Joseph Banks. Chief among these were Thomas Young and the two secretaries of the Admiralty (both members of the Board of Longitude), John Barrow and John Wilson Croker.[40] Such opponents of reform were happy to advance the cause of Peel as Davy's successor, and they viewed Wollaston as their chief adversary.[41] The opposition to his candidacy from the reform group was enough to cause Peel to withdraw from the contest, and he recommended that either the traditionalist Davies Gilbert or the reformer Wollaston be candidates in his stead.[42]

Wollaston, of course, had no interest in becoming president, for all the same reasons that had led to his standing aside for Davy in 1820, as well as (if any further reason were needed) his recent concerns for his health. Unable to find a man of science willing to stand as a candidate, Wollaston gave his grudging support instead to Gilbert, an effective advocate for science in parliament and a tolerable mathematician, but not one who would effect meaningful reform of the Society. With the support of Peel and the traditionalists and the acquiescence of Wollaston and the reformers, Davies Gilbert was elected unopposed to the presidency in 1827.

It is unfortunate that Wollaston did not play a more significant role in the 1827 election, because he was himself keen to see the Society become more scientific and he had the support of the reform-minded members who were counting on his scientific eminence to advance their cause. But his habitual aversion to overt political action kept him on the sidelines. Consequently, the reform movement in the Royal Society was derailed during Gilbert's three-year period as president, only to reappear with even more ferocity in the presidential election of 1830. Unfortunately, none of the senior trio of Wollaston, Davy, and Young lived to engage in that succession battle.

THE LAST YEAR

Wollaston continued his normal activities in the opening months of 1828, but an ominous entry by Warburton in his transcription of Wollaston's daybook notes that "all the entries from February in bad hand," suggesting that he was beginning to have problems with fine motor control.[43] It was in this month also that Sir Thomas Lawrence, then England's most famous portraitist, drew a sketch of the sixty-one-year-old Wollaston, shown in Figure 14.3, which depicts him with a much more angular face than earlier likenesses by Chantrey and Jackson, suggestive of some weight loss.[44] It appears Wollaston did his best to ignore such early symptoms of what he must have recognized to be harbingers of more serious brain disease, for he maintained his regular attendance at meetings of the Royal Society and the Board of Longitude, and made several trips to the countryside. It was in May while fishing for trout that he was again afflicted by numbness in his left arm. Details of this episode were later reported by William Thomas Brande, Davy's successor as professor of chemistry at the Royal Institution.

> In the month of May, 1828, Dr. Wollaston accompanied a friend to Stockbridge on a fishing excursion, and then expressed some alarm about a

Fig. 14.3. Sir Thomas Lawrence's Sketch of Wollaston, 1828.

> numbness in his left arm, which he said he had long called rheumatism, but which he now considered as a paralytic symptom; he was more especially apprehensive respecting it, in consequence of several of his family having suffered from that disease.[45]

Wollaston recovered sufficiently well to attend a meeting of the Board of Longitude on June 5, the last before it was dissolved by an act of Parliament on July 15. Then, before leaving for a short visit to the Isle of Wight in late July, Wollaston made appearances at two of his regular clubs, unaware that circumstances would mark both of those gatherings as his last. On Tuesday, June 17, he met with friends at perhaps his favorite grouping, the Chemical Club. On July 17, he attended the regular Thursday dinner of the Royal Society Club. Since that Club had decided to adjourn for the period from August to the end of October, Wollaston had no opportunity of attending another meeting, as his health confined him to his home when meetings resumed in November.

The short visit to the Isle of Wight was uneventful except for the second incident of a "fortification spectrum" appearing in Wollaston's visual field, and he returned to London in early August. Later that month, as entries in his daybook indicate, he began his normal round of sporting excursions, in spite of his deteriorating health. He went first to the Danesbury estate of William Blake and on August 30 he moved on to visit Sebright at Beechwood. A few days later he continued his shooting holiday at the Southhill estate of William Whitbread, where a successful outing seemed to buoy his spirits, according to one account.

> In the month of September, he went upon a shooting excursion to Mr. Whitbread's, in Bedfordshire, where he remained eight days, and enjoyed his sport: one evening he jokingly said, "For the first time in my life I count my game by the quarter of a hundred," having killed twenty-five head; he returned home in better health, but soon after the sensation in his arm became worse.[46]

Back in London on September 8, he found time to respond to Julia Hankey's earlier offer to join her and her family at Airdrie, Scotland, for some pheasant shooting, and replied that he could not fit such a trip into his busy schedule.[47] This was no lame excuse, for a week after writing he doggedly set out again for the sporting fields of Hunsdon (home of the MP Nicholson Calvert), Cassiobury (estate of the earl of Essex), and Beechwood before revisiting Blake at Danesbury on October 9. It was there that his final illness descended upon him.

> [In October] the sensation in his arm became worse, and whilst staying with Mr. Blake in Hertfordshire, he felt so unwell and uncomfortable as to be obliged to return home a day or two earlier than he intended. In the course of the following week, symptoms more decidedly alarming came on; the use of his arm was much impaired, and the muscles of the face and organs of speech were affected: his mental faculties, however, remained entire to the last.[48]

One cannot help but suspect that Wollaston wished to use as many of the useful weeks that remained to him for his cherished trips to the country for hunting, fishing, and conviviality because the tour of friends' estates in September 1828 was as ambitious as any he had previously undertaken.

That he was far from perfect health before and (especially) after the excursions was poignantly noted by Julia Hankey:

> When we left town early in August last [1828] we felt considerable uneasiness about the state of his health, he looked ill, was often depressed & (what we considered a very bad sign of his spirits) he spoke doubtfully of his shooting. . . . When we returned from Scotland in October I was much shocked at his appearance, I perceived a slight contraction at one side of his mouth, he walked feebly, he owned that he was ill, he complained of his eyes & of one hand, and there was an evident oppression on his spirits tho' he conversed cheerfully and expressed extreme pleasure at our return. We remained but one week in town during which he spent four evenings with us, he seemed to regret very much that my Uncle and I were to go to Brighton, he was pressed to accompany us & at first hesitated a little, but afterwards as if he recollected himself said "No, I have business here that must be done, don't ask me for I ought not to go." . . . We told him we should return in a fortnight, —"A fortnight!" he said, "that is a long time, I may be in Heaven in a fortnight." When he rose to go, he put his hand on my brother's shoulder to support himself & he could scarcely walk downstairs,—he took an affectionate leave of us all and the tears stood in his eyes; but altho' we saw evident marks of indisposition, altho' we saw that he struggled against the depression he felt, and that his manner struck us as remarkable, we none of us suspected that this was to be the last Evening he would ever pass with us. It was the last evening he ever spent away from his own house.[49]

PREPARATIONS FOR DEATH

By mid October, as the Hankey memoir makes painfully obvious, Wollaston's health was in rapid, and apparently irreversible, decline. Accordingly, he set about putting his personal affairs in order in the short time he had left. First he brought his will up to date and distributed his estate among all his brothers and sisters. He signed it on October 18 and had it witnessed by his three domestic servants.[50]

Wollaston turned his attention next to preparing experimental results for publication that were complete enough for dissemination. First up was a complete and detailed description of the chemical processes employed to purify platinum and the metallurgical steps needed to consolidate it into malleable metal. The paper was read (by one of the secretaries) to the Royal Society on November 20, and it earned for him the Society's highest honor, a Royal Medal, which was announced at the Anniversary Meeting of the Society a week later. Next, Wollaston turned his attention to the three papers discussed earlier in this chapter, one on a microscopic doublet (read November 27), one on the comparative brightness of the sun and two stars (read December 11), and one on a differential barometer (read February 5, 1829), together with a fourth (discussed in Chapter 2) on the water of the Mediterranean (read December 18). Wollaston wrote the drafts of a few of these papers himself while his handwriting was still legible, but some later sections were dictated to Warburton. After all his scientific work was wrapped up, Wollaston moved next to create a different type of legacy for each of the three English scientific societies to which he belonged.

On November 26, Wollaston wrote to the Royal Society to declare that he had transferred to it £2,000 of 3 percent Consols to initiate what he termed a "Donation Fund." His letter stipulated that the Society

> shall apply the said dividends from time to time in promoting experimental researches, or in rewarding those by whom such researches may have been made, . . . [and] to apply the said dividends to aid and reward any individual or individuals of any country, saving only that no person being a Member of the Council . . . shall receive or partake of such aid or reward . . . [and] not to hoard the said dividends parsimoniously, but to expend them liberally, and, as nearly as may be, annually. . . .[51]

Wollaston's hope that others would contribute to the fund was realized, although none of his contemporaries matched his generosity. By 1848, the fund (which continues today) had grown to a total of £4,843.[52] Only in the second half of the nineteenth century, however, did scientists begin to use it with any frequency to support their own research. On December 7, Wollaston made a second gift to the Royal Society, presenting it with all the platinum and palladium that remained in his possession, to be used in, or as a reward for, chemical experiments judged acceptable.[53] Both donation letters are in Warburton's handwriting, but each bears Wollaston's feeble signature.

Wollaston also initiated a Donation Fund at the Geological Society, to which he had been elected in 1813 and had served as Council member. By a letter dated December 8, he transferred £1,000 of 3 percent Consols to the Society, with terms for its use identical to those specified for the Royal Society fund.[54] The geologists decided in 1829 to use the first year's income from the fund to create a die for a gold medal termed the Wollaston Medal, to be given annually for outstanding achievement in geology.[55] From 1846 to 1860, the medal was fashioned out of palladium donated by Percival Norton Johnson, the son of the John Johnson who had imported platina from New Granada on Wollaston's behalf. The Wollaston Medal was first awarded in 1831 to William Smith, creator of the first geological map of England and Wales.

The third scientific bequest was to the Astronomical Society, which had come into existence in 1820 and had just elected him as a member. Wollaston had been proposed for membership in June 1828, and would by the normal course of events have been balloted for at the December meeting. However, due to the rapidly deteriorating state of his health, the Society acclaimed him as a member on November 14, foregoing even the formality of a ballot.[56] By a letter to the president of the Society, John Herschel, on December 8, Wollaston bequeathed to the Society the telescope made by Peter Dollond in 1771 for his father, which he had himself improved by adjustment of the triple object glass. In a separate memorandum, Wollaston expressed the wish that the Society "will not keep it useless, but lend it, or give it if they think proper, to any industrious and useful member of the Society, he not being at the time a member of the Council."[57]

By these bequests to scientific societies, Wollaston demonstrated his deep commitment to independent and democratic research. His wishes that the funds, and the instrument, be used for the betterment of science by members not on the Councils of the societies prevented institutional abuse of the donations. Moreover, his wish that the funds would act as seed money for subsequent donations reflected his desire that persons with aptitude but insufficient personal funds could pursue scientific research in the future. He had in mind, however dimly, the twentieth-century model of science funding, independent (to the largest extent possible) of political interference and open to all with talent. It is easy to see why he was so highly regarded by the reform-minded members of the Royal Society.

With his family, scientific and institutional interests taken care of, Wollaston turned his thoughts to his friends, and to many of them he prepared and presented mementos that held meaning for both donor and recipient. His first cousin, Elizabeth, wrote to Julia Hankey's aunt shortly after his death to say there was "not a soul who was ever kind to him to whom he has not sent some memorial before he died, and I understand that he has left nothing for his brothers and sisters to do but to divide his bequests amongst them & probably that was so arranged that it could be done without trouble."[58] Evidence for a few of these gifts exists in various letters and remembrances that circulated among his acquaintances. Some were small tokens of appreciation given to the children of friends, such as a small length of gold chain given to one of Edward Codrington's daughters.[59] Some were more relevant to shared experiences. To Mary Somerville, for example, whom he had instructed in crystallography, he gave his cabinet of wooden models of the natural crystals.[60] To both Maria Edgeworth and John Brinkley's wife, he sent a rhodium-tipped pen. Mrs. Brinkley's son-in-law later recounted the circumstances surrounding her gift:

> [Prior to his death, Wollaston] had long prepared little tokens of friendship for various persons, among the rest for Mrs. Brinkley a Rhodium pen. This was packed up with his peculiar neatness, contained directions, was sealed and addressed with his own hand, which must, from the goodness of the handwriting, have been done probably before the attack of paralysis. It was his wish to keep them until as near his death as possible, in order to show his friends how long he thought of them. And accordingly, long after he lost his speech, two days before his death, he gave directions with his pencil to have the pen sent to Mrs. Brinkley! It came in a frank [a free transit letter], along with the post which brought the account of his death![61]

One of the most meaningful gifts, not surprisingly, was sent to Julia Hankey. It was held back until the day of Wollaston's death, likely because she continued to visit him during the last days of his life. In 1815 Berzelius had agreed to tutor the young Prince Oscar of Sweden in chemistry and had asked Wollaston to purchase on his behalf a number of scientific instruments from London craftsmen. This Wollaston did and, in gratitude, Berzelius sent him a porphyry vase, a symbol of their mutual interest in mineralogy. This vase, together with a broach, was given to Julia.[62]

When Wollaston departed the Hankey home in late October with tears in his eyes, he knew that he was not going to recover from what he might by then have believed to be a brain tumor. From that point on, he worked diligently to put his affairs in order, assisted in the last few weeks of incapacitation by his brother George and Henry Warburton. As word of his failing health began to circulate, several visitors dropped by to enquire of his health, but his guardians allowed only a select few to meet with him. One of these was Admiral Edward Codrington, who was then battling with his superiors at the Admiralty over the international fallout from his naval triumph at the Battle of Navarino. Shortly after visiting Wollaston in late November, he wrote to his wife:

> Wollaston's appearance is very afflicting, and the disease seems to pervade the whole body more or less. His head is nevertheless quite clear, and he continues dictating to different people on scientific sub-

> jects about which his mind has been employed. . . . Be assured that . . . he will die the truly great man that his intimate friends have thought him. Die, however, of his present attack, I sadly fear he must, although some little oscillations between better and worse have occasionally excited hopes. I say *fear* when one might rather be expected to say *hope,* considering the physical dependence of one who has ever been the most physically as well as mentally independent of any man I ever knew.[63]

The judgment of Codrington, a man who had battle command of others in naval engagements and who had undoubtedly witnessed numerous acts of individual bravery, to assess Wollaston as the "most physically as well as mentally independent" of any man of his acquaintance, forces us to appreciate how self-reliant Wollaston was in all of his personal, social, and scientific endeavors. Such resolute individualism was both his strength and his weakness, and it must have sustained him in his final days and hours.

Even as his brain tumor was robbing him of his movement, his eyesight, and his powers of speech in December 1828, Wollaston refused to allow his unimpaired mental faculties to rest. As noted by one of his deathbed visitors, together with the attending surgeon, Benjamin Brodie,

> it was a matter of deep interest to us to observe his philosophical mind taking calm but careful note of its own decay—the higher faculties, which were little if at all impaired, occupied in testing by daily experiments of his own suggestion, the changes gradually taking place in the functions of the senses, the memory, and the voluntary power. Diagrams and figures drawn on a board before him were among the methods he thus employed. He had manifestly much interest, if not indeed a certain pleasure, in detecting the changes going on and in describing them to us. He would admit no interpretation of them save in reference to that final change which he constantly and calmly kept in view. It was a self-analysis of mind carried on to the last moments of life.[64]

It was almost as if Wollaston had decided to write one last paper on the perceptual decline of someone dying of brain cancer, to be published in *Philosophical Transactions,* of course. Remarkably, on the first day of winter, the day before his death, he left those gathered around him dumbfounded with one last display of mental acumen.

> [On December 21] his physician, conceiving all his senses were destroyed and his intellect gone, observed in the room that Dr. W. was dying, and

> could not understand. Wollaston when the physician left the room, to the surprise of all, made a sign for his pencil, and although quite blind, with some difficulty, but still with much of his usual precision, wrote down the numbers from 500 to 520 in their regular order, no doubt to show his memory and reflection were unimpaired. How like him! I suppose that he did not write from 1 to 20 lest it might be attributed to mere habit, his beginning with 500 showed reflection. Two hours before his death he wrote *end* - *near* - and between that and actual death he made several attempts to write—the mind survived the body—for his hand failed to trace the ideas; and most unfortunately the last notes of this great philosopher are illegible. It strikes me that he was endeavouring to convert his death into a grand philosophical experiment, to give data for determining the influence of the body on the mind, and to try whether it was possible for the latter to remain until the very end.[65]

Sometime during the day of December 22, 1828, William Hyde Wollaston, sixty-two years of age and in the company of friends, died at his home in Dorset St. Later that day, George had the unpleasant task of carrying the news to all of his sisters, who had assembled at Anna's and Henrietta's residence in Greenwich.[66] There he learned that they had decided to erect a monument in his memory, an initiative that never came to fruition, perhaps because of the death of Anna in 1829.

CHAPTER 15

Post Mortem and Legacy 1828–Present

> The seal of a superior genius is stamped on even the least of the works of the English philosopher [Wollaston], and from the outset of my career, I have paid homage to this great intellect.[1]

POST MORTEM

Having monitored so carefully the decline in his mental faculties in the last days of his life, it is not a surprise that Wollaston asked that an autopsy be conducted on his body. This was done on December 24 by the physicians William Babington and James Somerville, and the surgeon Benjamin Brodie. Not unexpectedly, they found a tumor the size of a "hen's egg" in the right optic thalamus of the brain, and, likely seeking an explanation for the origin of Wollaston's intellectual acuity, they found it "worthy of notice that the Brain generally was of a large size, but that the principal development of it was in the posterior lobes of the Cerebrum which were much larger in proportion than the anterior lobes."[2] After the autopsy was completed, Wollaston's remains were placed in the family vault in the cemetery of St. Nicholas's Church in Chislehurst. The vault, which contains the remains of his parents and a number of siblings, commemorates the great scientist interred there with the simplest of inscriptions, now just barely legible, "William Hyde Wollaston, 3rd son. He was born Aug. 6, 1766 and died Dec. 22, 1828." Like the epitaph, the funeral was a simple one, as his cousin Elizabeth related: "It was his express desire that not a single carriage should attend his funeral & that he should only be followed by his Brothers & his nephews."[3] Humility in life was followed by unpretentiousness in death.

How does one assess William Hyde Wollaston's contributions to the society in which he flourished and to the sciences to which he contributed?

It is important, I think, to gain an appreciation of his strengths and weaknesses as a human being and his impact upon others, emotionally and intellectually. Obviously, the most significant of his achievements, and the primary reason for writing this biography, is his scientific work. It is a fact that he achieved an international reputation in chemistry, linear optics, and mineralogy during his lifetime and was acknowledged to be one of the most acute experimental philosophers of his age. His place in the historical record, however, does not come close to matching the status he attained during an age of startling and foundational discovery. The reasons for his relative invisibility in modern accounts of scientific progress are many and varied. Some are due to the man himself, some are accidents of history, and some are consequences of inadequate historical scholarship. Before directing attention to his scientific legacy in the last section of this chapter, I will try to recapture the contemporary opinion of the man by citing some of the comments of those who knew him personally and professionally.

First are three comments made by contemporaries who had been subjected to Wollaston's unsettling methods of responding to questions. The first is by John Barrow, a man of Wollaston's age who served for many years as a secretary of the Admiralty and who interacted frequently with him at the Chemistry Club, the Royal Society Club, and the Board of Longitude, but was not otherwise close to him.

> With all his learning and variety of knowledge, scientific and practical, Wollaston was not the most ready to communicate it; nor was he always the most courteous when engaged in argument. He knew so much, that he could not always avoid betraying a consciousness of his own superior knowledge, and of the want of it in others. Still to those who understood him his manner of disputing was not disagreeable. He would rarely give an immediate or direct answer to a question, but generally respond by putting another analogous question, of an opposite tendency. Thus, for instance, I ask him to define the word heat; he replies, "Tell me how you define cold." I say, "I cannot." "Then I will do it for you—cold is the absence of heat." But having thus got him under weigh, a noble dissertation on heat would surely follow.[4]

Not surprisingly, many found such intellectual scientific exchanges intimidating. Not many people relish having their own knowledge tested when seeking the same from another. Even some who had strong scientific credentials and a healthy dose of self-assurance could be offended by some of Wollaston's mannerisms. One of those was Charles Babbage, the Cambridge mathematician of calculating engine fame who was a generation younger than Wollaston. He made reference to Wollaston's reflexive tendencies in his assessment of the man whom he admired.

> The most singular characteristic of Wollaston's mind was the plain and distinct line which separated what he knew from what he did not know; and this again, arising from his precision, might be traced to caution.
>
> It would, however, have been visible to such an extent in few except himself, for there were very few so perfectly free from vanity and affectation. To this circumstance may be attributed a peculiarity of manner in the mode in which he communicated information to those who sought it from him, which was to many extremely disagreeable. He usually, by a few questions, ascertained precisely how much the inquirer knew upon the subject, or the exact point at which his ignorance commenced, a process not very agreeable to the vanity of mankind; taking up the subject at this point, he would then very clearly and shortly explain it.[5]

However unsettling one's initial encounters with Wollaston might be, those who persevered often found ways to break down the intellectual barriers and to establish a congenial relationship. A good example of such a person was the Scottish chemist Thomas Thomson, who often prodded Wollaston for chemical information during their overlapping years in London and who found that their friendship improved over time.

> Few individuals ever enjoyed a greater share of general respect and confidence, or had fewer enemies, than Dr. Wollaston. He was at first shy and distant, and remarkably circumspect, but he grew insensibly more and more agreeable as you got better acquainted with him, till at last you formed for him the most sincere friendship, and your acquaintance ended in the warmest and closest attachment.[6]

The comments of Barrow, Babbage, and Thomson establish the public persona of Wollaston. He was exceedingly careful in his observations, reluctant to generalize, confident in his knowledge, irritatingly logical in discussion, and incapable of self-aggrandizement. This was Wollaston the natural philosopher, the one encountered at the Royal and Geological Society meetings, at scientific soirées, in government committees, and at the Board of Longitude. But among family and close friends he comported himself quite differently.

The most compelling assessment of this part of his character comes from the perceptive memoir of Julia Hankey:

> You [Warburton] can require no proofs of the warmth with which he loved his friends, of the infinite trouble he took to serve them, & the affectionate watchfulness with which he endeavoured to present their wishes. His timidity with strangers, his modesty & reserve in speaking of himself when the subjects could not be altogether avoided, the indignation he felt at the slightest deviation from honesty & truth & his anxiety for strict justice cannot be unknown to you. Perhaps you may have had less opportunity of observing how easily he was amused with most trifling conversation if carried on with mirth & good humour, with what keenness he entered into every species of game, displaying as much anxiety & amusement as the youngest of the party, and how well he was contented in the society of those who had no merit but that of meaning well. You may not have remarked that unpretending stupidity never offended him, whilst his irritation at presumptuous ignorance was excessive; that from his own rectitude and singleness of heart he sometimes gave people credit for more virtue than he afterwards found them to possess & that he was more easily imposed upon by an affectation of kindness than one would naturally have expected from so acute a mind. He supposed all men to be good till he had proved them to be bad & never spoke ill of any one if he could say a word in their favor or remain silent. He seldom conversed about people, he preferred speaking of things and never either intruded his own affairs on the notice of others or showed the slightest curiosity about theirs. If indeed he imagined people were speaking of matters in which he had no concern he frequently abstracted his mind so entirely from the conversation as to be totally ignorant of the subject of their discourse. No one better understood the art of keeping a secret for he avoided even the appearance of having one to keep.[7]

Julia's intelligent and informed memoir of Wollaston, written at Warburton's request, remained unexamined in the Bathurst family papers until L. F. Gilbert of University College, London, made a copy a century later. Thus, like much of the information Warburton assembled and guarded for his intended biography, Julia's comments have escaped the notice of historians, leading to a widespread and inaccurate judgment of Wollaston as an austere, aloof and even asocial character. Henry Warburton, as a failed biographer, has a lot to answer for.

Warburton, nonetheless, was the logical choice of Wollaston's friends to write his biography. He had similar scientific interests, belonged to many of the same Societies, traveled and fished with Wollaston, and had the intellect most thought essential for the task. Warburton certainly began his task with energy and enthusiasm. But it does not appear that he got much further than the composition of a few pages of biographical snippets covering Wollaston's years at Cambridge. Even more unfortunately, Warburton refused to give others access to the documents he collected and, most egregiously, he made no great effort to preserve them. Consequently, Wollaston's legacy began to dim with time. About the middle of the nineteenth century, those of Wollaston's colleagues who were still alive began to recognize that Warburton would never complete the task entrusted to him, and some short reminiscences began to appear. Extracts of the works by Henry Hasted, Thomas Thomson, and George Wilson have been cited in earlier chapters. How differently might Wollaston have been interpreted if Julia Hankey's memoirs, for example, had been more generally available.

Warburton's failure was exacerbated by his death in 1858, after which the Wollaston manuscripts disappeared. This was the situation when the University College chemist Lionel F. Gilbert spent many years in the first half of the twentieth century collecting materials for a book-length biography, which remained incomplete at the time of his death in 1955. Fortunately, Gilbert's materials have been preserved in the archives of the D. M. S. Watson Library of University College, London. When all the Warburton material was rediscovered in the Department of Mineralogy and Petrology of the University of Cambridge in 1949, it was too late for Gilbert to do much other than to identify its contents.[8] Ironically, the inadvertent storage of Warburton's collection in an unsuspected location probably saved it from fragmentation and dispersal, and possibly complete loss. The Wollaston manuscripts, including his laboratory notebooks, were transferred to the main Cambridge University Library in 1968; they have provided the foundation for this book.

THE LEGACY OF WILLIAM HYDE WOLLASTON

The three leading figures of English science in the first quarter of the nineteenth century were Humphry Davy, Thomas Young, and Wollaston (the Manchester chemist John Dalton, more celebrated today than any of those three, was too distant from the London scene to have as much contemporary impact). Their scientific lives were closely intertwined, as many examples included in this book attest. They all arrived in London around

1800 and, by most unfortunate coincidence, died within six months of each other: Wollaston in December 1828 at the age of sixty-two, Young in May 1829 at fifty-five, and Davy in June of the same year at fifty. Eulogies, not surprisingly, often drew comparisons among the three influential philosophers, each of whom had made significant contributions to the science of the time. One example is the address given by Davies Gilbert, president of the Royal Society, at the Anniversary Meeting on November 30, 1829, in which he drew a facile comparison that nonetheless rings true.

> Having characterized Davy by poetic genius, I would venture to ascribe minute accuracy, even in the merest trifles, as the distinction of Wollaston, and almost universal acquirements as the characteristic of Young. While in soundness of judgement combined with general ability of the highest class, no discriminations can be found.[9]

Gilbert's assessment reveals how generally England's scientific eminence was shared among Davy, Young, and Wollaston at the time of their deaths, and how detrimental to Wollaston's reputation it has been that only two of the three were commemorated after their deaths by complete biographies.

A biography is a story of an individual and the times in which he or she lived, and the impact that individual had on their generation and later ones. This book has followed Wollaston from his birth in 1766 to his death in 1828, with special attention to his many seminal contributions to scientific knowledge and applications. We know that he was a kind and generous man, with sincere empathy for the well-being of others. He had his faults, of course, emanating primarily from his resolute independence and his austere modes of technical communication. But his personal foibles did not diminish the fame he justly accumulated in his time: he was judged to be unique in the way he conducted his scientific studies, which ranged over a diverse set of disciplines and can be readily appreciated by scanning the list of his publications given in the Bibliography. He is remembered eponymously by Wollaston wires, the Wollaston prism for light splitting, the Wollaston doublet for microscopes, the Wollaston Medal and the silicate mineral Wollastonite, and his name was given to several geographical landmarks (such as the Wollaston Islands in both the Canadian Arctic and off the coast of Chile) by British naval explorers. His enviable record of scientific achievement places him among the most acute observational and experimental scientists of his age. But he lacked one attribute that scientists and historians value above all else in their pantheon of heroes. He refused to advance theories without overwhelming experimental justification, and that caution has caused his reputation to suffer.

A persistent reluctance to make sense of observational data by situating them in a broader explanatory framework prevented Wollaston from entering into the first rank of scientists. It is no exaggeration to conclude, as I have suggested in earlier chapters, that he could easily have been the first to discover what became known as Fraunhofer lines, or the polarization of light by Iceland spar, or isomorphism in crystals, or the structural consequences of atomic theory, or a comprehensive theory of vision, or even electromagnetic rotation. Boldness in thinking, even when tarnished by precipitous error, captivates those who seek out and immortalize the great figures in history. John Dalton was a bold thinker and an unexceptional experimenter, the methodological opposite of Wollaston. Today Dalton is known to everyone with an interest in the origins of modern science, Wollaston to almost no one. And we can't blame Warburton for all of this imbalance. It does not help Wollaston's cause that he placed such a complete blanket of secrecy around his commercial work or that he had such an aversion to personal acclaim. Not only did he not wish to have his accomplishments lauded by his colleagues, he actively and even uncharitably scolded many of them for mentioning his work. Had he been motivated to, or badgered to, publish a compilation of his micro techniques for the qualitative analysis of minerals and chemical substances, for example, his name could well have been indelibly connected to the rise of analytical chemistry. By so effectively shrouding the light of his own ingenuity and limiting the imaginative scope of his thinking, Wollaston reduced his visibility to twentieth-century historians who had no nineteenth-century biography to alert them to their oversight. As is the case in so many other areas of human achievement, scientists and historians gravitate to winners, those who however recklessly extrapolate from observation and experiment to comprehensive explanations. But there is an essential role in the evolution of modern science for those, like Wollaston, who carefully discover the fundamental consistencies in natural occurrences, which are the starting point for hypothesis formation and ultimately become the set of accepted observational data upon which theories ultimately rest.

Wollaston's life and times also have interest beyond the scientific. He lived in a period of astonishing agricultural, economic, political, industrial, and social change, collectively identified as the English Industrial Revolution. By dint of his own efforts and seizing the opportunities provided by good birth and family connections, Wollaston is a paradigmatic example of the individualism that so invigorated and characterized British society in the early nineteenth century.[10] From life as a provincial physician visiting patients on horseback, he rose to a life in the capital in which he conducted a profitable chemical business, became a scientific consultant to (and employee of) government, and played a leading role in one of the world's leading scientific societies. His was a case where curiosity-driven research led to important commercial applications, completely unfettered by governmental oversight or control, except for an irritating period when income was taxed. This was not entirely beneficial to a society that allowed him to establish a chemical-processing facility in a residential neighborhood, buy and sell large amounts of industrial chemicals, and dispose of waste products in outdoor drains. Nonetheless, he saw his products transform the gunnery and sulfuric acid industries. He moved confidently among scientists, craftsmen, artists, engineers, military men, politicians, aristocrats, field sportsmen, and anglers. His science had no barriers between what is now known as pure and applied science, nor between different disciplines, nor between social classes. His closest personal friends were drawn from all sides and periods of his life, of both genders and of a wide range of ages.

But a biography should focus on the individual, and it is appropriate to end this one with a short but appropriate memorial sent to Warburton by the Scottish manufacturing chemist Charles Macintosh, which well illustrates the close relationship Wollaston had with many of his contemporaries. Macintosh, the same age as Wollaston, had established a successful chemical business in Glasgow, culminating in the 1820 discovery of a process to coat fabrics with rubber (the ubiquitous Mackintosh (*sic*) raincoat). Despite their geographical separation, the two met frequently and became good friends.

> Since the commencement of our acquaintance, near thirty years ago, he [Wollaston] condescended to show me unremitting acts of kindness and friendship, flowing solely and spontaneously from his excellency of heart and disposition, quite regardless of the humble sphere I move in, in relation to science, or acquirement of any kind, when compared to him. He appeared to take pleasure in listening to my reports of proceedings going on in my various chemical manufactures; his friendly and valuable advice was always at my command; and he very often, in a good-humoured, jocular way, used to call me the great chemist, in allusion, of course, to the extended scale of my operations. I was always perfectly at my ease with the Doctor, trusting to the goodness of his heart for overlooking the weakness of my head. You know, as well as I do, that he was Argus-eyed;

> and that almost instinctively he discovered the weak point of any relation made to him. I have heard it said that on such occasions he was nothing loath to apply the probe with a very firm, unrelenting hand; but I never experienced this. God knows he had no occasion to stand upon ceremony with me, —and he did not; but I never had occasion to wince under his strictures, that he did not immediately apply a cordial balm, which made me admire, and like him more and more.[11]

William Hyde Wollaston was, and still is, a man worth knowing.

NOTES

PROLOGUE

1. Blomefield, *Chapters*, 42.

2. Maria Edgeworth to W. R. Hamilton, July 1830; quoted in Graves, *Life of Hamilton*, I, 382.

CHAPTER ONE

1. Wollaston Mss., Warburton's biographical notes, box 2, envelope B, item 4.

2. F. Wollaston, *Secret History*, 10.

3. H. W. Wollaston, *Wollaston Family*, gives essential family information.

4. Wollaston Mss., notebook **Daybook**, slip of paper inside back cover.

5. E. Heberden, *William Heberden*, 77.

6. Ibid., 78.

7. Thomson, *History of Chemistry*, II, 247. Elsewhere Thomson, never a reliable source for factual information, says more correctly that Wollaston "enjoyed uninterrupted health for many years"; Thomson, "Biographical Account," 136.

8. F. Wollaston, *Secret History*, 10.

9. Reproduced with the kind permission of Mrs. Daphne Wollaston.

10. Information is taken from entries for William Wollaston in *Biographia Britannica*, 6, 4302–4308, *The General Biographical Dictionary*, 240–246 and *oDNB*.

11. Biographical details are taken from F. Wollaston, *Secret History*, and H. W. Wollaston, *History*.

12. Information on Heberden is taken largely from E. Heberden, *William Heberden*, and the entry by the same author in *oDNB*.

13. H. W. Wollaston, *History*.

14. F. Wollaston, *Secret History*, 4.

15. Ibid.

16. Ibid., 4–5.

17. Ibid., 11.

18. Ibid., 54–55.

19. Ibid., 62–63. Emphases in the original.

20. Wollaston Mss., notebook **Daybook**, 97.

21. Gilbert Collection, filing box 4, file 3, enclosure B, Genealogy, 12.

22. Information on Charterhouse is taken largely from Combe, *History of the Colleges*, Anon., *Charterhouse* and Quick, *Charterhouse: A History*.

23. Ibid., Combe, 20.

24. Quick, *Charterhouse*, 33.

25. Arrowsmith, *Charterhouse Register*, 368. The oDNB and several other sources do not include Wollaston's years as a day scholar in their reports of his Charterhouse years.

26. Searby, *History of the University*, Vol III, 1.

27. Venn, *Caius College*, 155.

28. Searby, *History*, 99–100.

29. Venn, *Biographical History*, II, "William Hyde Wollaston," 106.

30. Ibid. Venn mistakenly lists Wollaston's first degree as a BA.

31. Gilbert Collection, Chronology of Wollaston's Life, box 2, item F, entry for 1783.

32. Wollaston Mss., notebook **Daybook**, 97; also Venn, *Biographical History*.

33. Winstanley, *Unreformed Cambridge*, 61.

34. Wollaston Mss., Warburton's biographical notes, box 2, envelope B, item 4.

35. Gilbert Collection, box 5, file 1, enclosure A, item 1, Chap.1, 5.

36. Hasted, *Reminiscences*, 3.

37. Wollaston Mss., Warburton's biographical notes, box 2, envelope B, item 4.

38. Ibid.

39. Ibid.

40. Hasted, *Reminiscences*, 2.

41. The leading sources on Tennant are Anon. [John Whishaw], "Smithson Tennant, Esq."; Wales, "Smithson Tennant"; McDonald, "Smithson Tennant"; and Usselman, "Smithson Tennant."

42. Wollaston Mss., box 2, notebook of Tennant's **Tours**. The Wollaston collection also contains much Tennant material, presumably because Warburton was a close friend of his and at one time planned to write a biography of him.

43. Anon. [Whishaw], *Tennant*, 5.

44. Perrin, "A Reluctant Catalyst."

45. Whishaw, *Tennant*, 5.

46. Wollaston Mss., Warburton's biographical notes, box 2, envelope B, item 4. Wollaston and Tennant first met about 1786.

47. Wollaston Mss., notebook **Daybook**, 97.

48. Information taken from the corporate records, the *Gesta Collegii*, of Caius College, available in the Gilbert Collection, Wollaston chronology, box 2, item F.

49. Ibid.

50. Venn, *Caius College*, 196.

51. *Gilbert Collection, Wollaston chronology, Gesta Collegii, box 2, item F.*

52. Winstanley, *Unreformed Cambridge*, 76.

53. Wollaston Mss., notebook **Daybook**, inside back cover.

54. A transcription of the manuscript is available in Crummer, "Introduction to the Study of Physic," and a sound analysis of the work is in E. Heberden, *William Heberden*, 31–39.

55. Wollaston Mss., Heberden's *An Introduction to the Study of Physic*, 1.

56. Ibid., 38.

57. Wollaston Mss., notebook **Daybook**, inside back cover.

58. The case for the pre-eminence of London's clinical instruction is advanced in Gelfand, "'Invite the philosopher,'' and Porter, "Medical lecturing."

59. Wollaston Mss., notebook **Daybook**, 94.

60. Warburton's "Notes of Wollaston's Life." The original manuscript was in the possession of the Herschel family in 1889. This version of Warburton's preliminary study of Wollaston can no longer be found. I thank Larry Schaaf for the extract.

61. Wollaston Mss., notebook **7**, back 2. This notebook contains medical entries in the pages numbered from back to front, and personal and scientific entries in pages numbered from front to back.

62. Wollaston Mss., notebook **Daybook**, 1.

63. Gilbert Collection, Wollaston chronology, *Gesta Collegii*, box 2, item F.

64. Royal Society Library, certificate of election of William Hyde Wollaston.

65. Wollaston to Isaac Pennington, April 22, 1793. Wollaston Mss, notebook **7**, 26.

66. Wollaston Mss., notebook **Daybook**, 0.

67. Wollaston Mss., notebook **7**, front 15–24.

68. Ibid., back 4.

69. Hasted, *Reminiscences*, 5.

70. Wollaston Mss., notebook **Daybook**, 2.

71. Ibid.

72. Information on the workings of the College is available in Clark, *A History*.

73. Royal College of Physicians, London, *Register Book*.

CHAPTER TWO

1. Hasted, *Reminiscences*, 3.

2. Ibid. The quotation is adapted from Shakespeare's *As You Like It*.

3. Anon., "Henry Hasted," *Gentleman's Magazine*.

4. Wollaston Mss., notebook **7**, front 1.

5. Ibid., 7.

6. Ibid., 11.

7. W. H. Wollaston, "Potash in Sea Water," 1820.

8. Wollaston Mss., notebook **7**, front 1.

9. W. H. Wollaston, "On Fairy-rings," 1807.

10. See, for example, Edwards, "Growth of Fairy Rings."

11. Hasted, *Reminiscences*, 3–4.

12. Caroline Wollaston to unknown recipient, date unknown. I thank Daphne Wollaston for permission to quote from the letter in her possession.

13. Wollaston Mss., notebook **Daybook**, facing p. 11.

14. I thank John Keyworth of the Museum and Historical Research Section of the Bank of England and the helpful staff at the Record Office in Roehampton for access to, and assistance with, the Wollaston accounts held by the Bank of England.

15. W. H. Wollaston, "Urinary Concretions," 1797.

16. For a broader discussion of nineteenth-century studies of calculi, and methods of treatment, see Coley, "Animal Chemists."

17. W. H. Wollaston, "Urinary Concretions," 1797, 390.

18. Ibid., 397.

19. Buckland, "Fossil Teeth and Bones," 187. William Prout, chemist and physician, reached similar conclusions on samples sent to him by Buckland; see Buckland, " Fossil Faeces," 237–238.

20. Pearson, "Urinary Concretions."

21. Fourcroy and Vauquelin, "Calculs Urinaires."

22. Marcet, *Calculous Disorders.*

23. Ibid., 60–61.

24. W. H. Wollaston, "Cystic Oxide," 1810.

25. Berzelius to Marcet, Feb. 6, 1818; Söderbaum, *Berzelius Bref*, III, 160.

26. Marcet to Berzelius, Apr. 3, 1818, ibid., 169.

27. Griffiths et al., "Cystine Stone."

28. Wellner & Meister, *Survey*, 912.

29. W. H. Wollaston, "Cystic Oxide," 1810, 230.

30. Wollaston Mss., notebook **Daybook**, 3.

31. Wollaston to Hasted, Nov. 3, 1797, Gilbert Collection, box 1 file 1. Typed extracts of the letters to Hasted were made by P. J. Hartog early in the twentieth century from the originals then held by D. O. Wollaston. The present location of the letters is not known.

32. Royal College of Physicians, Register Book.

33. Coley, "Alexander Marcet."

34. For a contextual analysis of her publications, see Bahar, "Jane Marcet."

35. Royal College of Physicians, Register Book, minutes for Apr. 12, 1802.

36. Clark, *Royal College*, 624.

37. Jones, *Mental Health Services*, 32.

38. Royal College of Physicians, Account of Receipts & Payments by the Treasurer of the College of Physicians under an Act of Parliament for regulating Madhouses.

39. Woodville, *Cow-Pox.*

40. Wollaston's marginal notes in Woodville, *Cow-Pox*, 38–39. The book, together with other tracts on inoculation belonging to Wollaston, is in the library of the Royal College of Physicians.

41. Royal College of Physicians, Register Book, minutes for Apr. 12, 1802.

42. Wollaston to Hasted, July 28, 1798, Gilbert Collection, box 1, file 1.

43. Ibid., May 3, [1798].

44. Ibid., Jan. 23, [1800].

45. Ibid., Aug. 30, 1801.

46. Ibid., Apr. 6, 1802.

47. Wollaston Mss., notebook **Daybook**, 5.

48. Thomson, *History of Chemistry*, II, 236.

49. For details see Gilbert, "Wollaston Mss," and Usselman, "Platinum Notebooks."

50. Wollaston Mss, notebooks $\mathbf{L}_1$ and $\mathbf{B}_2$.

51. Wollaston to Hasted, May 3, 1798, Gilbert Collection, box 1, file 1.

52. A good history of the Royal Society at Wollaston's time is M. Boas Hall, *All Scientists Now*.

53. Royal Society, *Journal Book*.

54. Wollaston Mss., notebook **Inkbook**.

55. Ibid., 39; also summarized on a slip of paper inside p. 43.

56. W. H. Wollaston, "On double Images," 1800, 249.

57. Vince, "Unusual Horizontal Refraction."

58. Wollaston, "Double Images," 1800, 241.

59. Ibid., 243.

60. See, for a modern explanation of the phenomena, Fraser and Mach, "Mirages."

61. Biot, "Réfractions Extraordinaires," 6–7. I have taken the English translation from Goodman, "William Hyde Wollaston," 203–204.

62. Bynum, "Hospitals and Career Structures."

63. Results for the election are taken from the Minutes Book of St. George's Hospital and are cited in a letter from George Peachey to L. F. Gilbert, Oct. 14, 1932, Gilbert Collection, Box 9, file 2, enclosure B.

64. Wollaston Mss., notebook **Daybook**, 5.

65. Wollaston to Hasted, postmarked Dec. 29, 1800, Gilbert Collection, box 1, file 1.

66. Thomson, *History of Chemistry*, II, 247.

CHAPTER THREE

1. Henry, *Elements of Experimental Chemistry*, 1, x.

2. A leading reference to the reception of Volta's discovery in England is Sudduth, "The Voltaic Pile."

3. Wollaston to Hasted, June 6, 1800, Gilbert Collection, box 1, file 1.

4. Hasted, *Reminiscences*, 7.

5. Pictet, *Voyage de trois mois*, 17–18.

6. Usselman, "Wollaston's Microtechniques," 25.

7. W. H. Wollaston, "Production and Agency of Electricity," 1801.

8. Davy, "Causes of the Galvanic Phenomena."

9. W. H. Wollaston, "Production and Agency of Electricity," 1801, 431.

10. Ibid.

11. Ibid., 432.

12. Ibid., 434. Like most of his peers, Wollaston used the terms "voltaic" and "galvanic" as near synonyms.

13. Pictet, *Voyage*, 17.

14. Cited in Knight, *Transcendental Part of Chemistry*, 43.

15. Donovan, *Present State of Galvanism*, 51.

16. W. H. Wollaston, "Production de l'électricité," 1821, 45.

17. Faraday, *Experimental Researches*, 1, 93.

18. Paris, *Sir Humphry Davy*, 1, 97–98.

19. W. H. Wollaston, "Elementary Galvanic Battery," 1815.

20. Usselman, "Wollaston's Microtechniques," 29.

21. Wollaston Mss., notebook **Daybook**, 16.

22. Gordon, "Wollaston, F.R.S."

23. Gilbert, "William Hyde Wollaston."

24. Wollaston to Hasted, Nov. 16, 1801; Gilbert Collection, box 1, file 1.

25. Wollaston to Hasted, Nov. 24, 1801; ibid.

26. Ibid.

27. Certificates of Admission, Royal Society, London.

28. Geikie, *Royal Society Club*.

29. Barrow, *Sketches of the Royal Society*, 9.

30. Minute Books of the Royal Society Club, Royal Society, London.

31. Jungnickel and McCormmach, *Cavendish*, 305.

32. Barrow, *Sketches*, 58.

33. There are several book-length biographies of Davy. I have drawn information from Hartley, *Humphry Davy*; Knight, *Humphry Davy*; and Fullmer, *Young Humphry Davy*.

34. Pictet, *Voyage*, 30.

35. Information on Young is drawn from Wood and Oldham, *Thomas Young*; Robinson, *The Last Man*; and Cantor, "Thomas Young."

36. Berman, *The Royal Institution*, 23–24.

37. Robinson, *The Last Man*.

38. Blomefield, *Chapters in My Life*, 42.

39. For a concise overview of English optics of this time, and Young's contribution to it, see Cantor, " 'Georgian' Optics."

40. Wollaston to Young, Nov. 20, 1800; Thomas Young Correspondence, Royal Society Library, letter Yo44. Emphasis in the original.

41. Wollaston to Young, Nov. 22, 1800, ibid., Yo45. Emphasis in the original.

42. Wollaston to Young, Monday 17 [Aug, 1801], ibid., Yo47.

43. Wollaston to Hasted, Aug. 30, 1801, Gilbert Collection, box 1, file 1.

44. Wollaston to Young, undated [ca. 1801], Royal Society Library, Yo50.

45. Babbage, *Reflections*, 203–205.

46. W. H. Wollaston, "Refractive and Dispersive Powers," 1802.

47. Ibid., 372.

48. Such inverted dispersion had previously been described in Blair, "Unequal Refrangibility," which Wollaston cited.

49. W. H. Wollaston, "Refractive and Dispersive Powers," 1802, 378.

50. King, "Optical Work of W. H. Wollaston," 15.

51. Wollaston, "Refractive and Dispersive Powers," 1802, 378.

52. David Brewster, "Report on Optics," delivered to the British Association for the Advancement of Science in 1832. Quoted in Watson, " Discovery of the Dark Lines in the Solar Spectrum," 497.

53. Charles Babbage, *Reflections*, 209–210.

54. Ibid., 210–211.

55. Young, "The Production of Colours," 395.

56. W. H. Wollaston, "Refractive and Dispersive Powers," 1802, 379–380.

57. Ibid.

58. Anon. [H. Brougham], *Edinburgh Review*, 1803, 98.

59. W. H. Wollaston, "Chemical Effects of Light," 1804.

60. For a helpful discussion of competing explanations of double refraction in the early nineteenth century, see Frankel, "Corpuscular Theory of Double Refraction." A more comprehensive and mathematically rigorous treatment is Buchwald, "Double Refraction." I have drawn heavily on Buchwald's conclusions in my assessment of Wollaston's contributions to the subject.

61. Buchwald, "Double Refraction," gives a detailed comparison of the experimental fit of each explanation.

62. Anon. [T. Young], "Review of Laplace's Memoir," 339.

63. W. H. Wollaston, "Iceland Crystal," 1802.

64. Ibid., 383.

65. Buchwald, "Double Refraction," 342–349.

66. Anon. [H. Brougham], *The Edinburgh Review*, 99.

67. Full details are available in Buchwald, "Double Refraction," 350–366.

68. See Frankel, "Double Refraction," for additional details on the French contributions to the topic.

69. Anon. [T.Young], "Review of Laplace's memoire," 338–339.

70. Ibid., 348.

71. W. H. Wollaston, "Quantity of Horizontal Refraction," 1803.

72. We now know that atmospheric refraction depends on how steep the temperature gradient of the air is and whether it is positive or negative. For a modern, well illustrated explanation, see Fraser and Mach, "Mirages."

73. Greenwich Observatory, Minutes of the Board of Longitude, VII, 23–24.

74. Anon., "Instruments intended for the Northern Expeditions."

75. Simms, *A Treatise*, 65.

76. The sketch of the dip sector is taken from Hall, *Voyage of Discovery*, Appendix, xxxii.

77. Ibid., xxxviii.

78. There is a more complete discussion of the use of Wollaston's instruments in Goodman, "William Hyde Wollaston," 204–209.

79. Bektas and Crosland, "The Copley Medal."

CHAPTER FOUR

1. McDonald & Hunt, *History of Platinum*, 173.

2. Ibid., 18.

3. Vallvey, "Smuggling of Spanish Platina."

4. Usselman, "Colombian Alluvial Platina," 256.

5. Watson, "A new Semi-Metal."

6. In this book, I will refer to the naturally-occurring, unpurified ore as platina, and the purified metal as platinum. This distinction was not generally made in the eighteenth and early nineteenth centuries.

7. McDonald & Hunt, *History of Platinum*, 48.

8. Ibid., 77.

9. Vallvey, "Smuggling of Spanish Platina."

10. F. J. H. Wollaston, *A Course of Chemical Lectures*, 25.

11. Tennant, "On the Action of Nitre," 221.

12. Wollaston Mss., notebook **Daybook**, 11, 17.

13. R. Knight, "A new and expeditious Process."

14. Ibid., 3.

15. The purchase was made on Feb. 18, 1800. Wollaston Mss., notebook **Daybook**, 17.

16. Wollaston Mss, notebook **J**, 17–18. Reproduced by kind permission of the Syndics of Cambridge University Library.

17. The abbreviation "gn." is used for grain(s), to avoid confusion between Wollaston's shortform "gr" and the abbreviation "g." for gram.

18. Kronberg, Coatsworth, and Usselman, "Mass Spectrometry as a Historical Probe."

19. Wollaston Mss., notebook $\mathbf{L_1}$, 4, lists Wollaston's share as £475 and Tennant's as £320.

20. Wollaston Mss., notebook **J**, 20, 29, 31.

21. Ibid., 31.

22. W. H. Wollaston, "Rendering Platina malleable," 1829.

23. Ibid., 1.

24. Wollaston Mss., notebook **O**, 22.

25. Ibid.

26. Ibid., 23.

27. Knight, "Expeditious process," 3.

28. Lavoisier, *Elements of Chemistry*, 492.

29. Wollaston Mss., notebook **D**, iii.

30. Ibid.

31. Wollaston Mss., notebook **G**, 3.

32. Ibid., 4.

33. Ibid., 8.

34. Wollaston Mss., notebook $\mathbf{L_1}$, 4 (Wollaston account) and T4 (Tennant account).

35. Wollaston Mss., notebook **G**, 23. Reproduced by kind permission of the Syndics of Cambridge University Library.

36. In Troy weights, still used to this day for precious metals, 1 pound consists of 12 ounces, 1 ounce of 20 pennyweights (dwt.), 1 pennyweight of 24 grains. Troy pounds and ounces were the same as their apothecary equivalents, and an ounce in each scale was made up of 480 gn.

37. Wollaston Mss., notebook **G**, 13.

38. Wollaston Mss., notebook $\mathbf{L_1}$, 4.

39. A photograph of the blade is available in Usselman, "Secret History of Platinum," 38–39.

40. For an analysis of Proust's platinum work see Garcia, "Early Studies of Platinum in Spain."

41. Proust, "Experiments on Platina."

42. Ibid., 122.

43. Wollaston Mss., notebook **G**, 8.

44. W. H. Wollaston "Rendering Platina Malleable," 1829.

45. McDonald & Hunt, *History of Platinum*, 173.

46. Wallach, *Briefwechsel*, 1, 253; quoted in Weekes and Leicester, *Discovery of the Elements*, 403.

47. Granville, *The Royal Society in the XIXth Century*, 70.

CHAPTER FIVE

1. J. Banks to R. Chenevix, Mar. 22, 1805, British Museum (Natural History).

2. Wollaston Mss., notebook **G**, facing p. 1.

3. Ibid., 11–12.

4. Ibid., 82.

5. Ibid., 12. Selenite is actually a calcium sulfate.

6. Wollaston Mss., notebooks **F**, 3–15 and **D**, 1–10.

7. Wollaston Mss., notebook **F**, 3.

8. Wollaston Mss., notebook **D**, 3.

9. W. H. Wollaston, "On the Discovery of Palladium," 1805, 320.

10. Ibid., 322.

11. Ibid., 322–323.

12. Wollaston Mss., notebook **O**, 66. Reproduced by kind permission of the Syndics of Cambridge University Library.

13. Details of the procedure are given in Usselman, "The Wollaston/Chenevix Controversy," 564.

14. Wollaston Mss., notebook **O**, 66+.

15. W. H. Wollaston, "Discovery of Palladium," 1805, 324. Wollaston considered a "triple salt" to be one that had a third constituent combined with the acidic and basic portions of normal two-component salts. This differs from the modern usage of the term.

16. Wollaston Mss., notebook **G**, 31.

17. W. H. Wollaston, "On a method of rendering Platina malleable," 1829, 7–8.

18. McDonald & Hunt, *History of Platinum*, 191.

19. A photograph of one of these palladium samples is in Kronberg et al, "Palladium and Rhodium," following p. 32.

20. Ibid., 23.

21. W. H. Wollaston, "Discovery of Palladium," 1805, 327.

22. Wollaston Mss., notebook **F**, 13.

23. Wollaston Mss., notebook **G**, 31.

24. Ibid., 32.

25. Wollaston Mss., notebook **F**, facing p.15.

26. For a more complete analysis of the palladium controversy, see Usselman, "The Wollaston/Chenevix Controversy."

27. Wollaston Mss., notebook $\mathbf{L_1}$, 18.

28. A photograph of the leaflet is reproduced in McDonald & Hunt, *History of Platinum*, 154.

29. Frondel, "Jacob Forster." Frondel suggests that the leaflet was distributed by Forster, which is plausible given Wollaston's desire for anonymity and secrecy.

30. Wollaston Mss., notebook **Forster Account**, 2.

31. Chenevix, "Enquiries concerning the Nature of a Metallic Substance," 291.

32. Ibid., 290.

33. Ibid., 307.

34. Royal Society of London, Minute Book of the Committee of Papers, entry for the meeting of June 16, 1803. Thomas Thomson in his 1830 *History of Chemistry* (2, 216) mistakenly writes that Wollaston was one of the secretaries of the Royal Society when Chenevix's paper was read. Others have repeated this error.

35. Wollaston Mss., notebook **Forster Account**, 2.

36. Wollaston Mss., notebook **G**, 53.

37. Wollaston, "On a new Metal," 1804, 429.

38. Dean and Usselman, "The 'Synthetic' Palladium of Richard Chenevix."

39. Nicholson, "Some account of a pretended new metal."

40. Chenevix, "D'une lettre . . . Au cit. Vauquelin."

41. Ibid., 337.

42. Wollaston Mss., notebook **Daybook**, 30, 98.

43. Banks to Chenevix, London, Mar. 22, 1805. British Museum (Natural History).

44. Royal Society Journal Book, Copley Award speech for Nov. 30, 1803.

45. Anon. [W. H. Wollaston], "Reward of Twenty Pounds," 1804.

46. W. Nicholson, "Prize."

47. Wollaston Mss., notebook **Forster Account**, 5.

48. Thomson, *A System of Chemistry*, 4, 785.

49. Collet-Descotils, "On the Cause of the different Colours."

50. Fourcroy and Vauquelin, "Expériences sur le platine brut."

51. Wollaston Mss., notebook **G**, fac. p. 43.

52. W. H. Wollaston, "On a new Metal," 1804, 420.

53. Ibid., 419.

54. Wollaston Mss., notebook **G**, 48.

55. Kronberg et al., "Palladium and Rhodium," 23.

56. W. H. Wollaston, "On a new Metal," 1804. The printed paper mistakenly lists June 24 as the date of reading.

57. Tennant, "Two metals."

58. W. H. Wollaston, "On a new Metal," 1804, 430.

59. Chenevix, "Platina and Mercury."

60. Royal Society Library, Journal Book, Minutes of Weekly Meetings (1793–1829), Jan. 31, 1805. The contents of the paper are summarized in the meeting minutes.

61. W. H. Wollaston, "Letter Concerning Palladium," 205.

62. W. H. Wollaston, "Discovery of Palladium," 1805.

63. Ibid., 317–318.

64. Wollaston Mss., bound collection of published papers.

65. Thomson, "Analysis of the Ore of Iridium."

66. W. H. Wollaston, "Discovery of Palladium," 1805, 329.

67. Wollaston Mss., notebook **O**, 41. See also Chaldecott, "Wollaston's Platinum Thermometer."

68. Wollaston Mss., notebook $\mathbf{L_1}$, 34.

69. Ure, *Dictionary of Chemistry*, 768.

70. See, for example, Reilly, "Richard Chenevix."

71. Banks to Chenevix, Mar. 22, 1805, British Museum (Natural History).

72. Vauquelin, "Le palladium et le rhodium," 170–171.

73. Wollaston Mss., notebook $\mathbf{L_1}$, 18.

74. Wollaston Mss., notebook **Forster Account**, 4–7.

75. Ibid., 7.

76. Wollaston Mss., notebook $\mathbf{L_1}$, T7.

77. Wollaston Mss., notebook **Forster Account**, 10.

78. Ibid., 11.

79. See Chaldecott, "Platinum and Palladium in Astronomy and Navigation."

80. Wollaston Mss., notebook **Forster Account**, 12.

81. Ibid.

82. Full details are given in Kronberg et al., "Palladium and Rhodium."

83. Chaldecott, "Platinum and Palladium," 95.

84. Wollaston Mss., notebook **Forster Account**, 17.

85. Wollaston Mss., notebook **G**, 50.

86. Wollaston Mss., notebook **D**, 47–48.

87. More information is given in Kronberg et al., "Palladium and Rhodium."

88. W. H. Wollaston, Notebook on **Rhodium**, Science Museum, 1.

89. W. H. Wollaston, Notebook on **Rhodium**, contains receipts from Cuttell for slicing the alloy.

90. Kronberg et al, "Palladium and Rhodium," 29.

91. W. H. Wollaston, Notebook on **Rhodium**, 5.

92. Wollaston Mss., notebook $\mathbf{B_2}$, 17–19.

93. Wollaston Mss., notebook **Daybook**, 78–79.

94. A photograph is reproduced in Kronberg et al, "Palladium and Rhodium."

95. Edward Codrington to William Codrington, May 5, 1822. I thank Mrs. Daphne Wollaston for permission to quote from the letter in her possession.

CHAPTER SIX

1. Hall, *Forty Etchings*, ii. Macadamised roads consisted of layers of sized stone constructed to the design of the Scot, John Macadam.

2. W. H. Wollaston, "Spectacle Glasses," 1804.

3. English patent No. 2752, Feb. 9, 1804.

4. Wollaston Mss., notebook **O**, 30–31.

5. Wollaston Mss., notebook **Daybook**, 32.

6. Jones, "Observations on Dr. Wollaston's Statements."

7. Ibid., *Phil. Mag.*, 67.

8. Von Rohr, "Meniscus Spectacle Lenses."

9. W. H. Wollaston, "Periscopic Spectacles," 1804.

10. Jones, "Examination of Dr. Wollaston's Experiment."

11. Wollaston Mss., notebook **Daybook**, fac. p. 44, fac. p. 56 and fac. p. 71.

12. W. H. Wollaston, "Periscope Camera Obscura and Microscope," 1812.

13. Ibid., 372.

14. Clay and Court, *History of the Microscope*, 74.

15. Jones, "Critical Observations."

16. Ibid., *J. Nat. Phil. Chem and Arts*, 101.

17. Ibid., 107.

18. W. H. Wollaston, "Report of Mons. Biot," 1813.

19. Ibid., *J. Nat. Phil. Chem and Arts*, 319–320.

20. For more information, see Goodman, "William Hyde Wollaston," 238–247.

21. Von Rohr, "Spectacle Lenses," 186.

22. Hasted, *Reminiscences*, 6.

23. English patent No. 2993, Dec. 4, 1806.

24. W. H. Wollaston, "The Camera Lucida," 1807.

25. Ibid., 344.

26. Hammond and Austin, *Camera Lucida*, 79.

27. Wollaston to Hasted, Sept. 26, 1806; Gilbert Collection, box 1, file 1.

28. Hooke, "A Contrivance."

29. See Hammond and Austin, *Camera Lucida*, 13–17.

30. Ibid., 25–27.

31. Wollaston Mss., notebook **Daybook**, fac. p. 44, fac. p. 56.

32. Sheldrake, "On the Camera Lucida."

33. Bate, "On the Camera Lucida," 147–148.

34. Taken from Benjamin (ed.), *Wrinkles and Recipes*, 195.

35. Wollaston Mss., notebook **Daybook**, 42.

36. See Schaaf, *Tracings of Light* and Warner, *Cape Landscapes*.

37. Herschel's sketch of the Temple of Juno.

38. W. H. Wollaston by Francis Chantrey. Reproduced with permission.

39. Lithograph of Wollaston from a sketch (made with a camera lucida) by Chantrey, drawn on Stone by R. J. Lane. Reproduced with permission.

40. Hall, *Travels in North America*.

41. Hall, *Forty Etchings*, ii.

42. Fox Talbot, *The Pencil Of Nature*, Brief Historical Sketch, 3.

43. For a more complete discussion, see Hammond and Austin, *Camera Lucida*.

44. Weld, *History of the Royal Society*, 2, 561.

45. Hall, *All Scientists Now*, 10.

46. W. H. Wollaston, "On the Force of Percussion," 1805.

47. For the scientific context, see G. E. Smith, "The *vis viva* dispute."

48. Cardwell, "The Concepts of Power, Work and Energy," 213.

49. W. H. Wollaston, "On the Force of Percussion," 1805, 22.

50. Ibid.

51. See, for added discussion of Wollaston's contributions to this topic, Cardwell, "The Concepts of Power, Work and Energy," 209–224.

52. Young, *A Course of Lectures*, 1, 59.

53. Anon. [John Playfair], *Edin. Rev., 128.*

54. Ewart, "On the measure of moving force."

55. Wollaston Mss., notebook **Daybook**, 43.

56. Wollaston to Hasted, Feb. 28, 1807, Gilbert Collection, box 1, file 1.

57. For information on several such clubs, see Averly, "The 'Social Chemists'" and Inkster, "Science and Society in the Metropolis."

58. There are a few pages of references to the Chemical Club in the Gilbert Collection, Box 1, File 3, Enclosure B.

59. J. Dalton to Rev. Mr. Johns, Dec. 27, 1809, quoted in Patterson, *John Dalton*, 171.

60. Söderbaum, *Jac. Berzelius Reseantechningar*, 38, entry for July 28, 1812.

61. Typed copies of fifty letters from Wollaston to Marcet, the originals of which are now in private hands, are in the Gilbert Collection, University College, London.

62. Marcet, "Various Dropsical Fluids," 358.

63. Most of the information on Howard is taken from Kurzer, "Edward Charles Howard."

64. Ibid., 138.

65. See Woodward, *The History of the Geological Society of London*; Davies, *The Geological Society of London*; and Rudwick, "The Foundation of the Geological Society of London."

66. Rudwick, ibid., 336.

67. Woodward, *Geological Society*, 274.

68. Payments from Wollaston to the secretary of the Geological Society, Leonard Horner, are recorded in the copy of Wollaston's banking account at Coutts & Co contained in the Gilbert Collection. I thank the Directors of Messrs. Coutts & Co for permission to quote from the account.

69. Leonard Horner to Wollaston, Aug. 14, 1811. I thank Mrs. Daphne Wollaston for permission to quote from the letter.

CHAPTER SEVEN

1. Streicher, "Some Remarks on the Wollaston Method," 16.

2. Chaldecott, "William Cary."

3. Wollaston Mss., notebook **I**, 2.

4. Banks to Chenevix, Mar. 22, 1805, British Museum (Natural History).

5. McDonald & Hunt, *History of Platinum*, 138–139.

6. Hall to Marcet, Mar. 18, 1805, National Library of Scotland MS 3813. Quoted in Chaldecott, "Fabrication of Platinum Vessels," 159–160.

7. Ibid., May 14, 1805. Quoted in Chaldecott, ibid., 163–164.

8. Thomson, "Biographical Account of Dr. Wollaston," 139.

9. Wollaston Mss., notebook **I**, i–ii.

10. Ibid., 2.

11. Ibid., 3.

12. Thorpe, *Essays in Historical Chemistry*, 558–559.

13. Chaldecott, "William Cary," 122, quotes 14s. 6d. as the retail price of unworked platinum, based on a comparison of overall weights sold by Cary and deposits into Wollaston's bank account. However, not all of Cary's remittances to Wollaston were made by bank deposit, especially in the early years, so this figure for the retail price of platinum is somewhat lower than the price of 16s. given here.

14. Wollaston Mss., notebook **D**, iii.

15. Wollaston Mss., notebook **F**, 10.

16. For more on financial details of the Wollaston-Tennant partnership, see Usselman, "Merchandising Malleable Platinum."

17. Wollaston Mss., notebook $\mathbf{L}_1$, 29.

18. Wollaston Mss., notebook **G**, -5. Reproduced by kind permission of the Syndics of Cambridge University Library.

19. Much of the information on the use of platinum in firearms is taken from Cottington, "An Early Industrial Application for Malleable Platinum."

20. Ibid., 79.

21. The Figure is taken from Cottington, ibid., and reproduced with permission of *Platinum Metals Review* and the Royal Armouries, Leeds, where the gun is located.

22. Gilbert, "Wollaston Mss. at Cambridge," 320.

23. Composite data compiled from Wollaston Mss., notebook **I**.

24. For more details on the methods of acid production, see A. and C. R. Aikin, *Dictionary of Chemistry and Mineralogy*, 2, 367–371; and J. G. Smith, *Heavy Chemical Industry in France*, 5–112.

25. Ibid., Smith, 98.

26. Gilbert, "Wollaston Mss.," 322.

27. Wollaston Mss., notebook **H**, 3.

28. Gilbert, "Wollaston Mss.," 323.

29. Wollaston Mss., notebook $\mathbf{L}_1$, T8.

30. Wollaston Mss., box 2, file B, item 2.

31. Wollaston Mss., notebook $\mathbf{L}_1$, T8.

32. Wollaston to Farmer, Nov. 30, 1809, Science Museum Library.

33. Wollaston Mss., notebook $\mathbf{L}_1$, 44.

34. Wollaston Mss., notebook **H**, 5.

35. Wollaston Mss., notebook $\mathbf{L}_1$, summary sheet inserted before p. 43.

36. Wollaston Mss., notebook **H**, 12.

37. Wollaston to Farmer, Dec. 17, 1809, Science Museum Library.

38. Wollaston Mss., notebook $\mathbf{L}_1$, summary sheet inserted before p. 43.

39. Ibid., 47 and T11.

40. Wollaston to Parkes, Jan., 1812, quoted in Hinde, "William Hyde Wollaston," 674.

41. Parkes, *A Chemical Catechism*, 370–371.

42. Wollaston Mss., notebook **H**, 4.

43. *English Cyclopaedia*, 6, 795–796.

44. Composite data from Wollaston Mss., notebook **H**.

45. Ibid., 44.

46. Data for the chart is taken from Wollaston Mss., notebooks **I** and **H**.

47. Some of the other uses to which bar platinum was put are described in Chaldecott, "William Cary."

48. McDonald, *The Johnsons of Maiden Lane*.

49. Usselman, "Colombian Alluvial Platina."

50. Data for the chart is taken from Wollaston Mss., notebooks **D**, $\mathbf{L_1}$, and $\mathbf{B_2}$.

51. Vallvey, "Smuggling of Spanish Platina."

52. Usselman, "Colombian Alluvial Platina," 259.

53. Vallvey, "Smuggling of Spanish Platina," 484.

54. Wollaston Mss., notebook $\mathbf{L_1}$, 1.

55. Ibid., 51.

56. Data for the chart is taken from Wollaston Mss., notebooks $\mathbf{L_1}$ and $\mathbf{B_2}$.

57. Composite data from Wollaston Mss., notebooks **D** and **F**.

58. Wollaston Mss., notebook **D**, 35.

59. Wollaston Mss., notebook **F**, fac.p. 17.

60. Wollaston Mss., notebook **D**, 14.

61. Ibid.

62. W. H. Wollaston, "Rendering Platina Malleable," 1829, 3.

63. Ibid.; also Wollaston Mss., notebook **D**, 14, 16.

64. Ibid., notebook **D**, 38.

65. Wollaston Mss., notebook $\mathbf{B_2}$, 4.

66. W. H. Wollaston, "Rendering Platina Malleable," 1829, 4–5. The press, minus the cradle, is now in the Whipple Museum of the History of Science in Cambridge, England.

67. Wollaston Mss., notebook **E**, 13.

68. Streicher, "Some Remarks on the Wollaston Method," 16.

69. Bank of England account ledger for William Hyde and Henry Septimus Hyde Wollaston. I thank John Keyworth of the Museum and Historical Research Section of the Bank of England and the helpful staff at the Record Office for their kind assistance.

70. Ibid.

CHAPTER EIGHT

1. W. H. Wollaston, "On Super-acid and Sub-acid Salts," 1808, 96.

2. Use of the term "organic" to denote compounds that originate in living organisms, or are produced from them, was in use in the early nineteenth century, and it is in that sense that the term is used here.

3. The abbreviations I use throughout this chapter for the organic acids and their salts are intended to focus attention on the acidic hydrogen atoms (as we now

understand them) present in those compounds. Such symbolism will make their combining proportions easier to understand, although the proton donor theory of acidity was not proposed until early in the twentieth century.

4. A. and C. R. Aikin, *Dictionary of Chemistry and Mineralogy*, 2, 192.

5. Wollaston Mss., notebook $\mathbf{L_1}$, 10.

6. Wollaston Mss., notebook **F**, 36.

7. Ibid., fac. p. 36.

8. Ibid., fac. p. 38.

9. Ibid.

10. Ibid., fac. p. 39.

11. Ibid., fac. p. 41, 41.

12. Wollaston Mss., notebook $\mathbf{L_1}$, 10.

13. Ibid., 30.

14. James, "Faraday as an Administrator," 122.

15. Wollaston Mss., notebooks $\mathbf{L_1}$, 44, and $\mathbf{B_1}$, 12.

16. Ibid., notebook $\mathbf{L_1}$.

17. Wollaston Mss., notebook **F**, fac. p. 41, 41.

18. Ibid.

19. Wollaston Mss., notebook $\mathbf{L_1}$, T6.

20. Ibid., T13.

21. Wollaston Mss., notebook **F**, 40.

22. All data is taken from Wollaston Mss., notebook $\mathbf{L_1}$.

23. Wollaston Mss., notebook **O**, 38.

24. Wollaston Mss., notebook $\mathbf{L_1}$, T4.

25. Usselman, "Smithson Tennant," 129.

26. Wollaston Mss., notebook $\mathbf{L_1}$, fac. p. T15.

27. Wollaston Mss., notebook **F**, fac. p. 44. Natural lime juice contains about 7 .5 oz. of citric acid per gallon, which suggests that the purchased lime juice had been previously concentrated, perhaps for use by sailors as a scurvy preventative.

28. Ibid., 44.

29. Wollaston Mss., notebook $\mathbf{L_1}$, 59.

30. Ibid., T15–T17.

31. Wollaston Mss., notebook **O**, 59.

32. Ibid.

33. Ibid.

34. Ibid., 60.

35. Wollaston Mss., notebook **F**, fac. p. 30.

36. Wollaston Mss., notebook **O**, 72.

37. Ibid., fac. p. 60.

38. Thomson, "On the Oxides of Lead."

39. See Freund, *Chemical Composition*, 153–157, for a succinct summary of Dalton's findings. Dalton's observations are placed in context in Thackray, *John Dalton*; and Rocke, *Chemical Atomism*.

40. Thomson, *A System of Chemistry*, 1807, 3, 424–429.

41. Ibid., 626.

42. Thomson, "On Oxalic Acid."

43. Ibid., 70.

44. Ibid., 73–74. See Usselman, "Multiple Combining Proportions," which discusses the non-replicability of Thomson's results.

45. Thomson, "Oxalic Acid," 87.

46. Thomson, *History*, 2, 293.

47. W. H. Wollaston, "On Super-acid and Sub-acid Salts," 1808.

48. Ibid., 96.

49. Ibid., 99.

50. Ibid.

51. Ibid., 101.

52. Ibid.

53. Ibid., 101–102.

54. Ibid., 102.

55. Whewell, *History of the Inductive Sciences*, 3, 150.

56. Berzelius, *Autobiographical Notes*, 172–173.

57. Berthollet, Introduction to *Système de chimie*.

58. Bérard, "Les Oxalates et Suroxalates alcalins." In a study to be published elsewhere, I will report a set of replication experiments which confirm the experimental implausibility of Thomson's results.

CHAPTER NINE

1. Herschel, *A Preliminary Discourse*, 354.

2. Wollaston Mss., notebook **Daybook**, 30.

3. Ibid.

4. Wollaston Mss., Smithson Tennant's notebook **Tours**, 1805 entry for tour in Ireland.

5. Wollaston Mss., notebook **Daybook**, 43.

6. Ibid., 49.

7. Ibid., 52, 98.

8. Bostock to Marcet, Sept. 16, 1809; typed copy in Gilbert Collection, box 4, file 1, enclosure B.

9. Geike, *Life of Sir Roderick Murchison*, 1, 129. No date for the event is given.

10. Haüy, *Sur la structure des crystaux*. Haüy's contributions are assessed in Burke, *Origins of the Science of Crystals*, and the relevance of Haüy's ideas to Wollaston's work is discussed in Goodman, "Problems in Crystallography."

11. Burke, *Origins of the Science of Crystals*, 88.

12. Carangeot, "Goniomètre."

13. W. H. Wollaston, "Refraction of Iceland Crystal," 1802.

14. Ibid., 385.

15. Wollaston Mss., notebook **Daybook**, 47.

16. The letters written by Wollaston to Clarke are now in the possession of Mr. Herb Obodda, who has very kindly given me access to them. I thank him for his generosity.

17. W. H. Wollaston, "Reflective Goniometer," 1809.

18. Ibid., 253–254.

19. Ibid., 256–257.

20. Wollaston Mss., Wollaston's copies of his own published papers.

21. Wollaston to Clarke, Feb. 2, 1810, H. Obodda library. Underlining in the original.

22. Wollaston to Allan, Feb. 12, 1812; copy in Gilbert Collection, box 1, file 2, enclosure A. Underlining in the original.

23. Buchwald, "Double Refraction from Huygens to Malus," 350.

24. Ibid., 354–357, has a full description and diagram of the instrument.

25. Ibid., 357.

26. Wollaston to Hasted, Jan. 26, (1811?); Gilbert Collection, box 1, file 1.

27. Haüy, *Tableau comparatif.*

28. Ibid., 122.

29. Wollaston to Hasted, June 12, 1809, Gilbert collection, box 1, file 1.

30. Goodman, "William Hyde Wollaston," 357–358.

31. Herschel, *A Preliminary Discourse*, 354.

32. W. H. Wollaston, "On the Primitive Crystals," 1812, 161.

33. See, for a sound analysis of Mitscherlich's discovery, Melhado, "Mitscherlich's Discovery of Isomorphism." Goodman, "William Hyde Wollaston," 97–100, discusses the measurement of crystal angles with the reflective goniometer made by some of Wollaston's contemporaries.

34. W. H. Wollaston, "Elementary Particles," 1813.

35. Ibid., 57.

36. Ibid., 53.

37. Dalton, *New System of Chemical Philosophy*, Part 1, 210–211.

38. Dalton to Rev. Mr. Johns, Dec. 27, 1809, quoted in Patterson, *John Dalton*, 171.

39. Dalton to Daubeny, quoted in Daubeny, *Introduction to the Atomic Theory*, 137.

40. Wollaston, "Elementary Particles," 1813, 58.

41. Ibid., 59.

42. Ibid., 60–61. The modern sodium chloride crystal lattice has the same structure as that proposed by Wollaston, but the constituent particles are ions of different sizes.

43. Goodman, "Problems in Crystallography," discusses the impact of Wollaston's Bakerian Lecture.

44. W. H. Wollaston, "Elementary Particles," 1813, 61.

45. Ibid., 53.

46. Beudant, "Recherches." Thomson's summary appeared as Beudant, "An Abstract of an Inquiry." See Melhado, "Isomorphism," for a thorough analysis of Beudant's paper.

47. Ibid., "An Abstract," 265.

48. W. H. Wollaston, "Observations on M. Beudant's Memoir," 1818.

49. Ibid., 284.

50. Ibid., 286.

51. W. H. Wollaston, "On the Octohedral Form of Iodine," 1815, and "The primitive Form of Bitartrate of Potash," 1817.

52. W. H. Wollaston, *Crystallographic Notes and Measurements.*

53. For a comprehensive discussion of Mitscherlich's research in mineralogy, see Melhado, "Mitscherlich's discovery." There is no evidence that Mitscherlich knew of Wollaston's work on structurally similar crystal forms, although he did use a reflecting goniometer to measure crystal angles.

54. See, for example, Schütt, "Der Entdeckung des chemischen Isomorphimus," and Morrow, "One Hundred and Fifty Years of Isomorphism."

55. LN, "Meionite," 31.

CHAPTER TEN

1. W. H. Wollaston, "Chemical Equivalents," 1814, 7.

2. W. H. Wollaston, "The Croonian Lecture," 1810.

3. Ibid., 2.

4. Ibid., 3.

5. Ibid., 3–4.

6. Helmholtz, "Über den Muskeltone," 208–209.

7. For a comprehensive modern discussion of muscle sounds, see Oster, "Muscle Sounds."

8. W. H. Wollaston, "On Platina and native Palladium from Brazil," 1809.

9. Hatchett, "A Metal Hitherto Unknown."

10. Ekeberg, "Of the Discovery of a New Substance."

11. See Weeks and Leicester, *Discovery of the Elements*, for a full account of the discovery of the two elements.

12. W. H. Wollaston, "Columbium and Tantalum," 1809.

13. Weeks and Leicester, *Discovery of the Elements*, 347–348.

14. W. H. Wollaston, "On a Method of Freezing at a Distance," 1813.

15. See B. A. Smith, "Wollaston's cryophorus," for a modern explanation of the theoretical basis for the device.

16. W. H. Wollaston, "Extremely Fine Wires," 1813, 114–115.

17. Wollaston Mss., notebook **C**, 20.

18. Ibid.

19. Wollaston, "Extremely Fine Wires," 1813, 115.

20. Ibid., 116.

21. Wollaston Mss., notebook **C**, 21.

22. See Coatsworth et al., "Artefact as Historical Document," for a more thorough discussion of Wollaston's wires.

23. Wollaston Mss., notebook **C**, 25.

24. Sacharoff et al., "Fabrication of Ultrathin Drawn Pt Wires."

25. Coatsworth et al., "Artefact as Historical Document," 101.

26. Ibid., 103.

27. Beamish et al., "The Platinum Metals," 319.

28. Wollaston, "Extremely Fine Wires," 1813, 117.

29. The draft manuscript is present in the Wollaston Mss., Box 2, file D, item 3. The table's data is given in Coatsworth et al., "Fine Platinum Wires," 97.

30. W. H. Wollaston, "On a Method of Rendering Platina Malleable," 1829, 7.

31. Lavoisier, *Elements of Chemistry*, 492.

32. H. W. Wollaston, *History of the Wollaston Family*, 116.

33. Wollaston Mss., notebook **Tours**, 23.

34. H. S. H. Wollaston, undated journal letter to his children. I thank Mrs. Daphne Wollaston for permission to quote from the letter.

35. Wilson, *Religio Chemici*, 295.

36. W. H. Wollaston, "Scale of Chemical Equivalents," 1814.

37. Ibid., 1–2.

38. Ibid., 2.

39. Ibid., 3.

40. Ibid., 18–22.

41. Wollaston to Berzelius, Aug. 12, 1812, library of The Royal Swedish Academy of Sciences.

42. Berzelius to Davy, undated [August or September 1812], in Soderbaum, *Jac. Berzelius Bref*, I, ii, 32. Translation is taken from Goodman, "William Hyde Wollaston," 25.

43. Berzelius and Marcet, "Experiments on the Alcohol of Sulphur," 192.

44. Marcet to Berzelius, Jun. 10, 1813, in Soderbaum, *Jac. Berzelius Bref*, 1, iii, 51.

45. The ongoing debate about the number of hydrogen and oxygen equivalents (or atoms) contained in water, whether two-to-one or one-to-one, was not satisfactorily resolved until much later in the nineteenth century. Since the values for many elements depended on the ratio chosen, several tables of atomic/equivalent weights of the nineteenth century have values half of the modern ones. See Knight, *Atoms and Elements*, for a general discussion of the problem. A thorough comparison of Wollaston's values with those of his contemporaries is given in Rocke, *Chemical Atomism*, especially 49–97.

46. W. H. Wollaston, "Scale of Chemical Equivalents," 1814, 14.

47. Reid, *Principles of Chemistry*, i, 59. A good description of the widespread adoption of Wollaston's scale is available in Goodman, "Wollaston and the Atomic Theory of Dalton."

48. Goodman, "William Hyde Wollaston," 43–61, discusses development of the scale in greater detail.

49. Wollaston, "Chemical Equivalents," 1814, 2–3.

50. Ibid., 4–5.

51. Ibid. Wollaston's calculation rests, of course, on the unstated (and unproven) assumption that equal volumes of gases contain equal numbers of particles.

52. Ibid., 7.

53. This conclusion has been emphasized in Rocke, *Chemical Atomism*, 61–66.

54. Prout, "Analysis of Organic Substances," 270. See Goodman, "William Hyde Wollaston," for several other examples.

55. Davy, *Six Discourses*, 128–129.

56. Anon., "Adjudgement of the Royal Medals," 15. Quoted in Goodman, "William Hyde Wollaston," 47.

57. W. H. Wollaston, "On the Finite Extent of the Atmosphere," 1822.

58. Ibid., 98.

59. Dalton, "On the constitution of the atmosphere." See Goodman, "Wollaston and the Atomic Theory," for more on the reception of Wollaston's paper.

60. Faraday, "A Limit to Evaporation."

61. Graham, "Finite Extent of the Atmosphere."

62. Whewell, *Philosophy of the Inductive Sciences*, 1, 436–438.

63. Wilson, *Religio Chemici*, 259–263.

64. Daubeny, *Introduction to the Atomic Theory*, 125.

CHAPTER ELEVEN

1. Berzelius to Gahn, Jan. 25, 1813, Söderbaum (ed.), *Jac. Berzelius Bref*, 9, 72.

2. W. H. Wollaston, "Animal Secretions," 1809.

3. See Goodman, "William Hyde Wollaston," 300–302, for a brief discussion of the contributions of others who built on Wollaston's idea.

4. W. H. Wollaston, "On the Non-Existence of Sugar," 1811.

5. Rollo, *Two Cases of the Diabetes Mellitus*.

6. Wollaston to Baillie, undated [1798 or 1799], Royal College of Surgeons of England Library, Hunter-Baillie Collection, vol. 3, 28, no. 23.

7. Wollaston, "Non-Existence of Sugar," 1811, 97–98.

8. Ibid., 99.

9. Bernard, "De l'origine du sucre. . . ."

10. Wollaston Mss., notebooks **M-Pend I**, **M-Pend II**, **M-Pend III** and **Pencil Tabs**.

11. Pond to Wollaston, Jan. 5, 1808, Wollaston Mss., letter inserted in notebook **Pencil Tabs**.

12. Wollaston Mss., notebook **M-Pend I**, 1.

13. Ibid., 93.

14. Wollaston Mss., notebook **M-Pend III**, 15.

15. MacArthur, "Davy's Differences," 208.

16. Berger to Marcet, Jun. 12, 1810; Bibliothèque Publique et Universitaire, Geneva, Fonds Marcet, MS fr. 4245. I thank Colin MacArthur for bringing this letter to my attention, and for the translation from the original French.

17. Wollaston to Marcet, Jun. 29, 1810, Gilbert Collection.

18. MacArthur, "Davy's Differences," 223, note 21.

19. Jones, *The Royal Institution*, 300. The date of Wollaston's election is taken from the minute books of the Royal Institution.

20. Coutts banking account, Gilbert Collection.

21. Jones, *The Royal Institution*, 303.

22. Wollaston to Thomas Harrison (signature portion only), secretary of the Royal Institution, undated; Royal Institution General Archives, 112 Signatures of Members and Correspondence.

23. Wollaston to Clarke, Mar. 19, 1813, H. Obodda library, emphasis in the original. Reprinted with permission.

24. See Usselman, "Smithson Tennant."

25. Warburton to Clarke, Mar. 31, 1813, copy in the Gilbert Collection, University College, London, Box 12, File 3, Enclosure B. Original in the H. Obodda library. Reprinted with permission.

26. Anon. [Whishaw], "The Late Smithson Tennant," 92.

27. Berzelius to Gahn, Jan. 25, 1813, Söderbaum (ed.), *Jac. Berzelius Bref*, 9, 73.

28. Ibid., 72.

29. Berzelius to Berthollet, [October, 1812], ibid., 1, 41–42; quoted in Weeks and Leicester, *Discovery of the Elements*, 411.

30. W. H. Wollaston, "Single-lens Micrometer," 1813.

31. The elections are mentioned in letters from Berzelius to Marcet, May 22, 1813 and Jan. 7, 1814, in Söderbaum (ed.), *Jac. Berzelius Bref*, 1, pt.3, 48, 89. Both diplomas are in the library of the Royal Society of London.

32. Children, "Large voltaic battery."

33. Wollaston Mss., notebook **C**, 28, 29; also mentioned in W. H. Wollaston, "Elementary Galvanic Battery," 1815, 211.

34. Children, "Large voltaic battery," 363.

35. Marcet to Berzelius, Mar. 29, 1815; in Söderbaum (ed.), *Jac. Berzelius Bref*, 1, pt.3, 122.

36. Children, "Large voltaic battery," 368.

37. Ibid., 373.

38. Wollaston to Hasted, undated [1808], Gilbert Collection, box 1, file 1.

39. Wollaston Mss., notebook **Daybook**, fac. p. 65.

40. Ibid., 65.

41. Wollaston to Hasted, July 8, 1814, Gilbert Collection, box 1, file 1.

42. Marcet to Berzelius, May 24, 1814; in Söderbaum (ed.), *Jac. Berzelius Bref*, 1, pt. 3, 98–99.

43. Marcet to Berzelius, ibid.; also July 1814, ibid., 1, pt. 3, 106.

44. Wollaston to Clarke, undated [summer 1814, by context]; H. Obodda library, cited with permission.

45. Wollaston to Hasted, Sept. 15, 1814, Gilbert Collection, box 1, file 1.

46. For more details, see Usselman, "Smithson Tennant."

47. Accounts of Tennant's death are given in Thomson, *History of Chemistry*, 2, 239–240, and in a letter from Marcet to Berzelius, Mar. 29, 1815, in Söderbaum (ed), *Jac. Berzelius Bref*, 1, pt. 3, 118–119.

48. Wollaston to Marcet, Feb. 25, 1815, Gilbert Collection.

49. Wollaston Mss., notebook $\mathbf{L}_1$, 65.2.

50. Ibid., 65.4.

51. Ibid., T17.

52. Wollaston to Hasted, Dec. 24, 1814, Gilbert Collection, box 1, file 1.

53. Wollaston to Unknown, Feb. 27, 1815, Gilbert Collection, box 1, file 2, enclosure A.

54. Wollaston Mss., **Daybook**, 67 and notebook **D**, 57.

55. Probate will of Francis Wollaston, National Archives, Kew, England.

56. Marcet to Berzelius, Nov. 7–9, 1815, in Söderbaum (ed.), *Jac. Berzelius Bref*, 1, pt.3, 128–129.

57. Wollaston Mss., notebook **Daybook**, 63.

58. D. R. Fisher, "Sebright, Sir John Saunders, seventh baronet (1767–1846)," oDNB.

59. Marcet to Berzelius, May 24, 1814, Söderbaum (ed.), *Jac. Berzelius Bref*, 1, pt. 3, 98.

60. Wollaston to Berzelius, Jan. 16, 1816, ibid., 7, 292.

61. Marcet to Berzelius, Sept. 24, 1813, ibid., 1, pt. 3, 73.

62. Wollaston to Hasted, Oct. 9, 1815, Gilbert Collection, box 1, file 1.

63. Marcet to Berzelius, Jan. 23, 1816, Söderbaum (ed.), *Jac. Berzelius Bref*, 1, pt. 3, 136.

64. Wollaston Mss., notebook **Tours**, 27.

65. Wollaston to Hasted, Oct. 9, 1815, Gilbert Collection, box 1, file 1.

CHAPTER TWELVE

1. Wollaston to Marcet, Aug. 5 or 6, 1820, Gilbert collection.

2. Tate, *Alcoholometry*, x.

3. Ashworth, "Between the Trader and the Public," 29–31.

4. Tate, *Alcoholometry*, xi.

5. Ibid., xiv.

6. Simmonds, *Alcohol*, 264.

7. Tate, *Alcoholometry*, xviii.

8. Blagden, "Proportioning the Excise."

9. Gilpin, "Tables."

10. Ashworth, "Between the Trader and the Public," 42.

11. A. Nicholas to Excise Commissioners, Jan. 25, 1803, TNA, CUST 148/17/4. Documents relevant to the hydrometer evaluations are grouped together (in no particular order) with the reference numbers CUST 148/17/1 through to CUST 148/17/6. I thank William Ashworth for bringing these documents to my attention and providing helpful advice in their interpretation.

12. Thomas Groves to Excise Commissioners, Jan. 19, 1803, TNA, CUST 148/17/5.

13. Report addressed to the Honorable Board of Excise concerning Hydrometers, May 25, 1803, TNA, CUST 148/17/2.

14. Ibid.

15. Wollaston to the Board of Excise, May 25, 1803, TNA, CUST 148/17/2.

16. Wollaston to the Board of Excise, Jun. 10, 1803, TNA, CUST 148/17/2.

17. Treasury Committee to Board of Excise, Jul. 15, 1803, TNA, CUST 148/17/5.

18. Wollaston to Board of Excise, Aug. 9, 1803, TNA, CUST 148/17/1.

19. Treasury to Thomas Burton, Nov. 23, 1803, TNA, CUST 148/17/5.

20. A. M. Bate to the Board of Excise, Sept. 5, 1806, TNA, CUST 148/17/1.

21. Contents of the books are summarized and evaluated in Tate, *Alcoholometry*, 14–32.

22. George Harrison to the Board of Excise, Oct. 25, 1806, TNA, CUST 148/17/1.

23. Wollaston to the Treasury, Sept. 9, 1807, ibid.

24. R. Bate to Excise, Dec. 15, 1807 and May 30, 1808, ibid.

25. Wollaston to Excise, Oct. 26, 1810, TNA, CUST 148/17/2.

26. Groves to Excise, Nov. 29, 1810, ibid.

27. Wollaston to Excise, Jan. 25, 1813, TNA, CUST 148/17/5.

28. Nicholas to Excise, Jan. 27, 1813, TNA, CUST 148/17/4.

29. *The Statutes of the United Kingdom of Great Britain and Ireland*, 1816, 56 George III, c. 140.

30. *The Statutes of the United Kingdom of Great Britain and Ireland*, 1818, 58 George III, c. 28.

31. Wollaston Mss., notebook **Daybook**, fac. p. 71.

32. See, for example, T. Porter, *Trust in Numbers*, and Ashworth, *Customs and Excise*.

33. Information on this committee is taken from the Minutes of the Council of the Royal Society of London, vol. 8, for Feb. 24, 1814.

34. A full description of Davy's invention and the controversy that surrounded it is available in D. Knight, *Humphry Davy*, 105–120.

35. Davy, "On the fire-damp of coal mines," 8.

36. Davy to J. G. Lambton, Nov. 21, 1817; quoted in Paris, *Sir Humphry Davy*, 2, 129.

37. Ibid., 130-131.

38. Hoppit, "Reforming Britain's Weights and Measures."

39. "Report from the Select Committee on Weights and Measures," July 1, 1814, House of Commons, Parliamentary Papers (1813–1814).

40. Ibid., 7.

41. Ibid., 11.

42. Hoppit, "Reforming Britain's Weights," 98–99.

43. Goodman, "William Hyde Wollaston," 6–7.

44. Hoppit, "Reforming Britain's Weights," 99.

45. "Report [initial and final] of the Commissioners appointed for inquiring into the mode of preventing the Forgery of Bank Notes," House of Commons, Parliamentary Papers (1819–1820). I thank Mr. John Keyworth, Curator, Bank of England Museum and Historical Research Section, for copies of these reports.

46. Johnson, "The Board of Longitude," 64.

47. Information drawn from the Act is taken from the National Maritime Museum, Board of Longitude Papers, 1714–1829, MRF/L/1, vol. I.

48. Ibid., MRF/L/3, vols. VII and VIII.

49. W. H. Wollaston, "Ramsden's Dividing Engine," 1822. A description of the improvements in the design of English dividing engines is available in Brooks, "The Circular Dividing Engine."

50. For a concise discussion of the navigational and scientific goals of the polar expeditions, see Levere, "Science and the Canadian Arctic." A more comprehensive treatment is available in the same author's *Science and the Canadian Arctic*.

51. Anon., "Instructions."

52. Ross, *Voyage of Discovery*, Appendix iii, 87–89, "Crimson-Coloured Snow and Meteoric Iron."

53. Bauer, "Colouring Matter of the Red Snow."

54. Ross, *Voyage of Discovery*, Appendix iii, 89.

55. Codrington to Wollaston, Feb. 23, 1827, in Bourchier (ed.), *Life of Admiral Sir Edward Codrington*, 1, 351.

56. W. H. Wollaston, "On the Water of the Mediterranean," 1829. Wollaston's analysis was a continuation of a more extensive analysis of global waters conducted by Marcet, who died before the Mediterranean waters reached England.

57. Codrington to Wollaston, Mar. 16, Apr. 1, Apr. 22, May 7 and Aug. 2, 1827; Bourchier (ed.), *Life of Admiral Sir Edward Codrington*, 1, 353–400.

58. Codrington to Wollaston, Oct. 29, 1827, ibid., 2, 97.

59. Codrington to Captain Spencer, Jul. 22, 1828, ibid., 2, 378.

60. Miller, "Between Hostile Camps."

61. Gilbert, "Presidency of the Royal Society."

62. Wollaston to Marcet, Aug. 5 or 6, 1820, Gilbert Collection.

63. Miller, "Between Hostile Camps."

64. Barrow, *Sketches of the Royal Society*, 51–52.

65. Todd, *Beyond the Blaze*, 213–214.

66. Ibid., 214.

67. Babbage to Whewell, May 15, 1820. Quoted in Miller, "Hostile Camps," 24.

68. Wollaston to Marcet, Aug. 5 or 6, 1820, Gilbert Collection.

69. Herschel, "Circumstances of Wollaston's Declining to Contest the Presidency of the R. Socy with Sir H. Davy. June 1820." The manuscript is transcribed in Gilbert, "Presidency of the Royal Society," 258–263.

70. Ibid., 259.

71. Ibid., 261–262.

72. Ibid., 262.

73. Davy to Wollaston, Jun. 25, 1820, quoted in Gilbert, "Presidency of the Royal Society," 267.

74. The official document is included with Wollaston Diplomas, MS 240, Royal Society, London. The appointment is dated Dec. 7, 1820.

75. Hall, *All Scientists Now*, 21.

76. Wollaston to Marcet, Dec. 4, 1820, Gilbert Collection.

CHAPTER THIRTEEN

1. Hawkins, *The Works of B. C. Brodie*, 3, 653.

2. Marcet to Horner, Apr. 15, 1816; quoted in Lyell (ed.), *Memoir of Leonard Horner*, 1, 93.

3. Roberts, *Memorials*, 2, 35.

4. There are two different typed copies of Julia Hankey's memoirs in the Gilbert Collection. The location of the originals is unknown. The shorter of the two (Gilbert Collection, box 4, file 2, enclosure A), hereafter referred to as "Hankey Memoir A," is in the form of a letter to Warburton dated Jun. 30, 1829. The longer (Gilbert Collection, box 1, file 3, enclosure A, item 19), hereafter "Hankey Memoir B," is undated and may have been written for distribution among her friends.

5. Hankey Memoir B, Gilbert Collection.

6. Ibid.

7. Hankey Memoir A, Gilbert Collection.

8. Hankey Memoir B, Gilbert Collection.

9. Warburton to Marcet, Sept. 6, 1816, Gilbert Collection, box 4, file 1, enclosure A, item 12.

10. Hankey Memoir A, Gilbert Collection.

11. Hankey Memoir B, Gilbert Collection.

12. Ibid.

13. Wollaston to Hasted, Apr. 7, 1817, Gilbert Collection, box 1, file 1.

14. Wollaston Mss., notebook **Daybook**, 70 (rev).

15. Wollaston Mss., notebook **Tours**, 23.

16. Wollaston Mss., notebook **Daybook**, 98.

17. Bickley (ed.), *The Diaries of Sylvester Douglas*, 2, 296.

18. Ibid., 302.

19. Marcet to Berzelius, Feb. 18, 1819; in Söderbaum (ed.), *Jac. Berzelius Bref*, 1, pt. 3, 184.

20. W. H. Wollaston, "Iodine," 1814.

21. W. H. Wollaston, "On the Cutting Diamond," 1816.

22. W. H. Wollaston, "Native Iron Found in Brazil," 1816.

23. Wollaston Mss., notebook $\mathbf{B_2}$, fac. p. 6.

24. Usselman, "Colombian Alluvial Platina," 266–267.

25. Wollaston Mss., notebook **I**, 58.

26. See Williams, *Michael Faraday*, 137–190, for a thorough technical discussion of Faraday's electromagnetic researches.

27. Stauffer, "Oersted's Discovery."

28. Babington to Marcet, Sept. 7, 1820, Gilbert Collection, box 4, file 1, encl. B.

29. Davy, "On the Magnetic Phenomena."

30. Anon. [W. H. Wollaston], "Electric and Magnetic Phenomena," 1820, 363.

31. Faraday, "Historical Statement," 288. Reprinted in Faraday, *Experimental Researches*, II, 159.

32. Faraday, ibid.; *Experimental Researches*, 160.

33. Faraday to Stodart, Oct. 8, 1821, in H. B. Jones, *Life and Letters of Faraday*, 1, 339–340.

34. Faraday to Wollaston, Oct. 30, 1821, ibid., 344.

35. Wollaston to Faraday, Oct. 30, or Nov. 1, 1821, ibid., 344–345.

36. Faraday, "Historical Statement."

37. H. B. Jones, *Life and Letters of Faraday*, 1, 349.

38. Faraday, *Experimental Researches*, 159, 161.

39. Davy, "On a New Phenomenon."

40. Ibid., 158–159.

41. H. B. Jones, *Life and Letters of Faraday*, 1, 346.

42. Ibid., 1, 379.

43. Faraday to Wollaston, May 30, 1823, ibid., 347–348.

44. Ibid., 349.

45. Warburton to Faraday, Jul. 8, 1823, ibid., 351–352.

46. Fullmer and Usselman, "Faraday's Election," 18.

47. W. H. Wollaston, "Cutting Rock Crystal," 1820.

48. W. H. Wollaston, "A Triple Object-Glass," 1822.

49. W. H. Wollaston, "On Metallic Titanium," 1823.

50. W. H. Wollaston, "Magnetism of Metallic Titanium," 1823.

51. Anon., [W. H. Wollaston], "Dr. Wollaston's Method."

52. Ibid.

53. W. H. Wollaston, "On Sounds Inaudible by Certain Ears," 1820.

54. Bunbury to Hasted, May 5, 1849, in Hasted, *Reminiscences*, 4.

55. W. H. Wollaston, "Semi-Decussation of the Optic Nerves," 1824.

56. Ibid., 224–225.

57. J. Taylor, *Le Méchanisme ou le nouveau traité de l'anatomie du globe de l'oeil...* (Paris: Michel-Estienne David, 1738), reprinted in Rucker, "Semidecussation of the Optic Nerves," 167.

58. W. H. Wollaston, "Semi-Decussation," 1824, 227.

59. See Rucker, "Semidecussation," for a review of prior thought on the matter.

60. W. H. Wollaston, "Semi-Decussation," 1824, 228.

61. Plant, "Fortification Spectra of Migraine."

62. Wollaston Mss., notebook **Tours**, inserted slip of paper.

63. Airy, "Hemiopsy." See also Levene, "Symptomatology of Migraine."

64. I thank Dr. Gordon T. Plant for first suggesting migraine to me as a cause of Wollaston's hemianopsia, and Dr. Paul Cooper for expert advice.

65. Hawkins, *B. C. Brodie*, 3, 653.

66. Brande, *Manual of Chemistry*, 103.

67. Maria Edgeworth to Mrs. Edgeworth, Jan. 16, 1822, in Colvin (ed.), *Maria Edgeworth: Letters from England*, 321.

68. Patterson, *Mary Somerville*, 12.

69. More information on Somerville is available in Chapman, *Mary Somerville*, and Brock, "The Public Worth of Mary Somerville."

70. Somerville, *Personal Recollections*, 128–129.

71. Ibid., 130.

72. W. Somerville to C. R. Weld, Apr. 20, 1848; in Weld, *History of the Royal Society*, 2, 313–314.

73. Pencil sketch of W. H. Wollaston by J. Jackson.

74. McCormack, "Edgeworth, Maria (1768–1849)," *oDNB*.

75. Maria Edgeworth to Mrs. Edgeworth, Jan. 16, 1822, in Colvin (ed.), *Maria Edgeworth: Letters from England*, 321.

76. Ibid., Apr. 4, 1822, 383.

77. Ibid., Apr. 6, 1822, 387–388.

78. Ibid., Jun. 3, 1822, 405.

79. Harriet Edgeworth to Mrs. Ruxton, Jun. 16, 1822. I thank Christina Colvin for this extract from the letter.

CHAPTER FOURTEEN

1. Bourchier (ed.), *Sir Edward Codrington*, 2, 437–438.

2. William to Henry Wollaston, Apr. 10, 1823. A handwritten copy is in the possession of the Wollaston family, and a typed copy is in the Gilbert Collection, box 1, file 2, encl A.

3. Wollaston Mss., notebook **Daybook**, 63.

4. Lockhart, *Life of Sir Walter Scott*, 6, 218–219.

5. Wollaston Mss., notebook **Daybook**, 93.

6. Davy, *Salmonia*, 329.

7. Knight, "Davy's Salmonia," 212.

8. Maxwell (ed.), *Houghton Fishing Club*, 2.

9. Ibid., 17, 18, 24.

10. Prescott, "Royal Institution's Visual Collections," 70. The bust bears the date 1830, which is more than a year after Wollaston's death.

11. F. Chantrey bust of William Hyde Wollaston.

12. W. H. Wollaston, "Eyes in a Portrait," 1824.

13. Ibid., 250–251.

14. Quoted in Weld, *History of the Royal Society*, 2, 313.

15. Wollaston to Hasted, Oct. 6, 1823, Gilbert Collection, box 1, file 1.

16. Wollaston Mss., notebook **Daybook**, 79.

17. Register Book of the Royal College of Physicians, entry for Feb. 13, 1824.

18. Usselman, "Faraday's Use of Platinum."

19. James, "Faraday's work on optical glass," 298.

20. Ibid.

21. Wollaston Mss., notebook **F**, loose sheet; see also Usselman, "Faraday's Use of Platinum," 177–179.

22. Ibid., 179.

23. W. H. Wollaston, "The Light of the Sun," 1829.

24. Ibid., 24.

25. W. H. Wollaston, "A Microscopic Doublet," 1829.

26. Ibid., 10.

27. Ibid., 12.

28. Chevalier, *Des Microscopes*, 55, 60.

29. Goodman, "Wollaston," 258–265.

30. W. H. Wollaston, "On a Differential Barometer," 1829.

31. Wollaston Mss., notebook **Daybook**, 86.

32. Wollaston to Herschel, [Oct. 6, 1826], Royal Society of London, HS 18, 286.

33. Babbage, *Passages*, 186–187.

34. Wollaston to Hasted, Mar. 27, 1828, Gilbert Collection, box 1, file 1.

35. Geikie, *Life of Sir Roderick Murchison*, 1, 123.

36. Blomefield [Jenyns], *Chapters in My Life*, 38.

37. Wollaston to Hasted, Oct. 30, 1827, Gilbert Collection, box 1, file 1.

38. D. M. Knight, *Humphry Davy*, 151.

39. The issues at stake in the Royal Society presidential elections of 1827 and 1830 are discussed in Miller, "The Royal Society of London 1800–1835," 297–382.

40. Ibid., 306.

41. Croker to Peel, Nov. 17, 1827, Croker Papers, William L. Clements Library, University of Michigan., Letter Book, 393–397.

42. Todd, *Beyond the Blaze*, 238.

43. Wollaston Mss., notebook **Daybook**, fac. p. 88.

44. T. Lawrence portrait of William Hyde Wollaston.

45. Brande, *Manual of Chemistry*, 103.

46. Ibid.

47. Wollaston to Julia Hankey, Sept. 8, 1828, Gilbert Collection, box 1, file 2, encl. A.

48. Brande, *Manual of Chemistry*, 103.

49. Hankey Memoir A, Gilbert Collection.

50. Probate will of William Hyde Wollaston, The National Archives, Kew, Richmond, Surrey, UK.

51. Wollaston to the Royal Society, Nov. 26, 1828, original in the Archives of the Royal Society, MC 1, 161; reprinted in Weld, *History of the Royal Society*, 2, 446–447.

52. Weld, *History of the Royal Society*, 2, 448.

53. Wollaston to the Royal Society, Dec. 7, 1828, original in the Archives of the Royal Society, MC 1, 162.

54. Wollaston to the Geological Society, Dec. 8, 1828, reprinted in *Proceedings of the Geological Society*, 1, 1834, 110.

55. Woodward, *History of the Geological Society of London*, 90.

56. Dreyer et al., *History of the Royal Astronomical Society*, 1, 45–46.

57. Extracts from the Astronomical Society of London, Monthly Reports, Feb. 13, 1829; copy in the Gilbert Collection, box 1, file 3, encl. A, item 16.

58. E. W. [Elizabeth Wollaston] to Christine Alexander, [January] 1829, copy in Gilbert Collection, box 1, file 3, encl. A, item 15.

59. The chain, enclosed in an envelope, which declares it to be a gift from Wollaston "a few days before his death" to J. B. C. (Jane-Barbara Codrington), is now in the possession of the Wollaston family.

60. Somerville, *Personal Recollections*, 129.

61. Robert Graves to George Kiernan, Dec. 27, 1828, in Graves, *Life of Hamilton*, 382–383.

62. Note included with the letter from E. W. to C. Alexander, ref. 58.

63. Bourchier (ed.), *Life of Admiral Sir Edward Codrington*, 2, 437–438.

64. Holland, *Recollections of Past Life*, 214.

65. Graves to Kiernan, ref. 61, 383.

66. E. W. to C. Alexander, ref. 58.

CHAPTER FIFTEEN

1. Chevalier, *Des Microscopes*, 14.

2. W. Babington, B. Brodie, and J. Somerville, "Appearances Observed on Inspecting the Body of the Late William Hyde Wollaston," Dec. 24, 1828, Wollaston Mss., box 2, envelope B.

3. E. W. [Elizabeth Wollaston] to Christine Alexander, [January] 1829, copy in Gilbert Collection, box 1, file 3, encl. A, item 15.

4. Barrow, *Sketches of the Royal Society*, 58.

5. Babbage, *Reflections on the Decline of Science in England*, 205.

6. Thomson, *History of Chemistry*, 2, 249–250.

7. Hankey Memoir A, Gilbert Collection.

8. Gilbert, "Wollaston Mss. at Cambridge."

9. D. Gilbert, "Anniversary Address," 41.

10. See, for example, Morrell, "Individualism and the Structure of British Science in 1830."

11. C. Macintosh to H. Warburton, Feb. 25, 1829, in Macintosh, *Biographical Memoir*, 94.

BIBLIOGRAPHY

MANUSCRIPTS

Bank of England Archive. Threadneedle St., London. W. H. Wollaston accounts.

British Museum (Natural History), London. Dawson-Turner Collection of the correspondence of Joseph Banks.

Geological Society of London. W. H. Wollaston. *Notebook of Crystallographic Notes and Measurements*, 1813–1828.

L. F. Gilbert Collection. D. M. S. Watson Library. University College, London. References to the collection correspond to entries in the Finder's Guide compiled by M. C. Usselman, which accompanies the collection.

The National Archives (TNA) of the UK. Public Record Office. Sikes Hydrometer Holdings, CUST 148/17/1-6.

National Maritime Museum. Greenwich Observatory. Minutes of the Board of Longitude.

Royal College of Physicians, London. *Register Book of the Royal College of Physicians*.

Royal Society of London Library. London,. Wollaston papers.

———. Thomas Young correspondence.

Science Museum, London.

Wollaston Manuscripts, Cambridge University Library, Mss. Add. 7736. References to the collection correspond to entries in the Finder's Guide compiled by M. C. Usselman, which accompanies the collection.

DICTIONARIES AND ENCYCLOPEDIAS

Biographia Britannica. London: John Nichols, 1735–1826.

Dictionary of Scientific Biography. New York: Charles Scribner's Sons, 1970–1978.

English Cyclopaedia. London: Charles Knight, 1854–1862.

The General Biographical Dictionary. London: Alex Chalmers, 1817.

Oxford Dictionary of National Biography. Oxford: Oxford University Press, 2004.

BOOKS AND ARTICLES

Aikin, A. and C. R. *A Dictionary of Chemistry and Mineralogy.* 2 vols. London: W. Phillips, 1807.

Airy, George B. "The Astronomer Royal on Hemiopsy." *Phil. Mag.* 30 (1865): 19–21.

Anon. *Charterhouse, Its Foundation and History.* London: M. Sewell, 1849.

———. "Henry Hasted." *The Gentleman's Magazine* 39 (1853): 98–99.

———. "On Fairy Rings." *The Gentleman's Magazine* 62 (1792): 209–211.

———. "On the Recent Adjudgement of the Royal Medals by the President and Council of the Royal Society." *Quart. J. Sci.* 1 (1827): 15.

——— [Henry Brougham]. *The Edinburgh Review* 2:3 (1803: Apr.): 97–98.

——— [Henry Kater, William Hyde Wollaston, Humphry Davy and Henry Englefield]. "Instructions for the Adjustments and Use of the Instruments Intended for the Northern Expeditions." *Quart. J. Sci. Arts* 5 (1818): 202–233.

——— [John Playfair]. *Edin. Rev.* 12 (1808): 120–130.

——— [John Whishaw]. "Some Account of the Late Smithson Tennant, Esq." *Ann. Phil.* 6 (1815): 1–11, 81–100.

——— [T. Young]. "Review of Laplace's Memoir 'Sur la Loi de la Réfraction extraordinaire.'" *Quart. Rev.* 2 (1809): 337–348.

——— [W. H. Wollaston]. "Reward of Twenty Pounds for the Artificial Production of Palladium." *J. Nat. Phil. Chem. and Arts,* 7 (1804): 75.

——— [W. H. Wollaston]. "On the Connexion of Electric and Magnetic Phenomena." *Quart. J. Sci.* 10 (1820): 363.

——— [W. H. Wollaston]. "Dr. Wollaston's Method of Detecting Magnesia, on the Smallest Scale." *Ann. Phil.* 25 (1823): 155–156.

Arrowsmith, R. L. (ed.). *Charterhouse Register, June 1769 – May 1872.* Godalming: 1964.

Ashworth, William J. "'Between the Trader and the Public': British Alcohol Standards and the Proof of Good Government." *Tech & Cult.* 42 (2001): 27–50.

———. *Customs and Excise: Trade, Production and Consumption in England, 1640–1845.* Oxford: Oxford University Press, 2003.

Averly, Gwen. "The 'Social Chemists': English Chemical Societies in the Eighteenth and Early Nineteenth Century." *Ambix* 33 (1986): 99–128.

Babbage, Charles. *Reflections on the Decline of Science in England, and on Some of its Causes.* London: Fellowes, 1830.

———. *Passages from the Life of a Philosopher.* London: Longman, Green, et al., 1864.

Bahar, Saba. "Jane Marcet and the limits to public science." *BJHS* 34 (2001): 29–49.

Barrow, Sir John. *Sketches of the Royal Society and Royal Society Club.* London: John Murray, 1849.

Bate, R. B. "On the Camera Lucida." *J. Nat. Phil. Chem. and Arts* 24 (1809): 146–150.

Bauer, Francis. "On Some Experiments on the Fungi Which Constitute the Colouring Matter of the Red Snow Discovered in Baffin's Bay." *Phil. Trans.* 110 (1820): 165–173.

Beamish, F. E., W. A. E. McBryde, and R. R. Barefoot. "The Platinum Metals." In *Rare Metals Handbook.* Second edition. Edited by C. A. Hampel, 304–335. London: Reinhold Publishing, 1961.

Bektas, M. Yakup, and Maurice Crosland. "The Copley Medal: The Establishment of a Reward System in the Royal Society, 1731–1839." *Notes Rec. R. Soc. Lond.* 46 (1992): 43–76.

Benjamin, Park (ed.). *Wrinkles and Recipes, Compiled from the Scientific American.* Second edition, revised. New York: John Wiley & Sons, 1875.

Bérard, Jacques Étienne. "Observations sur les Oxalates et Suroxalates alcalins." *Annales de Chimie* 73 (1810): 263–289; reprinted in *Journal of Natural Philosophy, Chemistry and the Arts* 31 (1812): 20–33.

Berman, Morris. *Social Change and Scientific Organization: The Royal Institution, 1799–1844.* Ithaca: Cornell University Press, 1978.

Bernard, Claude. "De L'Origine du Sucre. . . ." *Arch. Gén. Méd.* 18 (1848): 313.

Berthollet, C. L. Introduction to T. Thomson, *Système de Chimie.* Third edition. Translated by J. Riffault. Paris: Ve Bernard, 1809.

Berzelius, Jöns Jacob. *Autobiographical Notes.* Baltimore: Williams & Wilkins, 1934.

Berzelius, J. J., and Alexander Marcet. "Experiments on the Alcohol of Sulphur, or Sulphuret of Carbon." *Phil. Trans.* 103 (1813): 171–199.

Beudant, F. S. "Recherches tendantes à déterminer l'importance relative des formes cristallines et de la composition chimique dans la détermination des espèces minérales." *Annales des mines,* 2 (1817): 1–29. The English version appeared as "An Abstract of an Inquiry into the Relative Importance of the Crystalline Form and Chemical Composition in Determining the Species of Mineral Bodies." *Ann. Phil.* 11 (1818): 262–271.

Bickley, Francis (ed.). *The Diaries of Sylvester Douglas.* 2 vols. London: Constable & Co., 1928.

Biot, J. B. "Recherches sur les Réfractions Extraordinaires." *Mem. Acad. Sci.* 10 (1810): 1–266.

Blagden, Charles. "Report on the Best Method of Proportioning the Excise upon Spirituous Liquors." *Phil. Trans.* 80 (1790): 321–345.

Blair, Robert. "Experiments and Observations on the Unequal Refrangibility of Light." *Trans. Edin. Roy. Soc.,* 3 (1794): 3–76.

Blomefield [Jenyns], Leonard. *Chapters in My Life.* Bath: privately printed, 1889.

Boas Hall, Marie. *All Scientists Now: The Royal Society in the 19th Century.* Cambridge: Cambridge University Press, 1984.

Bourchier, J. B. (ed.). *Memoir of the Life of Admiral Sir Edward Codrington.* 2 vols. London: Longmans, Green & Co., 1873.

Brande, William Thomas. *A Manual of Chemistry.* Fourth edition. London: John Murray, 1836.

Brock, Claire. "The Public Worth of Mary Somerville." *BJHS* 39 (2006): 255–272.

Brooks, John. "The Circular Dividing Engine: Development in England 1739–1843." *Ann. Sci.* 49 (1992): 101–135.

Buchwald, Jed Z. "Experimental Investigations of Double Refraction from Huygens to Malus." *Arch. Hist. Exact Sci.* 21(1980): 311–373.

Buckland, W. "Account of an Assemblage of Fossil Teeth and Bones." *Phil. Trans.* 112 (1822): 171–236.

———. "On the Discovery of Coprolites, or Fossil Faeces." *Trans. Geol. Soc.* 3 (1829): 223–238.

Burke, John G. *Origins of the Science of Crystals.* Berkeley: University of California Press, 1966.

Bynum, W. F. "Physicians, Hospitals and Career Structures in Eighteenth-Century London." In *William Hunter and the Eighteenth-Century Medical World.* Edited by W. F. Bynum and Roy Porter, 105–128. Cambridge: Cambridge University Press, 1985.

Cantor, Geoffrey. "The Historiography of 'Georgian' Optics." *Hist. Sci.* 16 (1978): 1–21.

———. "Thomas Young." *Oxford Dictionary of National Biography.* Oxford: Oxford University Press, 2004.

Carangeot, A. "Goniomètre, ou Mesure-angle."*Observ. sur la Physique* 22 (1783): 193–197.

Cardwell, D.S.L. "Some Factors in the Early Development of the Concepts of Power, Work and Energy." *Brit. J. Hist. Sci.* 3 (1967): 209–224.

Chaldecott, John A. "Contributions of Fellows of the Royal Society to the Fabrication of Platinum Vessels: Some Unpublished Manuscripts." *Notes Rec. Roy. Soc. Lon.* 22 (1967): 155–172.

———. "Wollaston's Platinum Thermometer." *Plat. Met. Rev.* 16 (1972): 57–58.

———. "William Cary and His Association with William Hyde Wollaston." *Plat. Met. Rev.* 23 (1979): 112–123.

———. "Platinum and Palladium in Astronomy and Navigation: The Pioneer Work of Edward Troughton and William Hyde Wollaston." *Plat. Met. Rev.* 31 (1987): 91–100.

Chapman, Alan. *Mary Somerville and the World of Science.* Bristol: Canopus Pubs., 2004.

Chenevix, Richard. "Enquiries concerning the Nature of a Metallic Substance Lately Sold in London, as a New Metal, under the Title of Palladium." *Phil. Trans.* 93 (1803): 290–320.

———. "D'une lettre de M. CHENEVIX, . . . Au cit. Vauquelin." *Ann. Chim.* 46 (1803): 333–337.

———. "On the Action of Platina and Mercury upon Each Other." *Phil. Trans.* 95 (1805): 104–130.

Chevalier, Charles. *Des Microscopes et de leur Usage.* Paris: Ed. Proux et Co, 1839.

Children, John G. "An Account of Some Experiments with a Large Voltaic Battery." *Phil. Trans.* 105 (1815): 363–374.

Clark, Sir George. *A History of the Royal College of Physicians of London.* 2 vols. Oxford: Clarendon Press, 1964.

Clay, Reginald S., and Thomas H. Court. *The History of the Microscope.* London: Chas. Griffin & Co., 1932.

Coatsworth, L. L., B. I. Kronberg, and M. C. Usselman. "The Artefact as Historical Document. Part 1: The Fine Platinum Wires of W. H. Wollaston." *Hist. Tech.* 6 (1981): 91–111.

Coley, N. G. "Alexander Marcet (1770–1822): Physician and Animal Chemist." *Med. Hist.* 12 (1968): 394–402.

———. "Animal Chemists and Urinary Stone." *Ambix* 18 (1971): 69–93.

Collet-Descotils, H. V. "On the Cause of the different Colours of the Triple Salts of Platina, and on the Existence of a new Metallic Substance in that Metal." *J. Nat. Phil. Chem. and Arts,* 8 (1804): 118–126.

Colvin, Christina (ed). *Maria Edgeworth: Letters from England, 1813–1844.* Oxford: Clarendon Press, 1971.

Combe, William. *The History of the Colleges of Winchester, Eton, and Westminster; with the Charter-house, etc.* London: R. Ackermann, 1816.

Cottington, Ian E. "An Early Industrial Application for Malleable Platinum: Its Use in Flintlock Firearms." *Plat. Met. Rev.* 25 (1981): 74–81.

Crummer, Leroy. "An Introduction to the Study of Physic by William Heberden, M.D." *Ann. Med. Hist.* 10 (1928): 349–367.

Dalton, John. *A New System of Chemical Philosophy.* Part 1. Manchester: R. Bickerstaff, 1808.

———. "On the Constitution of the Atmosphere." *Phil. Trans.* 116 (1826): 174–188.

Daubeny, Charles. *An Introduction to the Atomic Theory.* London: John Murray, 1831. 2nd ed. Oxford: Oxford University Press, 1850.

Davies, Gordon L. Herries. *Whatever Is Under the Earth: The Geological Society of London 1807 to 2007.* London: Geological Society, 2007.

Davy, Humphry. "Notice of Some Observations on the Causes of the Galvanic Phenomena." *J. Nat. Sci .Phil. & Arts* 4 (1800): 337–342.

———. "On the Fire-Damp of Coal Mines, and on Methods of Lighting the Mines so as to Prevent Its Explosion." *Phil. Trans.* 106 (1816): 1–22.

———. "On the Magnetic Phenomena Produced by Electricity; in a Letter . . . to W. H. Wollaston." *Phil. Trans.* 111 (1821): 7–19.

———. "On a New Phenomenon of Electro-Magnetism." *Phil. Trans.* 113 (1823): 153–159.

———. *Six Discourses Delivered before the Royal Society.* London: John Murray, 1827.

———. *Salmonia.* 2nd ed. London: John Murray, 1829.

Dean, Philip A.W., and Melvyn C. Usselman. "The 'Synthetic' Palladium of Richard Chenevix: A Verdict on the Chemist and the Chemistry." *Ambix* 26 (1979): 100–115.

Donovan, Michael. *Essay on the Origin, Progress, and Present State of Galvanism.* Dublin: Hodges and McArthur, 1816.

Dreyer, J. L. E., et al. *History of the Royal Astronomical Society, 1820–1920.* London: Royal Astronomical Society, 1923.

Edwards, P. J. "The Growth of Fairy Rings of *Agaricus Arvensis* and Their Effect upon Grassland Vegetation and Soil." *J. Ecology* 72 (1984): 505–513.

Ekeberg, Anders G. "Of the Properties of the Earth Yttria, . . . and of the Discovery of a New Substance of a Metallic Nature (Tantalum)." *J. Nat. Phil. Chem. & Arts* 3 (1802): 251–255.

Ewart, Peter. "On the measure of moving force." *Memoirs of the Manchester Literary & Philosophical Society*, 2nd ser., 2 (1813): 105–258.

Faraday, Michael. "On the Existence of a Limit to Evaporation." *Phil. Trans.* 116 (1826): 484–493.

———. "Historical Statement Respecting Electro-Magnetic Rotation." *Quart. J. Sci.* 15 (1823): 288–292. Reprinted in Michael Faraday, *Experimental Researches in Electricity*, Volume 2. London: Richard and John Edward Taylor, 1844: 159–162.

———. *Experimental Researches in Electricity*, Volume 1. London: Richard and John Edward Taylor, 1839.

Fourcroy, A. F. de, and L. N. Vauquelin. "Mémoire sur l'analyse des calculs urinaires humains, . . ." *Ann. Chim.* 32 (1799): 216–222.

———. "Expériences sur le platine brut, sur l'existence de plusieurs métaux, et d'une espèce nouvelle de métal dans cette mine." *Ann. Chim.* 49 (1804): 177–183.

Frankel, Eugene. "The Search for a Corpuscular Theory of Double Refraction: Malus, Laplace and the Prize Competition of 1808." *Centaurus* 18 (1974): 223–245.

Fraser, Alistair B., and William H. Mach. "Mirages." *Sci. Am.* 234 (1976): 102–111.

Freund, Ida. *The Study of Chemical Composition.* Dover Publications: New York, 1968; reprint of 1904 edition published by Cambridge University Press.

Frondel, Clifford. "Jacob Forster (1739–1806) and His Connections with Forsterite and Palladium." *Mineralogical Mag.* 38 (1972): 545–550.

Fullmer, June Z. *Young Humphry Davy: The Making of an Experimental Chemist.* Philadelphia: American Philosophical Society, 2000.

Fullmer, June Z., and Melvyn C. Usselman. "Faraday's Election to the Royal Society: A Reputation in Jeopardy." *Bull. Hist. Chem.* 11 (1991): 17–28.

Garcia, R. Moreno. "Early Studies of Platinum in Spain: The Contribution Made by Joseph Louis Proust." *Plat. Met. Rev.* 37 (1993): 102–107.

Geike, Archibald. *Life of Sir Roderick Murchison.* 2 vols. London: John Murray, 1875.

———. *Annals of the Royal Society Club.* London: Macmillan and Co, 1917.

Gelfand, Toby. " 'Invite the Philosopher, as well as the Charitable': Hospital Teaching as Private Enterprise in Hunterian London." In *William Hunter and the Eighteenth-Century Medical World.* Edited by W. F. Bynum and Roy Porter, 129–151. Cambridge: Cambridge University Press, 1985.

Gilbert, Davies. "Anniversary Address of the Royal Society," Nov. 30, 1829. Printed in *Phil. Mag.* 7 (1830): 33-42.

Gilbert, L. F. "William Hyde Wollaston." *Chem. Ind.* (Jan. 5, 1952): 17.

———. "W. H. Wollaston Mss at Cambridge." *Notes Rec. Roy. Soc. Lon.* 9 (1952): 311–332.

———. "The Election to the Presidency of the Royal Society in 1820." *Notes Rec. Roy. Soc. Lon.* 11 (1955): 256–279.

Gilpin, George. "Tables for Reducing the Quantities by Weight, in Any Mixture of Pure Spirit and Water. . . ." *Phil. Trans.* 84 (1794): 275–382.

Goodman, D. C. "William Hyde Wollaston and His Influence on Early Nineteenth Century Science." D. Phil. Thesis, University of Oxford, 1965.

———. "Problems in Crystallography in the Early Nineteenth Century." *Ambix* 16 (1969): 152–166.

———. "Wollaston and the Atomic Theory of Dalton." *Hist. Stud. Phys. Sci.* 1 (1969): 37–59.

Gordon, W. T. "William Hyde Wollaston, F.R.S." *Nature,* July 21, 1934, 86–87.

Graham, Thomas. "On the Finite Extent of the Atmosphere." *Phil. Mag.* 1 (1827): 107–109.

Granville, A. B. *The Royal Society in the XIXth Century.* London: printed for the author, 1836.

Graves, Robert Percival. *Life of Sir William Rowan Hamilton.* 3 vols. Dublin: Hodges, Figgis & Co., 1882–1891.

Griffiths, P. D., R. G. Huntsman, and C. G. A. Thomas. "The First Cystine Stone?" *Brit. Med. J.* 4 Jan. 1964, 53.

Hall, Basil. *Account of a Voyage of Discovery to the West Coast of Corea, . . .* London: John Murray, 1818.

———. *Travels in North America in the Years 1827 and 1828.* Edinburgh: Robert Cadill, 1829.

———. *Forty Etchings: From Sketches Made with The Camera Lucida, in North America, in 1827 and 1828.* Edinburgh: Robert Cadill, 1829.

Hall, Marie Boas. *All Scientists Now: The Royal Society in the 19th Century.* Cambridge: Cambridge University Press, 1984.

Hammond, John H., and Jill Austin. *The Camera Lucida in Art and Science.* Bristol: Adam Hilger, 1987.

Hartley, Sir Harold. *Humphry Davy.* London: Thomas Nelson and Sons, 1966.

Hasted, Henry. "Reminiscences of Dr. Wollaston." *Proc. Bury and West Suffolk Archeological Institute* 1 (1849): 121–134; also privately printed as *Reminiscences of a Friend,* undated (ca.1850). There is a copy of the *Reminiscences* booklet in the Gilbert Collection. Page references to Hasted's work are to the booklet.

Hatchett, Charles. "An Analysis of a Mineral Substance from North America, Containing a Metal Hitherto Unknown." *Phil. Trans.* 92 (1802): 49–66.

Haüy, René Just. *Essai d'une théorie sur la structure des crystaux.* Paris: 1784.

———. *Tableau Comparatif des Résultats de la Cristallographie et de l'Analyse Chimique, relativement à la Classification des Minéraux.* Paris: Courcier, 1809.

Hawkins, Charles. *The Works of B. C. Brodie.* 3 vols. London: Longman, Green, et al., 1865.

Heberden, Ernest. *William Heberden: Physician of the Age of Reason.* London: Royal Society of Medicine, 1989.

Helmholtz, Hermann. "Über den Muskeltone." *Z. Gesammt. Naturwissenschaften* 34 (1869): 208–209.

Henry, William. *The Elements of Experimental Chemistry.* 11th ed. 2 vols. London, 1829.

Herschel, John F. W. *A Preliminary Discourse on the Study of Natural Philosophy.* London: Longman, Rees et al., 1831.

Hinde, P. T. "William Hyde Wollaston: The Man and His 'Equivalents.'" *J. Chem. Ed.* 43 (1966): 673–676.

Holland, Henry. *Recollections of Past Life.* London: Longmans, Green, and Co., 1872.

Hooke, Robert. "A Contrivance to Make the Picture of Any Thing Appear on a Wall . . . in the Midst of a Light Room in the Day-Time; . . ." *Phil. Trans.* 3 (1668): 741–743.

Hoppit, Julian. "Reforming Britain's Weights and Measures, 1660–1824." *Engl. Hist. Rev.* 108 (1993): 82–104.

Inkster, Ian. "Science and Society in the Metropolis: A Preliminary Examination of the Social and Institutional Context of the Askesian Society of London, 1796–1807." *Ann. Sci.* 34 (1977): 1–32.

James, Frank A. J. L. "Michael Faraday's Work on Optical Glass." *Phys. Educ.* 26 (1991): 296–300.

———. "Running the Royal Institution: Faraday as an Administrator." In *The Common Purposes of Life.* Edited by Frank A. J. L. James, 119–146. Burlington: Ashgate, 2002.

Johnson, Peter. "The Board of Longitude 1714–1828." *J. Brit. Astron. Assoc.* 99 (1989): 63–69.

Jones, [Henry] Bence. *The Life and Letters of Faraday.* 2 vols. London: Longmans, Green and Co., 1870.

———. *The Royal Institution: Its Founder and Its First Professors.* London: Longmans, Green, and Co., 1871.

Jones, Kathleen. *A History of the Mental Health Services.* London and Boston: Routledge & Kegan Paul, 1972.

Jones, William. "Observations on Dr. Wollaston's Statements Respecting an Improvement in the Form of Spectacle-Glasses. *Phil. Mag.* 18 (1804): 65–71; also *J. Nat. Phil. Chem. and Arts,* 7 (1804): 192–198.

———. "An Examination of Dr. Wollaston's Experiment on his Periscopic Spectacles." *Phil. Mag.* 18 (1804): 273–275; also *J. Nat. Phil. Chem. and Arts* 8 (1804): 38–40.

———. "Critical Observations, on Dr. Wollaston's Stated Improvement of the Camera Obscura and Microscope in the Application of the Meniscus, and Two Plano-Convex Lenses; . . ." *Phil. Mag.* 41 (1813): 247–253; also *J. Nat. Phil. Chem. and Arts* 34 (1813): 100–107.

Jungnickel, Christa, and Russell McCormmach. *Cavendish: the Experimental Life.* Lewisburg: Bucknell, 1999.

King, H. C. "The Life and Optical Work of W. H. Wollaston." *Brit. J. Physiol. Optics* 11 (1954): 10–31.

Knight, David M. *Atoms and Elements.* London: Hutchinson & Co., 1967.

———. *The Transcendental Part of Chemistry.* Folkestone: Wm Dawson & Son Ltd., 1978.

———. "Davy's Salmonia." In *Science and the Sons of Genius: Studies on Humphry Davy.* Edited by Sophie Forgan, 201–230. London: Science Reviews Ltd., 1980.

———. *Humphry Davy: Science and Power.* Oxford: Blackwell, 1992.

Knight, Richard. "A New and Expeditious Process for Rendering Platina Malleable." *Phil. Mag.* 6 (1800): 1–3.

Kronberg, B. I., L. L. Coatsworth, and M. C. Usselman. "The Artifact as Historical Document. Part 2: The Palladium and Rhodium of W. H. Wollaston." *Ambix* 28 (1981): 20–35.

———. "Mass Spectrometry as a Historical Probe: Quantitative Answers to Historical Questions in Metallurgy." *Archaeological Chemistry - III,* ACS Advances in Chemistry Series No. 205. Edited by J. B. Lambert, 295–310. Washington: American Chemical Society, 1984.

Kurzer, Frederick. "The Life and Work of Edward Charles Howard." *Ann. Sci.* 56 (1999): 113–141.

Lavoisier, Antoine. *Elements of Chemistry.* Translated by Robert Kerr. Edinburgh: William Creech, 1790.

Levene, John R. "Sir G. B. Airy, F.R.S. (1801–1892) and the Symptomatology of Migraine." *Notes Rec. Roy. Soc. Lon.* 30 (1975): 15–23.

Levere, Trevor H. "Science and the Canadian Arctic, 1818–76, from Sir John Ross to Sir George Strong Nares." *Arctic* 41 (1988): 127–137.

———. *Science and the Canadian Arctic.* Cambridge: Cambridge University Press, 1993.

LN. "Meionite." *Nouv. Dict. d'Hist. Naturelle* 20 (1818): 28-31.

Lockhart, John Gibson. *The Life of Sir Walter Scott.* 10 vols. Edinburgh: T. and A. Constable, 1902.

Lyell, Katherine M. (ed.). *Memoir of Leonard Horner.* 2 vols. London: Women's Printing Society, 1890.

MacArthur, C. W. P. "Davy's Differences with Gay-Lussac and Thenard: New Light on Events in Paris and on the Transmission and Translation Of Davy's Papers in 1810." *Notes Rec. Roy. Soc. Lon.* 39 (1985): 207–228.

Macintosh, George. *Biographical Memoir of the late Charles Macintosh.* Glasgow: W. G. Blackie & Co., 1827.

Marcet, Alexander. "A Chemical Account of Various Dropsical Fluids." *Med.- Chir. Trans.* 2 (1811): 340–381.

———. *An Essay on the Chemical History and Medical Treatment of Calculous Disorders.* London: Longman, Hurst, Rees, Orme and Brown, 1817.

Maxwell, Herbert (ed.). *Chronicles of the Houghton Fishing Club: 1822–1908.* London: Edward Arnold, 1908.

McDonald, Donald. "Smithson Tennant, F.R.S. (1761–1815)." *Notes Rec. Roy. Soc. Lon.* 17 (1962): 77–94.

———. *The Johnsons of Maiden Lane.* Martins Publishers Ltd: London, 1964.

McDonald, Donald, and Leslie B. Hunt. *A History of Platinum and its Allied Metals.* London: Johnson Matthey, 1982.

Melhado, Evan M. "Mitscherlich's Discovery of Isomorphism." *Hist. Stud. Phys. Sci.* 11 (1980): 87–123.

Miller, David Philip. "The Royal Society of London 1800–1835: A Study in the Cultural Politics of Scientific Organization." PhD Dissertation, University of Pennsylvania, 1981.

———. "Between Hostile Camps: Sir Humphry Davy's Presidency of the Royal Society of London, 1820–1827." *Br. J. Hist. Sci.* 16 (1983): 1–47.

Mordan, F. *Pens, Inks, & Inkstands.* London:W. Kent, 1858.

Morrell, J. B. "Individualism and the Structure of British Science in 1830." *Hist. Stud. Phys. Sci.* 3 (1971): 183–204.

Morrow, Scott I. "One Hundred and Fifty Years of Isomorphism." *J. Chem. Ed.* 46 (1969): 580–583.

Nicholson, W. "Some Account of a Pretended New Metal Offered for Sale, and Examined by Richard Chenevix,." *J. Nat. Phil. Chem. and Arts* 5 (1803): 136–139.

———. "Prize for the Artificial Composition of Palladium." *J. Nat. Phil. Chem. and Arts* 7 (1804): 159.

Oster, Gerald. "Muscle Sounds." *Sci. Am.* 250 (March, 1984): 108–114.

Paris, John Ayrton. *The Life of Sir Humphry Davy.* 2 vols. London: Colburn and Bentley, 1831.

Parkes, Samuel. *A Chemical Catechism.* 8th ed. New York, 1818.

Patterson, Elizabeth C. *John Dalton and the Atomic Theory.* New York: Anchor Books, 1970.

———. *Mary Somerville and the Cultivation of Science, 1815–1840.* The Hague: Martinus Nijhoff Publishers, 1983.

Pearson, George. "Experiments and Observations, Tending to Show the Composition and Properties of Urinary Concretions." *Phil. Trans.* 88 (1798): 15–46.

Perrin, C. E. "A Reluctant Catalyst: Joseph Black and the Edinburgh Reception of Lavoisier's Chemistry." *Ambix* 29 (1982): 141–176.

Pictet, M. A. *Voyage de trois mois en Angleterre, en Ecosse et en Irlande pendant l'été de l'an IX.* Geneva, 1802.

Plant, Gordon T. "The Fortification Spectra of Migraine." *Brit. Med. J.*, Dec. 20/27 1986, 1613–1617.

Porter, Roy. "Medical Lecturing in Georgian London." *Brit. J. Hist. Sci.* 28 (1995): 91–99.

Porter, Theodore. *Trust in Numbers: The Pursuit of Objectivity in Science and Public.* Princeton: Princeton University Press, 1995.

Prescott, G. M. "Forging Identity: The Royal Institution's Visual Collections." In *The Common Purposes of Life.* Edited by Frank A. L. James, 59–96. Burlington: Ashgate, 2002.

Proust, Louis. "Experiments on Platina." *Phil. Mag.* 11 (1801): 44–55, 118–128.

Prout, William. "Some Observations on the Analysis of Organic Substances." *Ann. Phil.* 6 (1815): 269–273.

Quick, Anthony. *Charterhouse: A History of the School.* London: James and James, 1990.

Reid, David Boswell. *Academical Examinations on the Principles of Chemistry.* Edinburgh: 1825.

Reilly, Desmond. "Richard Chenevix (1774–1839) and the Discovery of Palladium." *J. Chem. Ed.* 32 (1955): 37–39.

Roberts, Charles. *Memorials of the Earl of Stirling and of the House of Alexander.* 2 vols. Edinburgh: William Paterson, 1877.

Robinson, Andrew. *The Last Man Who Knew Everything.* New York: Pi Press, 2006.

Rocke, Alan J. *Chemical Atomism in the Nineteenth Century.* Columbus: Ohio State University Press, 1984.

Rohr, Moritz von. "Meniscus Spectacle Lenses." *Brit. J. Physiological Optics* 6 (1932): 183–187.

Rollo, John. *An Account of Two Cases of the Diabetes Mellitus.* 2 vols. London: C. Dilly, 1797.

Ross, John. *A Voyage of Discovery . . . for the Purpose of Exploring Baffin's Bay.* London: John Murray, 1819.

Rucker, C. Wilbur. "The Concept of a Semidecussation of the Optic Nerves." *Arch. Ophthal.* 59(2) (1958): 159–171.

Rudwick, M. J. S. "The Foundation of the Geological Society of London; Its Scheme for Co-Operative Research and Its Struggle for Independence." *Brit. J. Hist. Sci.* 1 (1963): 325–355.

Sacharoff, A. C., R. M. Westervelt, and J. Bevk. "Fabrication of Ultrathin Drawn Pt Wires by an Extension of the Wollaston Process." *Rev. Sci. Instrum.* 56 (1985): 1344–1346.

Schaaf, Larry J. *Tracings of Light: Sir John Herschel and the Camera Lucida- Drawings from the Graham Nash Collection.* Friends of Photography, 1989.

Schütt, Hans-Werner. "Zum Prioritätsproblem der Entdeckung des Chemischen Isomorphimus." *Physis*, 15 (1974): 5–22.

Searby, Peter. *A History of the University of Cambridge. Vol. 3: 1750–1870.* Cambridge: Cambridge University Press, 1997.

Sheldrake, T. "On the Camera Lucida." *J. Nat. Phil. Chem and Arts* 23 (1809): 372–377.

Simmonds, Charles. *Alcohol, Its Production, Properties, Chemistry, and Industrial Applications.* London: Macmillan, 1919.

Simms, Frederick W. *A Treatise on the Principal Mathematical Instruments Employed in Surveying, Levelling, and Astronomy* . . . 8th ed. London: Troughton and Simms, 1850.

Smith, B. A. "Wollaston's Cryophorus—Precursor of the Heat Pipe." *Phys. Educ.* 15 (1980): 310–314.

Smith, George E. "The *Vis Viva* Dispute: A Controversy at the Dawn of Dynamics." *Physics Today* 59 (Oct. 2006): 31–36.

Smith, John Graham. *The Origins and Early Development of the Heavy Chemical Industry in France.* Oxford: Clarendon Press, 1979.

Söderbaum, H. G. (ed.). *Jac. Berzelius Reseantechningar.* Stockholm: Norstedt & Söner, 1903.

———. *Jac. Berzelius Bref.* Uppsala: Almqvist & Wiksells, 1913.

Somerville, Martha. *Personal Recollections, from Early Life to Old Age, of Mary Somerville.* Boston: Roberts Brothers, 1874.

Stauffer, Robert C. "Speculation and Experiment in the Background of Oersted's Discovery of Electromagnetism." *Isis* 48 (1959): 33–50.

Streicher, J. S. "Some Remarks on the Wollaston Method of Producing Compact Platinum from Platinum Sponge Powder." In *Powder Metallurgy.* Edited by John Wulff, 16–17. Cleveland: American Society for Metals, 1942.

Sudduth, William M. "The Voltaic Pile and Electro-chemical Theory in 1800." *Ambix* 27 (1980): 26–35.

Talbot, W. H. Fox. *The Pencil of Nature.* London: 1844.

Tate, Francis G. H. *Alcoholometry.* London: H. M. Stationary Office, 1930.

Tennant, Smithson. "On the Action of Nitre upon Gold and Platina." *Phil. Trans.* 87 (1797): 219–221.

———. "On Two Metals, Found in the Black Powder Remaining after the Solution of Platina." *Phil. Trans.* 94 (1804): 411–418.

Thackray, Arnold. *John Dalton: Critical Assessments of His Life and Science.* Cambridge: Harvard University Press, 1972.

Thomson, Thomas. *A System of Chemistry.* 2nd ed. 4 vols. Edinburgh: Bell & Bradfute, 1804; 3rd ed. 5 vols. Edinburgh: Longman, Hurst, Rees & Orme, 1807.

———. "On the Oxides of Lead." *J. Nat. Phil. Chem and the Arts* 8 (1804): 280–293.

———. "On Oxalic Acid." *Phil. Trans.* 98 (1808): 63–95.

———. "Account of Dr. Wollaston's Scale of Chemical Equivalents." *Ann. Phil.* iv (1814): 176–177.

———. "Analysis of the Ore of Iridium." *Ann. Phil.*, n.s., 11 (1826): 17–19.

———. *The History of Chemistry.* 2nd ed. 2 vols. London: Colburn and Bentley, 1830.

———. "Biographical Account of Dr. Wollaston." *Proc. Phil. Soc. Glasgow* 3 (1850): 135–144.

Thorpe, T. E. *Essays in Historical Chemistry.* 2nd ed. London: Macmillan and Co., 1902.

Todd, A. C. *Beyond the Blaze: A Biography of Davies Gilbert.* Truro: D. Bradford Barton, 1967.

Ure, Andrew. *A Dictionary of Chemistry.* 3rd ed. London: T. Tegg, 1827.

Usselman, Melvyn C. "The Platinum Notebooks of William Hyde Wollaston." *Plat. Met. Rev.* 22 (1978): 100–106.

———. "The Wollaston/Chenevix Controversy over the Elemental Nature of Palladium: A Curious Episode in the History of Chemistry." *Ann. Sci.* 35 (1978): 551–579.

———. "William Wollaston, John Johnson and Colombian Alluvial Platina: A Study in Restricted Industrial Enterprise." *Ann. Sci.* 37 (1980): 253–268.

———. "Michael Faraday's Use of Platinum in His Researches on Optical Glass." *Plat. Met. Rev.* 27 (1983): 175–181.

———. "Wollaston's Microtechniques for the Electrolysis of Water and Electrochemical Incandescence." In *Electrochemistry, Past and Present, ACS Symposium Series 390*. Edited by J. T Stock and M. V. Orna, 20–31. Washington: American Chemical Society, 1989.

———. "Merchandising Malleable Platinum: The Scientific and Financial Partnership of Smithson Tennant and William Hyde Wollaston." *Plat. Met. Rev.* 33 (1989): 129–136.

———. "Multiple Combining Proportions: The Experimental Evidence." In *Instruments and Experimentation in the History of Chemistry.* Edited by F. L. Holmes and T. H. Levere, 243–271. Cambridge: MIT Press, 2000.

———. "A Secret History of Platinum." *Chem. Brit.* (Dec. 2001): 38–40.

———. "Smithson Tennant: The Innovative and Eccentric Eighth Professor of Chemistry." 113–137, In *The 1702 Chair of Chemistry at Cambridge: Transformation and Change.* Edited by Mary D. Archer and Christopher D. Haley. Cambridge: Cambridge University Press, 2004.

Vallvey, Luis Fermín Capitán. "Export and Smuggling of Spanish Platina in the Eighteenth Century." *Ann. Sci.* 53 (1996): 467–487.

Vauquelin, L. N. "Mémoire sur le Palladium et le Rhodium." *Ann. Chim.* 88 (1813): 167–198.

Venn, John. *Biographical History of Gonville and Caius College 1349–1897.* 2 vols. Cambridge: Cambridge University Press, 1897.

———. *Caius College.* London: F. E. Robinson & Co., 1901.

Vince, S. "The Bakerian Lecture. Observations upon an Unusual Horizontal Refraction of the Air; etc." *Phil. Trans.* 89 (1799): 13–23.

Wales, A. E. "Smithson Tennant, 1761–1815." *Nature,* 192 (1961): 1224–26.

Wallach, O. *Briefwechsel zwischen J. Berzelius und F. Wöhler.* Vol. 1. Leipzig: Verlag von Wilhelm Engelmann, 1901.

Warner, Brian. *Cape Landscapes: Sir John Herschel's Sketches, 1834–1838.* Cape Town: University of Cape Town Press, 2006.

Watson, E. C. "William Hyde Wollaston and the Discovery of the Dark Lines in the Solar Spectrum." *Am. J. Phys.* 20 (1952): 496–498.

Watson, William. "Several Papers Concerning a New Semi-Metal, Called Platina." *Phil. Trans.* 46 (1751): 584–596.

Weeks, Mary Elvira, and Henry M. Leicester. *Discovery of the Elements.* 7th ed. Easton: J. Chem. Ed., 1968.

Weld, Charles Richard. *A History of the Royal Society.* 2 vols. London: John Parker, 1848; repr. New York: Arno Press, 1975.

Wellner, D., and A. Meister. "A Survey of Inborn Errors of Amino-Acid Metabolism and Transport in Man." *Reviews of Biochemistry* 50 (1981): 911–968.

Whewell, William. *History of the Inductive Sciences.* 3 vols. London: John Parker, 1837.

———. *The Philosophy of the Inductive Sciences.* 2 vols. London: John W. Parker, 1847.

Williams, L. Pearce. *Michael Faraday: A Biography.* New York: Simon and Schuster, 1971.

Wilson, George. *Religio Chemici.* MacMillan & Co: London and Cambridge, 1862.
Winstanley, D. A. *Unreformed Cambridge.* Cambridge: Cambridge University Press, 1935.
Wollaston, Francis. *The Secret History of a Private Man.* London: privately printed, 1795.
Wollaston, Francis John Hyde. *A Plan of a Course of Chemical Lectures.* Cambridge: J. Archdeacon and J. Burges, 1794.
Wollaston, Henry Woods. *History of the Wollaston Family.* London: privately printed, 1960.
Wollaston, William Hyde. "On Gouty and Urinary Concretions." *Phil. Trans.* 87 (1797): 386–400.
———. "On Double Images Caused by Atmospherical Refraction." *Phil. Trans.* 90 (1800): 239–254.
———. "Experiments on the Chemical Production and Agency of Electricity." *Phil. Trans.* 91 (1801): 427–434.
———. "A Method of Examining Refractive and Dispersive Powers, by Prismatic Reflection." *Phil. Trans.* 92 (1802): 365–380.
———. "On the Oblique Refraction of Iceland Crystal." *Phil. Trans.* 92 (1802): 381–386.
———. "The Bakerian Lecture. Observations on the Quantity of Horizontal Refraction; with a Method of Measuring the Dip at Sea." *Phil. Trans.* 93 (1803): 1–11.
———. "On an Improvement in the Form of Spectacle Glasses." *Phil. Mag.* 17 (1803): 327–329; *J. Nat. Phil. Chem. and Arts* 7 (1804): 143–146.
———. "Experiment Showing the Advantage of Periscopic Spectacles." *Phil. Mag.* 18 (1804): 165–166; also *J. Nat. Phil. Chem. and Arts* 7 (1804): 241–242.
———. "On a New Metal, Found in Crude Platina." *Phil. Trans.* 94 (1804): 419–430.
———. "On Certain Chemical Effects of Light." *J. Nat. Phil. Chem. and Arts* 8 (1804): 293–297.
———. "Letter Concerning Palladium." *J. Nat. Phil. Chem. and Arts* 10 (1805): 204–205.
———. "On the Discovery of Palladium; with Observations on other Substances Found with Platina." *Phil. Trans.* 95 (1805): 316–330.
———. "The Bakerian Lecture on the Force of Percussion." *Phil. Trans.* 95 (1805): 13–22.
———. "Description of the Camera Lucida." *Phil. Mag.* 27 (1807): 343–347.
———. "On Fairy-Rings." *Phil. Trans.* 97 (1807): 133–138.
———. "On Super-Acid and Sub-Acid Salts." *Phil. Trans.* 98 (1808): 96–102.
———. "Description of a Reflective Goniometer." *Phil. Trans.* 99 (1809): 253–258.
———. "On Platina and Native Palladium from Brazil." *Phil. Trans.* 99 (1809): 189–194.
———. "On the Agency of Electricity on Animal Secretions." *Phil. Mag.* 33 (1809): 488–490.
———. "On the Identity of Columbium and Tantalum." *Phil. Trans.* 99 (1809): 246–252.
———. "The Croonian Lecture." *Phil. Trans.* 100 (1810): 1–15.
———. "On Cystic Oxide, a New Species of Urinary Calculus." *Phil. Trans.* 100 (1810): 223–230.
———. "On the Non-Existence of Sugar in the Blood of Persons Labouring under Diabetes Mellitus." *Phil. Trans.* 101 (1811): 96–105.
———. "On the Primitive Crystals of Carbonate of Lime, Bitter-Spar, and Iron-Spar." *Phil. Trans.* 102 (1812): 159–162.

———. "On a Periscope Camera Obscura and Microscope." *Phil. Trans.* 102 (1812): 370–377.

———. "A Method of Drawing Extremely Fine Wires." *Phil. Trans.* 103 (1813): 114–118.

———. "Description of a Single-Lens Micrometer." *Phil. Trans.* 103 (1813): 119–122.

———. "Letter from W. H. Wollaston, M.D. Sec. R. S. together with a Report of Mons. Biot, of the Imperial Institute of France." *Phil. Mag.* 42 (1813): 387–390; also *J. Nat. Phil. Chem and Arts,* 36 (1813): 316–321.

———. "On a Method of Freezing at a Distance." *Phil. Trans.* 103 (1813): 71–74.

———. "On the Elementary Particles of Certain Crystals." *Phil. Trans.* 103 (1813): 51–63.

———. "A Synoptic Scale of Chemical Equivalents." *Phil. Trans.* 104 (1814): 1–22.

———. "Iodine." *Ann. Phil.* 3 (1814): 313–314.

———. "Description of an Elementary Galvanic Battery." *Ann. Phil.* 6 (1815): 209–211.

———. "On the Octohedral Form of Iodine." *Ann. Phil.* 5 (1815): 237–238.

———. "Observations and Experiments on the Mass of Native Iron Found in Brazil." *Phil. Trans.* 106 (1816): 281–285

———. "On the Cutting Diamond." *Phil. Trans.* 106 (1816): 265–269.

———. "Determination of the Primitive Form of Bitartrate of Potash." *Ann. Phil.* 10 (1817): 37–38

———. "Observations on M. Beudant's Memoir 'Sur la Détermination des Espèces Minérales.'" *Ann. Phil.* 11 (1818): 283–286.

———. "On the Discovery of Potash in Sea Water." *Edinb. Phil. Journ.* 2 (1820): 325–326.

———. "On the Methods of Cutting Rock Crystal for Micrometers." *Phil. Trans.* 110 (1820): 126–131.

———. "On Sounds Inaudible by Certain Ears." *Phil. Trans.* 110 (1820): 306–314.

———. "Expériences sur la Production de l'électricité et sur son action chimique." *Ann. Chim.* 16 (1821): 45–53.

———. "On the Concentric Adjustment of a Triple Object-glass." *Phil. Trans.* 112 (1822): 32–37.

———. "On the Finite Extent of the Atmosphere." *Phil. Trans.* 112 (1822): 89–98.

———. "Report on the Present State of Ramsden's Dividing Engine." *Quart. J. Sci. Arts* 12 (1822): 381–388.

———. "On Metallic Titanium." *Phil. Trans.* 113 (1823): 17–22.

———. "On the Apparent Magnetism of Metallic Titanium." *Phil. Trans.* 113 (1823): 400–401.

———. "On Semi-Decussation of the Optic Nerves." *Phil. Trans.* 114 (1824): 222–231.

———. "On the Apparent Direction of Eyes in a Portrait." *Phil. Trans.* 114 (1824): 247–256.

———. "A Description of a Microscopic Doublet." *Phil. Trans.* 119 (1829): 9–13.

———. "On a Differential Barometer." *Phil. Trans.* 119 (1829): 133–136.

———. "On a Method of Comparing the Light of the Sun with That of the Fixed Stars." *Phil. Trans.* 119 (1829): 19–27.

———. "On a Method of Rendering Platina Malleable." *Phil. Trans.* 119 (1829): 1–8.

———. "On the Water of the Mediterranean." *Phil. Trans.* 119 (1829): 29–31.

Wood, A., and F. Oldham. *Thomas Young, Natural Philosopher.* Cambridge: Cambridge University Press, 1954.

Woodville, William. *Reports of a Series of Inoculations for the Variolae Vaccine or Cow-Pox.* London: Royal College of Physicians, 1799.

Woodward, Horace B. *The History of the Geological Society of London.* Geological Society: London, 1907; repr. Arno Press: New York, 1978.

Young, Thomas. "An Account of Some Cases of the Production of Colours, Not Hitherto Described." *Phil. Trans.* 92 (1802): 387–397.

———. *A Course of Lectures on Natural Philosophy and the Mechanical Arts.* 2 vols. London: Taylor and Walton, 1845; reprint of the 1807 edition.

PATENTS (GREAT BRITAIN)

Wollaston, William Hyde. English patent No 2752, "An Improvement in Spectacles by the Application of Concavo-Convex Glasses to Them." February 9, 1804.

———. English patent No 2993, "Drawing Apparatus." December 4, 1806.

INDEX